THE NON-SLEEPING UNIVERSE

Proceedings of two conferences on: 'Stars and the ISM'
held from 24–26 November, 1997 and on:
'From Galaxies to the Horizon' held from 27–29 November, 1997 at
the Centre for Astrophysics of the University of Porto, Portugal

Edited by

M.T.V.T. LAGO
University of Porto, Portugal

and

A. BLANCHARD
University of Strasbourg, France

Reprinted from *Astrophysics and Space Science*
Volume 261, Nos. 1–4, 1998/1999

SPRINGER-SCIENCE+BUSINESS MEDIA, B.V.

A C.I.P. Catalogue record for this book is available from the Library of Congress.

ISBN 978-94-010-5923-7 ISBN 978-94-011-4497-1 (eBook)
DOI 10.1007/978-94-011-4497-1

Printed on acid-free paper

TABLE OF CONTENTS

Preface ix

List of Sponsors xi

Committees xii

List of Participants xiii

NON-SLEEPING UNIVERSE: STARS AND THE ISM

INVITED PAPERS (I) *CONTRIBUTED PAPERS* (C) *AND POSTERS* (P)

I – *Asteroseismology, Stellar Rotation and Activity*

J. CHRISTENSEN-DALSGAARD / Asteroseismology (I) 1

I.W. ROXBURGH and M.J. THOMPSON / Space Asteroseismology on Later-Type Stars (P) 13

F.P. PIJPERS / Solar Rotation Inversions from Splittings and from A-Coefficients (P) 17

I.W. ROXBURGH for the COROT TEAM / COROT: Seismology of Stars (P) 19

I.W. ROXBURGH and S.V. VORONTSOV / On the Diagnostic Properties of Low Degree Acoustic Modes (P) 21

M.J. THOMPSON / What We have Learned from Helioseismology (I) 23

G. GIAMPIERI, A. POLNAREV, I.W. ROXBURGH and S.V. VORONTSOV / The Effect of Solar Oscillations on Space Gravitational Wave Experiments (P) 35

I.W. ROXBURGH / Helioseismic Constraints on the Solar Core (P) 37

I.W. ROXBURGH / Line Profiles of Solar P-Modes (P) 39

M.J. MONTEIRO, J. CHRISTENSEN-DALSGAARD and M.J. THOMPSON / Detection of the Lower Boundary of Stellar Convective Envelopes from Seismic Data (P) 41

I.W. ROXBURGH / Convective Overshooting and Mixing in Stellar Evolution (I) 43

K.I. MARCHENKOV, I.W. ROXBURGH and S.V. VORONTSOV / Non Linear Inversion for the Hydrostatic Structure of the Solar Interior (P) 51

H.P. SINGH, I.W. ROXBURGH and K.L. CHAN / A Parametric Study of 3-D Simulation of Penetrative Convection (P) 53

I.W. ROXBURGH / Steady Convection in Deep Compressible Layers (P) 55

I.W. ROXBURGH / Helioseismic Constraints on Solar Structure and the Solar Neutrino Problem (P) 57

D.R. SODERBLOM / Stellar Rotation and Activity (I) 59

L.J.R. MACHADO, M.T.V.T. LAGO and J.J.G. LIMA / Rotation of T Tauri Stars Revisited (P) 69

A.C. CAMERON / Stellar Tomography (I) 71

C.E. PARNELL / Our New View of the Solar Corona from Yohkoh and SOHO (I) 81

D.H. BROOKS, H.P. SUMMERS, R.A. HARRISON, J. LANG and A.C. LANZAFAME / EUV Spectral Variability and Non-Equilibrium Ionisation in the 'Quiet' Sun (C) 91

V.S. TAMAZIAN, J.A. DOCOBO and N.D. MELIKIAN / Flare Activity of Visual Binary COU 14 (C) 95

R. PALLAVICINI, A. MAGGIO, A. ORTOLANI, G. TAGLIAFERRI and S. COVINO / X-Ray Spectroscopy of Active Stars with ASCA and BeppoSAX (C) 101

G. MICELA, S. SCIORTINO, F.R. HARNDEN Jr. and R. ROSNER / X-Ray Variability and Rotation in the Pleiades Cluster (P) 105

II – *Time Variability in Young Stars, Stellar Disks and Winds*

M.T.V.T. LAGO, J.F. GAMEIRO, V.M. COSTA, L. MACHADO and A. PEDROSA / The Signatures of Structured Winds in T Tauri Stars (I) 107

L.P. CARRASQUEIRO, J.F. GAMEIRO and M.T.V.T. LAGO / A T Tauri Star Database (P) 117

V.M. COSTA and M.T.V.T. LAGO / Tauri Stars: The X-Ray Connection (P) 119

S. SCIORTINO, F. DAMIANI, G. MICELA and F. FAVATA / A Deep X-Ray Survey of the PMS Population of the Upper Sco-Cen Association (P) 121

A.I.G. de CASTRO, M. FRANQUEIRA, N. HUÉLAMO and E. VERDUGO / The Ultraviolet Variability of the T Tauri Stars (C) 123

N. HUÉLAMO, A.I.G. de CASTRO and M. FRANQUEIRA / X-Rays and UV Emission Lines in T Tauri Stars (P) 127

J.F. GAMEIRO and M.T.V.T. LAGO / Spectral Variability of T Tauri Stars in the Optical (C) 129

K.W. SMITH, G.F. LEWIS, I.A. BONNELL, P.S. BUNCLARK and J.P. EMERSON / Rapid Spectral Variability of DR Tau (P) 133

S.A. LAMZIN / The Structure of the Shock Wave in the Case of an Accretion onto Low Mass Young Stars (C) 137

P.J.V. GARCIA and E. THIÉBAUT / Extended (15 × 6 AU) Hα Emission around T Tau N (P) 141

J.M. OLIVEIRA, Y.C. UNRUH and B.H. FOING / Balmer Line Profiles Variations in SU Aurigae from the MUSICOS 96 Multisite Campaign (P) 143

S.A. LAMZIN, M. TEODORANI, L. ERRICO, A.A. VITTONE, F. GIOVANNELLI and C. ROSSI / Is it Possible to Observe Radiation of FUORs Boundary Layer? (P) 145

D.F.M. FOLHA and J.P. EMERSON / Near Infrared Lines as Diagnostics of Accretion in T Tauri Stars (C) 147

C. SAUTY, K. TSINGANOS and E. TRUSSONI / MHD Models for Stellar Winds (I) 151

S.A. LAMZIN, M. TEODORANI, L. ERRICO, A.A. VITTONE, E.A. KOLOTILOV, A.S. MIROSHNICHENKO and R.V. YUDIN / On the Source of Optical Variability and Outflow in Z CMa System (P) 161

A. PEDROSA and M.T.V.T. LAGO / Models in T Tauri Stars Tested through Hα (C) 163

S.A. LAMZIN, V. SHEVCHENKO, K. GRANKIN and S. MELNIKOV / On the Eclipses in GW Ori Double System (P) 167

III – *Jets and Outflows in the ISM*

M.D. SMITH / The Evolution of Young Stars, Protostellar Jets & Bipolar Outflows – a Unification Scheme (I) 169

A.J.L. FERNANDES and P.W.J.L. BRAND / Excitation of H_2 in Stellar Outflows: Shocks Versus Fluorescence (C) 179

M.D. SMITH / Molecular Outflows from High-Mass Protostars (P) 183

A.P. TSIVILEV, S. CORTIGLIONI, M.R. CIONI, M. ORSINI, E. PALAZZI and G.G.C. PALUMBO / A Microwave Jet Inside the Rosette Nebula? (P) 185

E. TOMMASI, D. LORENZETTI, T. GIANNINI, B. NISINI and F. PALLA / ISO-LWS Observations of NGC7129 Star Forming Region (C) 187

M.D. SMITH and M.M. MAC LOW / Supersonic Ambipolar Diffusion; Estimating the Magnetic Field Strength in Protostellar Outflows (P) 191

E.E. BLOEMHOF / Proper Motions of the H_2O Masers in W49N: The Velocity Variance/Covariance Matrix (P) 193

M.M. MAC LOW, M.D. SMITH, R.S. KLESSEN and A. BURKERT / The Decay of Supersonic and Super-Alfvénic Turbulence in Star-Forming Clouds (P) 195

I. SALAMANCA / The 'Seyfert-Like' Supernova SN 1997ab (C) 197

M.A. de AVILLEZ / Chimneys and Worms in the Galaxy (C) 201

J.A. DOCOBO, C. PRIETO and J.F. LING / The Influence of Mass Loss in the Eccentricity of Double Star Orbits (P) 205

T.C. TEIXEIRA, J.P. EMERSON and M.E. PALUMBO / CO Cocktails: Stirred ... but Not Shaken (P) 209

A. MAGALHÃES and J. CARNEIRO / The University of Porto Radio Spectrograph (P) 211

I.W. ROXBURGH / After Dinner Poster Paper (I) 213

NON-SLEEPING UNIVERSE: FROM GALAXIES TO THE HORIZON

INVITED PAPERS (I), *CONTRIBUTED PAPERS* (C) *AND POSTERS* (P)

I –*Gamma-Ray Bursts, the Hubble Deep Field and AGN's*

C. MEEGAN / Gamma-Ray Bursts: Where are We Now? (I) 215

O. BERTOLAMI / Axion-Dilaton Coupling and Gamma-Ray Bursts (C) 225

H. ZIAEEPOUR / Initial Data Set for Cosmology (P) 229

F. PASCUAL-SÁNCHEZ / On the Hypotheses of a New Model of Galaxy Formation (P) 231

R.D. BLANDFORD / Recent Results of the Hubble Deep Field (I) 233

J. DENNETT-THORPE / Relativistic Motion on Kilo-Parsec Scales in Powerful Radio Sources (C) 241

R.D. BLANDFORD / Recent Results on Active Galactic Nuclei (I) 245

A. ORR and A.N. PARMAR / Soft X-Ray Absorption in AGN Seen by BeppoSAX/LECS (C) 253

P. AUGUSTO / The Active CD Galaxy of Abell 2390 (P) 257

M.E. FILHO and M.J. MARCHÃ / The 200 mJy Sample at 1.4 GHz (P) 259

P. AUGUSTO, P.N. WILKINSON and I.W.A. BROWNE / The Population of KPC-Scale Flat-Spectrum Radio Sources (C) 261

II – *CMB Anisotropies: Present Status and Future Developments*

A. BLANCHARD, J.G. BARTLETT, M. DOUSPIS and M. LEDOUR / What do the CB Fluctuations Tell Us? (I) 265

A.J.C. da SILVA and P. CRAWFORD / CMB Angular Power Spectrum and the Cosmological Parameters (P) 275

D. BARBOSA, J.G. BARTLETT and A. BLANCHARD / Constraining Ω_0 with Sunyaev-Zel'dovich Observations (C) 277

III – *Cosmological Models, Structure Formation and Dark Matter*

A.R. LIDDLE / Inflation and the Cosmic Microwave Background (I) 281

P.T.P. VIANA and A.R. LIDDLE / Structure Formation with Scalar Fields (C) 291

P.V. MONIZ / Origin of Structure in SQU (C) 295

P. AGUIAR and P. CRAWFORD / Axially Symmetric Cosmological Models with Perfect Fluid and Cosmological Constant (P) 299

R. DORAN and P. CRAWFORD / Could the Observable Universe be Inside of a Black Hole? (P) 301

R.R. CALDWELL, R. DAVE and P.J. STEINHARDT / Quintessential Cosmology (I) 303

C.J.A.P. MARTINS / Wiggly Cosmic Strings (C) 311

P.P. AVELINO, E.P.S. SHELLARD, J.H.P. WU and B. ALLEN / Structure Formation Seeded by Cosmic Strings (P) 315

C.J.A.P. MARTINS and E.P.S. SHELLARD / Superconducting Strings and Magnetic Fields (P) 317

P.P. AVELINO, R.R. CALDWELL and C.J.A.P. MARTINS / The Promise of Structure Formation with Cosmic Strings in an Open Universe (P) 319

C.J.A.P. MARTINS and E.P.S. SHELLARD / Evolution of Superconducting Strings (P) 321

J.P.M. de CARVALHO and P.P. AVELINO / Cosmic Strings with Self-Interacting HDM (P) 323

C.J.A.P. MARTINS and E.P.S. SHELLARD / Vortons: Dark Matter from Cosmic Strings (P) 325

J.P. MIMOSO and A. NUNES / General Relativity as an Attractor to Scalar-Tensor Gravity Theories (C) 327

R. SADAT / High-Redshift X-Ray Clusters and Ω_0 (C) 331

M.J. CARR, P.P. AVELINO and J.P.M. de CARVALHO / Non-Linear Gravitational Clustering Using the Zel'dovich Approximation (P) 335

F.P. PIJPERS / Measuring Time Delays for Gravitationally Lensed Variable Quasars (C) 337

L.J. GOICOECHEA, E. MEDIAVILLA, A. OSCOZ, M. SERRA and J. BUITRAGO / Gravitational Lenses and the Hubble Constant: Present and Future (C) 341

R.D. BLANDFORD / Summary (I) 345

PREFACE

The "Non-Sleeping Universe" was a conference conceived to commemorate the opening of the new building of the Centro de Astrofísica da Universidade do Porto (CAUP). It seemed appropriate to include the two major areas of work at CAUP: Stellar Astrophysics and Cosmology. For the organizers the most obvious threading line would be activity. Therefore the choice of the name and contents.

The conference lasted for one week, the first three days dealing mainly with the non-sleeping stars, from Asteroseismology to Activity and Flows, involving both young stars and the interstellar medium. It included twenty five oral and thirty four poster presentations and involved 52 participants. The second part of the conference, covered another three days and addressed "non-sleepiness" at galactic scales and beyond. From gamma-ray bursts to the cosmic background radiation and cosmological models. It included twenty three oral and sixteen poster presentations and involved 46 participants.

Activity was also the environment dominant note. Only that this time of painters and builders rushing to finish the brand new building...

In any case it was a remarkable week with high level of excitement and interchange as expected from a Non-Sleeping Universe. Even if the Port wine in the cellars were the conference(s) dinner(s) took place seemed quite indifferent to all activity, dormant (...or not) in the amazingly large oak casks.

This was the first of a series of conferences/workshops that we intend to carry on a regular basis as a way to make the most out of our new and beautiful installations that we share with the town Planetarium. In fact, the building is a three partners adventure: the University of Porto, the Municipality and the Ministry for Science and Technology, who shared the costs. It is also the result of a long term dream of CAUP which shares its regular activities between research, education in Astronomy and public outreaching programs.

Astrophysics and Space Science is the original source of publication of his article. It is recommended that this article is cited as: *Astrophysics and Space Science* **261**: ix, 1999.

SPONSORS

- Junta Nacional de Investigação Científica e Tecnológica (JNICT)
- Centro de Astrofísica da Universidade do Porto (CAUP)
- Câmara Municipal do Porto
- Instituto do Vinho do Porto

Astrophysics and Space Science is the original source of publication of his article. It is recommended that this article is cited as: *Astrophysics and Space Science* **261**: xi–xii, 1999.

SCIENTIFIC ORGANIZING COMMITTEE

Paul Barr (ESA), John Barrow (Sussex), Alexander Boksenberg (Cambridge), Peter Brand (Edinburgh), Giuseppina Fabbiano (Harvard-Smithsonian) Gerald Fishman (NASA), Riccardo Giacconi (ESO), Jean Heyvarets (Strasbourg), Teresa Lago (Porto), Anthony Lasemby (Cambridge), Paulo Macedo (Porto), Françoise Praderie (Paris)

LOCAL ORGANIZING COMMITTEE

Pedro Augusto, Pedro P. Avelino, Jorge Paulo M. de Carvalho, Amadeu J. L. Fernandes, Jorge Filipe Gameiro, João J.A. Lima, Mário João Monteiro, António Pedrosa

LIST OF PARTICIPANTS

Joana Almeida
Centro de Astrofísica
Universidade do Porto
Rua das Estrelas
4150 Porto
PORTUGAL
joanafa@astro.fcma.up.pt

Miguel Avillez
Departamento de Matemática
Universidade de Évora
R. Romão Ramalho, 59
7000 Évora
PORTUGAL
pancho@galaxy.ce.uevora.pt

Jean-Loup Bertaux
Service d' Aeronomie du CNRS
BP3
91371 Verrieres-le-Buisson
FRANCE
jean_loup.Bertaux@aerov.jussieu.fr

Eric Bloemhof
California Institute of Technology
MS 105-24
Pasadena, CA 91125
UNITED STATES OF AMERICA
eeb@astro.caltech.edu

Vítor Bonifácio
Rua D. António Ferreira Gomes 10
3380 Ovar
PORTUGAL

Astrophysics and Space Science is the original source of publication of his article. It is recommended that this article is cited as: *Astrophysics and Space Science* **261**: xiii–xxxiii, 1999.

David H. Brooks
Centro de Astrofísica
Universidade do Porto
Rua das Estrelas
4150 Porto
PORTUGAL
brooks@astro.up.pt

Jorge Carneiro
Observatório Astronómico
Faculdade de Ciências
Universidade do Porto
PORTUGAL
jfcarne@oa.fc.up.pt

Andrew Collier-Cameron
School of Physics and Astronomy
University of St. Andrews
North Haugh
St Andrews, Fife
KY16 9SS
SCOTLAND
acc4@st-andrews.ac.uk

Luís P. Carrasqueiro
Centro de Astrofísica
Universidade do Porto
Rua das Estrelas
4150 Porto
PORTUGAL
lpc@stro.up.pt

J. Christensen-Dalsgaard
TAC and Institut for Fysik og Astronomi
Aarhus Universitet
DK-8000 Aarhus C
DENMARK
jcd@obs.aau.dk

Stefano Cortiglioni
Dept. of Astronomy
University of Bologna
I. Te. S. R. E. - C.N.R.
Via P. Gobetti, 101
40129 Bologna
ITALY
stefano@tesre.bo.cnr.it

Vítor Costa
Centro de Astrofísica
Universidade do Porto
Rua das Estrelas
4150 Porto
PORTUGAL
vcosta@astro.up.pt

José A. Docobo Durantez
Observatorio Astronomico Ramon Maria Aller
Universidade de Santiago de Compostela
Galicia
SPAIN
oadoco@usc.es

Amadeu Fernandes
Centro de Astrofísica
Universidade do Porto
Rua das Estrelas
4150 Porto
PORTUGAL
amadeu@astro.up.pt

João M. Fernandes
Observatório Astronómico da Universidade de Coimbra
Almas de Freire, Santa Clara
3040 Coimbra
PORTUGAL
jmfernan@gemini.ci.uc.pt

Bernard H. Foing
ESTEC/ESA
Solar System Division
PB 299
2200 AG Noordwijk
NETHERLANDS
foing@do.estec.esa.nl

Daniel Folha
Centro de Astrofísica
Universidade do Porto
Rua das Estrelas
4150 Porto
PORTUGAL
dfmf@astro.up.pt

Jorge F. Gameiro
Centro de Astrofísica
Universidade do Porto
Rua das Estrelas
4150 Porto
PORTUGAL
jgameiro@astro.up.pt

Paulo Garcia
CRAL - Observatoire Astronomique de Lyon
9, av. Charles Andre
65561 St. Genis Laval Cedex
FRANCE
pgarcia@avae.univ-lyon1.fr

Riccardo Giacconi
European Southern Observatory
Karl-Schwarzschild-Strasee 2
85748 Garching bei Munchen
GERMANY
rgiaccon@eso.org

Ana I. Gomez de Castro
Dept. de Astronomía y Geodesia, Facultad de C Matematicas
Univ. Complutense de Madrid
Ciudad Universitaria s/n
28040 Madrid
SPAIN
aig@eucmos.sim.ucm.es

Nuria Huélamo
Dept. de Astronomía y Geodesia, Facultad de C Matematicas
Univ. Complutense de Madrid
Ciudad Universitaria s/n
28040 Madrid
SPAIN
nuria@orion.mat.ucm.es

Maria Teresa Vaz Torrão Lago
Centro de Astrofísica
Universidade do Porto
Rua das Estrelas
4150 Porto
PORTUGAL
mtlago@astro.up.pt

Sergei Lamzin
Sternberg Astronomical Institute, Moscow V-234
Universitetskij Prospect, 13
119899 Moscow
RUSSIA
lamzin@sai.msu.su

João Lima
Centro de Astrofísica
Universidade do Porto
Rua das Estrelas
4150 Porto
PORTUGAL
jlima@astro.up.pt

Leandro Machado
Centro de Astrofísica
Universidade do Porto
Rua das Estrelas
4150 Porto
PORTUGAL
leandro@astro.up.pt

António Magalhães
Observatório Astronómico
Faculdade de Ciências
Universidade do Porto
Monte da Virgem
4430 Vila Nova de Gaia
PORTUGAL
asmagal@oa.fc.up.pt

Giuseppina Micela
Osservatorio Astronomico di Palermo
Piazza del Parlamento 1
90134 Palermo
ITALY
giusi@oasa.astropa.unipa.it

Mário J. Monteiro
Centro de Astrofísica
Universidade do Porto
Rua das Estrelas
4150 Porto
PORTUGAL
mjm@astro.up.pt

Joana Oliveira
Solar System Division
ESTEC/ESA
NETHERLANDS
joana@do.estec.esa.nl

Seathrun O'Tuairisg
Centro de Astrofísica
Universidade do Porto
Rua das Estrelas
4150 Porto
PORTUGAL
sea@astro.fcma.up.pt

Roberto Pallavicini
Osservatorio Astronomico di Palermo
Piazza del Parlamento 1
90134 Palermo
ITALY
pallavic@oapa.astropa.unipa.it

Clare Parnell
School of Mathematics
University of St. Andrews
North Haugh
St. Andrews, Fife
KY16 9SS
UNITED KINGDOM
clare@dcs.st-and.ac.uk

António Pedrosa
Centro de Astrofísica
Universidade do Porto
Rua das Estrelas
4150 Porto
PORTUGAL
apedrosa@astro.up.pt

Nuno V. Peixinho
Observatório Astronómico
Universidade de Lisboa
Tapada da Ajuda
1300 Lisboa
PORTUGAL

Frank P. Pijpers
Theoretical Astrophysics Center
Institute for Physics and Astronomy
Aarhus Universit
DENMARK
DK-8000 Aarhus C
fpp@obs.aau.dk

Filipe Pires
Centro de Astrofísica
Universidade do Porto
Rua das Estrelas
4150 Porto
PORTUGAL
pires@astro.up.pt

Sofia Randich
Osservatorio Astrofisico di Arcetri
Largo E. Fermi, 5
50125 Firenze
ITALY
randich@arcetri.astro.it

Ian Roxburgh
Astronomy Unit - Maths
Queen Mary & Westfield College
Mile End Road
London E1 4NS
ENGLAND
I.W.Roxburgh@qmw.ac.uk

Oliver Ryan
Centro de Astrofísica
Universidade do Porto
Rua das Estrelas
4150 Porto
PORTUGAL
ryan@astro.fcma.up.pt

Isabel Salamanca
Royal Greenwich Observatory
Madingley Road
Cambridge CB3 0EZ
UNITED KINGDOM
isabel@ast.cam.ac.uk

Christophe Sauty
D.A.E.C.
Observatoire de Paris
92 195 Meudon Cedex
FRANCE
christophe.sauty@daec.obspm.fr

Salvatore Sciortino
Osservatorio Astronomico di Palermo
Piazza del Parlamento 1
90134 Palermo
ITALY
sciorti@oapa.astropa.unipa.it

Kester Smith
Dept. of Physics
Queen Mary & Westfield College
Mile End Road
London E1 4NS
ENGLAND
k.w.smith@qmw.ac.uk

Michael Smith
Astronomisches Institut
University of Wuerzburg
Am Hubland
97074 Wuzburg
GERMANY
smith@astro.uni-wuerzburg.de

David Soderblom
Space Telescope Science Institute
3700 San Martin Dr.
Baltimore
MD 21218
UNITED STATES OF AMERICA
drs@stsci.edu

Vakhtang Tamazian
Obs. Astron. Ramon Maria Aller
University of Santiago de Compostela
SPAIN
oatamaz@usc.es

Teresa C. Teixeira
Theoretical Astrophysics Center
Institute for Physics and Astronomy
Aarhus University
DENMARK

A Michael J. Thompson
Astronomy Unit
School of Mathematical Sciences
Queen Mary & Westfield College
Mile End Road
London E1 4NS
ENGLAND
m.j.thompson@qmw.ac.uk

Elisabetta Tommasi
ISO Science Operation Centre
ESA - VILSPA
Villafranca del Castillo
P.O. Box 50727
28080 Madrid
SPAIN
etommasi@iso,vilspa.esa.es

Andrew Ulmer
Max-Planck Institut fur Astrophysik
Karl Schwarzschild-Str. 1
Postfach 1523
85740 Garching
GERMANY
aulmer@mpa-garching.mpg.de

Hans Zinnecker
Astrophysikalisches Institut Postdam
An der Sternwarte 16
D-14482 Potsdam
GERMANY
hzinnecker@aip.de

"NON-SLEEPING UNIVERSE: FROM GALAXIES TO THE HORIZON"

Rui Agostinho
Faculdade de Ciências
Universidade de Lisboa
Av. Prof. Gama Pinto, 2
1699 Lisboa Codex
PORTUGAL
ruiag@milkyway.cii.fc.ul.pt

Paulo Aguiar
Faculdade de Ciências
Universidade de Lisboa
Av. Prof. Gama Pinto, 2
1699 Lisboa Codex
PORTUGAL
paguiar@cosmo.cii.fc.ul.pt

Joana Almeida
Centro de Astrofísica
Universidade do Porto
Rua das Estrelas
4150 Porto
PORTUGAL
joanafa@astro.up.pt

Pedro Augusto
Universidade da Madeira
Funchal
PORTUGAL

Pedro Avelino
Centro de Astrofísica
Universidade do Porto
Rua das Estrelas
4150 Porto
PORTUGAL
pedro@astro.up.pt

Domingos Barbosa
Observatoire Astronomique de Strasbourg
11, rue de l' université
67000 Strasbourg
FRANCE
barbosa@wirtz.u-strasbg.fr

Jose L. Bermejo
Universidad de Barcelona
C/Catalunya, 68-3-2
08820 Barcelona
SPAIN
joseluis@tresnet.com

Orfeu Bertolami
Instituto Superior Técnico
Universidade de Lisboa
Av. Rovisco Pais
1096 Lisboa
PORTUGAL
orfeu@cosmos.ist.utl.pt

Alain Blanchard
Observatoire Astronomique de Strasbourg
11 Rue de l'Université
67000 Strasbourg
FRANCE
blanchar@astro.u-strasbg.fr

Roger Blandford
California Institute of Technology
Theoretical Astrophysics, 130-33
Pasadena, CA 991125
UNITED STATES OF AMERICA
rdb@tapir.caltech.edu

Robert Caldwell
Department of Physics and Astronomy
University of Pennsylvania
209 South 33rd Street
Philadelphia, PA 19104
UNITED STATES OF AMERICA
caldwell@dept.physics.upenn.edu

Alexandre Canavezes
Observatoire Astronomique de Strasbourg
11 Rue de l'Université
67000 Strasbourg
FRANCE
alex@astro.up.pt

Michael Carr
Dunsink Observatory
Dublin Institute of Advanced Studies
Castleknock
Dublin
IRELAND
mcarr@dunsink.dias.ie

Alexandre Cartaxo
Centro de Física Teórica
Universidade de Coimbra
R. Eng. Júlio A. Vieira, 53
S. M. Bispo
3040 Coimbra
PORTUGAL
cartaxo@ci.uc.pt

J. Paulo Carvalho
Centro de Astrofísica
Universidade do Porto
Rua das Estrelas
4150 Porto
PORTUGAL
mauricio@astro.up.pt

Paulo Crawford
Centro de Física Nuclear
Universidade de Lisboa
Av. Prof. Gama Pinto, 2
1699 Lisboa Codex
PORTUGAL
crawford@cosmo.cii.fc.ul.pt

Jane Dennett-Thorpe
Observatório Astronómico de Lisboa da Universidade de Lisboa
Tapada da Ajuda
1300 Lisboa
PORTUGAL
jdt@sintra.cc.fc.ul.pt

Rosa Doran
Faculdade de Ciências
Universidade de Lisboa
Av. Prof. Gama Pinto, 2
1699 Lisboa Codex
PORTUGAL
rdoran@cosmo.cii.fc.ul.pt

Mercedes Filho
Departamento de Física
Universidade de Lisboa
Av. Prof. Gama Pinto, 2
1100 Lisboa
PORTUGAL
mercedes@milkyway.cii.fc.ul.pt

Carlos Frenk
Physics Department
Durham University
South Road
Durham DH1 3LE
UNITED KINGDOM
c.s.frenk@durham.ac.uk

Luis Goicoechea
Departamento de Fisica Moderna
Universidad de Cantabria
Avda. de los Castros s/n
39005 Santander, Cantabria
SPAIN
goicol@besaya.unican.es

João de Jesus
Centro de Física Nuclear
Universidade de Lisboa
Av. Prof. Gama Pinto, 2
1699 Lisboa Codex
PORTUGAL
jjesus@alf2.cii.fc.ul.pt

Anthony Lasenby
Mullard Radio Astronomy Observatory
Cavendish Lab
Madingley Road
Cambridge, CB3 OHE
UNITED KINGDOM
A.N.Lasenby@mrao.cam.ac.uk

Andrew Liddle
Astronomy Centre
University of Sussex
Brighton BN1 9QH
UNITED KINGDOM
A.Liddle@sussex.ac.uk

Paulo G. Macedo
Centro de Astrofísica
Universidade do Porto
Rua das Estrelas
4150 Porto
PORTUGAL
pgmacedo@astro.up.pt

Carlos Martins
DAMTP
University of Cambridge
Silver Street
Cambridge CB3 9EW
UNITED KINGDOM
C.J.A.P.Martins@damtp.cam.ac.uk

Charles Meegan
NASA/Marshall Space Flight Center
Huntsville
Alabama
UNITED STATES OF AMERICA
meegan@sslles.msfc.nasa.gov

José P. Mimoso
Faculdade de Ciências
Universidade de Lisboa
Av. Prof. Gama Pinto, 2
1699 Lisboa Codex
PORTUGAL
jpmimoso@alf1.cii.fc.ul.pt

Paulo Moniz
DAMTP
University of Cambridge
Silver Street, Cambridge CB3 9EW
UNITED KINGDOM
prlvm10@damtp.cam.ac.uk

Astrid Orr
Astrophysics Division
Space Science Department
ESA / ESTEC
Postbus 299
2200 AG Noordwijk
THE NETHERLANDS
aorr@astro.estec.esa.nl

Seathrun O'Tuairisg
Centro de Astrofísica
Universidade do Porto
Rua das Estrelas
4150 Porto
PORTUGAL
sea@astro.up.pt

J.-Fernando Pascual-Sanchez
Fundamental Applied Math Dept
Faculty of Sciences
University of Valladolid
47011 Valladolid
SPAIN
passa@vax631.cpd.uva.es

Nuno V. Peixinho
Observatório Astronómico
Universidade de Lisboa
Tapada da Ajuda
1300 Lisboa
PORTUGAL

Frank P. Pijpers
Theoretical Astrophysics Center
Institute for Physics and Astronomy
Aarhus University
DENMARK
fpp@obs.aau.dk

Filipe Pires
Centro de Astrofísica
Universidade do Porto
Rua das Estrelas
4150 Porto
PORTUGAL
pires@astro.up.pt

Ana M. Rodrigues
Faculdade de Ciências
Universidade de Lisboa
Av. Prof. Gama Pinto, 2
1699 Lisboa Codex
PORTUGAL
anamarg@alf1.cii.fc.ul.pt

Alberto Roldan
Universitat de Barcelona
C/Napoles, 98-6-3
08013 Barcelona
SPAIN
94116794@xaee.ub.es

Oliver Ryan
Centro de Astrofísica
Universidade do Porto
Rua das Estrelas
4150 Porto
PORTUGAL
ryan@astro.up.pt

Rachida Sadat
Observatoire Astronomique de Strasbourg
11, rue de l' université
67000 Strasbourg
FRANCE
sadat@wirtz.u-strasbg.fr

Marc G. Sanchez
Universitat de Barcelona
C/Mossen Cinto Verdagrer, 53-1r
Sant Joan Despi
08970 Barcelona
SPAIN
95233121@xaee.ub.es

António J. Silva
Centro de Física Nuclear
Universidade de Lisboa
Av. Prof. Gama Pinto, 2
1699 Lisboa Codex
PORTUGAL
tosilva@alf1.cii.fc.ul.pt

Albert Stebbins
NASA/Fermilab
Box 500
Batavia, IL 60510
UNITED STATES OF AMERICA
stebbins@fnal.gov

Ismael Tereno
Centro de Física Nuclear
Universidade de Lisboa
Av. Prof. Gama Pinto, 2
1699 Lisboa Codex
PORTUGAL
ismael@cosmo.cii.fc.ul.pt

Andrew Ulmer
Max-Planck-Institut fuer Astrophysik
Karl Schwarzschild-Str. 1
Postfach 1523
85740 Garching
GERMANY
aulmer@mpa-garching.mpg.de

Pedro Viana
Centro de Astrofísica
Universidade do Porto
Rua das Estrelas
4150 Porto
PORTUGAL
viana@astro.up.pt

Simon White
Max-Planck-Institut fuer Astrophysik
Karl Schwarzschild-Str. 1
Postfach 1523
85740 Garching
GERMANY
swhite@mpa-garching.mpg.de

ASTEROSEISMOLOGY

JØRGEN CHRISTENSEN-DALSGAARD
Teoretisk Astrofysik Center, Danmarks Grundforskningsfond, and Institute for Fysik og Astronomi, Aarhus Universitet, DK-8000 Aarhus C, Denmark

1. Introduction

Frequencies of oscillation have two major advantages as diagnostics of stellar properties: they can be measured with great precision; and they are simply related to the properties of the star. A crude measure of the characteristic pulsation periods of a star is provided by the dynamical time scale $t_{\rm dyn} = [R^3/(GM)]^{1/2}$ where R and M are the stellar radius and mass, and G is the gravitational constant (for the Sun $t_{\rm dyn} \simeq 26\,{\rm min}$). Thus the frequency of a single (identified) mode of oscillation essentially provides a measure of the mean density $\bar{\rho} \propto M/R^3$ of the star. To obtain further information, more modes are required. In many pulsating stars, the spectrum of oscillation is in fact very rich.

The general goal of asteroseismology is obvious: to use data on stellar oscillations to improve our understanding of stellar structure and evolution. Indeed, given the limited observational basis for making detailed quantitative tests of the theory of stellar evolution, precise asteroseismic determination of just basic stellar parameters such as mass and radius for a range of stars would be of substantial value. More extensive data might provide information about the properties of the stellar interiors and allow investigation of departures from the standard picture. The computed models obviously depend on the micro-physics of the stellar interiors. Thus asteroseismic investigations offer the possibility of testing the physical properties of matter in stars, in particular the opacity (for a review, see *e.g.* Christensen-Dalsgaard 1995), and equation of state (*e.g.* Christensen-Dalsgaard & Däppen 1992). Potentially important non-standard effects are diffusion of elements through slow mixing in the radiative parts of stellar interiors; overshoot from convective cores; mass loss; and the effects of the detailed treatment of near-surface convection. In addition, it may be

Astrophysics and Space Science is the original source of publication of his article. It is recommended that this article is cited as: *Astrophysics and Space Science* **261**: 1–12, 1999.

possible to determine properties of the internal rotation of stars and find evidence for large-scale internal magnetic fields.

A mode of stellar oscillation is characterized by a spherical harmonic depending on the degree l and the azimuthal order m. Spherically symmetric, or radial, oscillation has $l = 0$, whereas for higher degree l is roughly a measure of the ratio between the stellar circumference and the average horizontal wavelength λ_h of the mode, $l \simeq 2\pi R/\lambda_h$; m measures the number of nodes along the equator. For each (l, m) the possible modes are characterized by their radial order n. Which modes are observable amongst the infinite number possible in a given star obviously depends on the mechanisms responsible for the excitation and amplitude limitation of the oscillations.

By far the most detailed information has been obtained in the case of the Sun (see Thompson, these Proceedings), where very detailed investigations of the internal structure and rotation have been possible. However, the Sun only samples a single, relatively simple stage of evolution in a rather unremarkable star. Thus it is crucially important that the solar information be complemented by information about other stars. Here I provide a brief overview of the properties of stellar oscillations and discuss some examples of the applications to asteroseismology. A much more detailed review was given by Brown & Gilliland (1994), while an extensive overview of stellar pulsation was provided by Gautschy & Saio (1995, 1996).

2. Observations of stellar oscillations

The observations are faced with two problems: to obtain sufficiently long series of continuous data; and to reach an adequate signal-to-noise ratio. An obvious and demanding goal is to reach, for other solar-like stars, the amplitude level of the solar oscillations: $15\,\mathrm{cm\,s^{-1}}$ in radial velocity and 5×10^{-6} in relative intensity. I note that stellar observations are generally restricted to modes of degree $l \lesssim 3$; the only exception is Doppler imaging in rapidly rotating stars (*e.g.* Kennelly 1995). A detailed review of the observational problems was given recently by Frandsen (1997).

Any diagnostic of the stellar atmosphere should in principle reflect the underlying oscillation. However, it is evident that the amplitude of the oscillation and the noise properties of the signal depend critically on the mode of observation. The simplest technique is probably to measure broad-band continuum intensity. This has the advantage of making full use of the light of the star and hence provides a low photon noise, even for observations of comparatively faint stars with modest telescopes. However, scintillation in the Earth's atmosphere is a serious problem for signals of an amplitude less than about 10^{-4}. A very ambitious attempt to detect solar-like oscillations in stars of the open cluster M67, involving week-long observations using

CCD photometry with some of the largest telescopes in the World, yielded a detection limit of around 2×10^{-5}, substantially higher than the solar amplitudes (Gilliland *et al.* 1993).

Doppler observations are less sensitive to atmospheric effects. If only a single spectral line is used, photon noise is a serious limitations except for the brightest stars. This problem can be overcome by combining data for many lines (assumed to oscillate in phase). The most serious limitation, however, is probably spectrograph stability; using fibre-fed echelle spectrographs it has been possible to reach a detection level of around $0.5\,\mathrm{m\,s^{-1}}$ (Brown *et al.* 1991, 1997; Edmonds & Cram 1995).

Differential intensity measurements, *e.g.* of the ratio of line to continuum intensity, can also be arranged to be essentially insensitive to scintillation, while photon noise can be reduced by combining data for several lines. A promising variant of this technique involves measurement of the equivalent width of the hydrogen Balmer lines (Kjeldsen *et al.* 1995); here it appears possible to get close to the level of the solar signal for relatively bright stars and moderately large telescopes (see also section 5.1).

Long series of observation are required to resolve the often dense spectrum of oscillations, to increase the signal-to-noise ratio in the spectrum and to improve the precision of the frequency determination. Observations lasting several weeks are quite common, and in the solar case the longest continuous time series span several years. Gaps in the data, such as are inevitable in single-site observations over several days from intermediate latitudes, cause sidebands in the spectra which greatly complicate the interpretation. To overcome this, campaigns have been established to coordinate observations from several sites (see below).

In photometric observations from space the dominant noise source is expected to be photon noise. This permits detection of solar-like amplitudes in relatively bright stars with very modest telescopes. Furthermore, space observations can be arranged to provide uninterrupted coverage. Such observations are, however, still in the future.

3. Properties of stellar oscillations

3.1. FREQUENCIES OF OSCILLATION

Apart from the simple scaling with the dynamical frequency $\omega_{\mathrm{dyn}} = t_{\mathrm{dyn}}^{-1}$ the frequencies are largely determined by the behaviour in the stellar interior of two characteristic frequencies: the Lamb frequency $S_l = [l(l+1)]^{1/2} c/r$, where c is the adiabatic sound speed and r is the distance to the centre; and the buoyancy frequency N which is determined by the difference between the adiabatic and actual density gradient. Modes with frequencies above both S_l and N in most of the star behave essentially as standing sound waves (or p modes); modes with frequencies below S_l and N are essentially standing gravity waves (or g modes). However, there may, particularly in evolved stars, be modes of mixed character, behaving as p modes in some region of the star and as g modes in others. Such modes add substantial complication, but also diagnostic richness, to the possible spectrum of stellar oscillations.

For high-order p modes of low degree the cyclic frequencies are approximately given by $\nu_{nl} \simeq \Delta\nu(n+l/2+\alpha+1/4)+\epsilon_{nl}$, where $\Delta\nu = \left(2\int_0^R \mathrm{d}r/c\right)^{-1}$ and ϵ_{nl} is a small correction term. Thus the spectrum is characterized by the large frequency separation $\Delta\nu \simeq \nu_{nl} - \nu_{n-1\,l}$ and the small frequency separation $\delta\nu = \nu_{nl}-\nu_{n-1\,l+2}$. The latter is determined largely by the conditions in the central parts of the star and hence is sensitive to the evolutionary state, while $\Delta\nu$ essentially scales as ω_{dyn}. This may be utilized by representing stellar evolution sequences in an asteroseismic Hertzsprung-Russell diagram (*e.g.* Christensen-Dalsgaard 1993) where $\delta\nu$ is plotted against $\Delta\nu$: a measurement of $(\Delta\nu, \delta\nu)$ in principle determines the mass and evolutionary state of a star. In practice, this assumes that all other relevant parameters of the star (such as chemical composition) are known. However, it is still the case that the pair $(\Delta\nu, \delta\nu)$ provides very useful information about the properties of a star (*e.g.* Brown *et al.* 1994).

For high-order g modes the periods are approximately uniformly spaced: $\Pi_{nl} \simeq [\Pi_0/l(l+1)](n + l/2 + \alpha_g)$, where the basic period spacing Π_0 is determined by the buoyancy frequency.

Frequencies of a spherically symmetric star are independent on the azimuthal order m of the mode. This degeneracy is lifted by rotation: slow rotation induces a frequency splitting that is approximately uniform in m, with a separation in most cases approximately equal to the average internal rotation rate. For more rapid rotation, the departure of stellar structure from spherical symmetry causes additional frequency shifts, of order $(\Omega/\omega_{\mathrm{dyn}})^2$. Here the effects of rotation might be comparable to the separation between adjacent multiplets (n, l), giving rise to a fairly complex spectrum of oscillations.

3.2. EXCITATION OF OSCILLATIONS

Roughly speaking, pulsations of an isolated star may be excited in two different ways: by extracting mechanical energy from the generation or transport of energy in the star; or by being driven stochastically by convection. In the former case, the star must function as a heat engine, where heat is added at compression and extracted at expansion. This is true of the perturbation in the nuclear energy-generation rate. In fact, low-order, low-degree g modes in evolving solar models have been found to be excited by this mechanism in a few cases (*e.g.* Christensen-Dalsgaard, Dilke & Gough 1974), although effects of such instability do not appear to have been observed. In general, however, the amplitude in the regions of nuclear burning is too small, compared with the amplitude closer to the surface, for the nuclear driving to dominate over the damping elsewhere. Radiative transport may lead to excitation if the perturbation in opacity is such that radiation is held back at compression and released at expansion (Eddington 1926). To produce net driving, the effect must arise in a region of localized opacity increase (or bump) in the transition region from nearly adiabatic to strongly nonadiabatic pulsation (*e.g.* Zhevakin 1953; Cox & Whitney 1958). As a result, the mechanism is effective only when the star satisfies certain specific constraints on its global parameters such as luminosity and effective temperature; this leads to well-defined instability regions in the Hertzsprung-Russell diagram. It should also be noted that at relatively low effective temperature convection dominates the energy transport in the region where instability might arise. This apparently in many cases leads to stability (*e.g.* Baker & Gough 1979; Houdek *et al.* 1998).

In the Sun all modes appear to be intrinsically stable, and the excitation is probably due to stochastic forcing of the oscillations by the turbulent convection in the near-surface region. Here convection reaches close to sonic speeds, leading to efficient generation of acoustic noise which excites the normal modes of the star. Early estimates of the resulting amplitude (*e.g.* Goldreich & Keeley 1977) indicated that the energy in one mode of pulsation is roughly the energy in one convective eddy with a timescale corresponding to the pulsation period; this gives amplitudes corresponding approximately to those observed in the Sun and has been used to estimate amplitudes in other, solar-like stars (Christensen-Dalsgaard & Frandsen 1983). More detailed calculations employing a time-dependent non-local mixing-length theory has given similar results (Houdek *et al.* 1998). It should also be noted that the excitation has been studied in considerable detail by means of extensive hydrodynamical simulations of the near-surface convection (Stein & Nordlund 1991; Bogdan, Cattaneo & Malagoli 1993; Nordlund & Stein 1998).

4. Which stars pulsate?

Pulsating stars are found throughout the HR diagram, as illustrated schematically in Fig. 1. "Classical" pulsating stars are largely confined to the so-called instability strip; in this region modes are excited through the opacity mechanism in the second helium ionization zone. The asteroseismically most interesting stars in this region are the δ Scuti stars, discussed below; we note also that a related class of longer-period pulsators (the γ Doradus

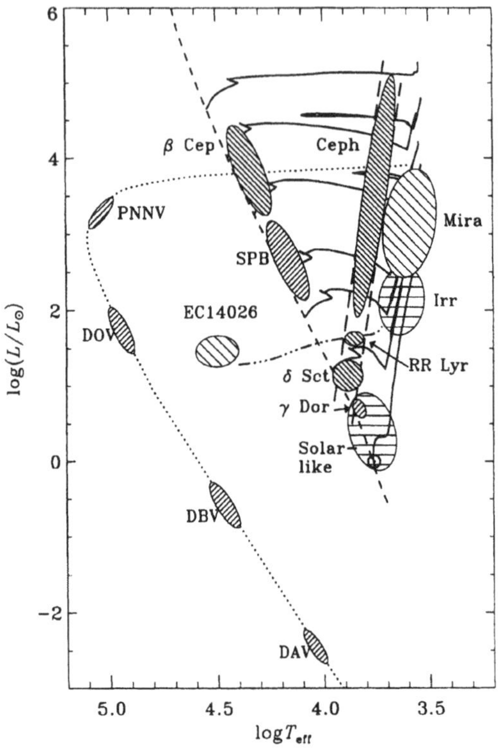

Figure 1. Schematic Hertzsprung-Russell diagram illustrating the location of several classes of pulsating stars (see also text). The dashed line shows the zero-age main sequence, the continuous curves are selected evolution tracks, at masses 1, 2, 3, 4, 7, 12 and 20 $M_\odot$, the dot-dashed line is the horizontal branch and the dotted curve is the white-dwarf cooling curve.

stars; *e.g.* Balona, Krisciunas & Cousins 1994) has recently been discovered. The β Cephei stars are main-sequence stars where p-mode pulsations are driven by the opacity mechanism operating in an opacity bump associated with iron-group elements (Cox *et al.* 1992; Moskalik & Dziembowski 1992); they typically show several modes and hence may provide important information about the main-sequence evolution of relatively massive stars (*e.g.* Dziembowski & Jerzykiewicz 1996). The same is true of the slowly-pulsating B (SPB) stars (*e.g.* Waelkens *et al.* 1998), where the same mechanism leads to the excitation of g modes with periods of the order of days (*e.g.* Dziembowski, Moskalik & Pamyatnykh 1993); data on these stars may yield valuable probes of conditions near convective cores. Solar-like pulsations, excited stochastically by convection, are expected for main-sequence stars with fairly vigorous near-surface convection; it is possible that the same mechanism is responsible for the irregular variations seen in many red-giant stars (*e.g.* Edmonds & Gilliland 1996). Finally, the EC 14026 variables and the pulsating white dwarfs are discussed below.

5. Applications of asteroseismology

Even the determination of two oscillation periods for double-mode Cepheids had interesting consequences: analysis of the periods showed a conflict with stellar masses obtained from evolution calculations, which was only resolved through the introduction of new opacity tables (*e.g.* Moskalik, Buchler & Marom 1992; Christensen-Dalsgaard & Petersen 1995). However, only with the availability of several modes may we probe details of stellar interiors. Here I briefly consider some classes of stars where this may be possible; a more detailed review was presented by Dziembowski (1997), while Thompson (these Proceedings) considers the case of the Sun.

5.1. SOLAR-LIKE STARS

There is no reason to believe that the Sun is peculiar, compared with other stars of similar parameters. Thus one would expect to see oscillations similar to those of the Sun in solar-like stars. Also, as pointed out by Thompson (these Proceedings), observations of just the low-degree modes in the Sun have substantial diagnostic potential. Detecting the oscillations in distant stars is difficult, however, because of the expected very low amplitudes. Indeed, despite many attempts over the past two decades no definite identification of solar-like oscillation in another star has been made (for a recent review, see Kjeldsen & Bedding 1997). Nonetheless, there have been interesting indications of the presence of significant signals. For example, Pottasch, Butcher & van Hoesel (1992) found hints of oscillations, with approximately the expected large frequency separation $\Delta\nu$, in α Cen A, while Brown *et al.* (1991) reported the possible detection of p modes in Procyon.

Perhaps the most convincing observations were obtained by Kjeldsen *et al.* (1995) in Balmer-line equivalent widths, for the subgiant star η Boo. They detected oscillations in a fairly extended frequency range, at an amplitude corresponding roughly to predictions based on stochastic excitation. The frequencies satisfied the p-mode asymptotic relation with a large frequency spacing of $40.3\,\mu$Hz. By comparing with models (Christensen-Dalsgaard, Bedding & Kjeldsen 1995) this led to a estimate of the luminosity of the star which was subsequently confirmed by HIPPARCOS data. Furthermore, the small frequency separation $\delta\nu$ computed for the model also agreed with the observations. Finally, both observed and model frequencies showed departures from the simple pattern expected for purely

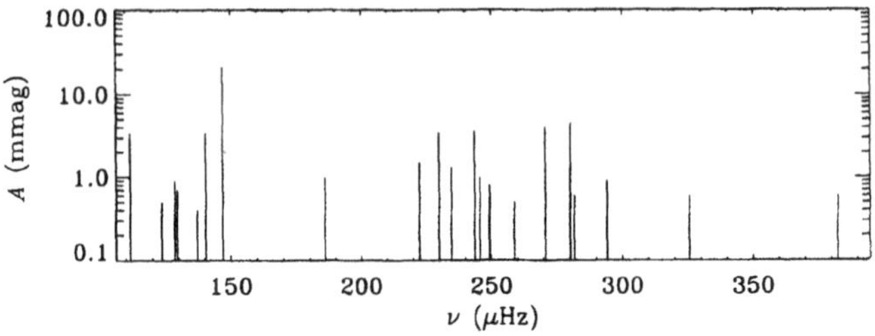

Figure 2. Schematic amplitude spectrum for the δ Scuti star FG Vir, based on frequencies and amplitudes (in V magnitude) obtained from extensive multi-site observations by Breger *et al.* (1998).

acoustic modes. In the model, these corresponded to mixed modes, with substantial amplitude in the region of strong hydrogen gradient in and just outside the shell-burning source. If confirmed, this behaviour would provide a very interesting diagnostic of the properties of this region. It should be noted, however, that radial-velocity observations by Brown *et al.* (1997) showed no evidence for the oscillations claimed by Kjeldsen *et al.* Thus further observations are needed to decide the reality of these oscillations.

5.2. δ SCUTI STARS

Apart from the Sun, the richest spectra of oscillation in main-sequence stars have been observed for the δ Scuti stars; a review of the pulsation properties of these stars was given by Guzik (1998). They have masses around $2\,M_\odot$ and are within or just after the main-sequence phase. As a result, they potentially allow the study of the properties of present or past convective cores, including possible overshoot (*e.g.* Dziembowski & Pamjatnykh 1991; Audard, Provost & Christensen-Dalsgaard 1995). Stability calculations generally show a substantial number of modes to be excited, including g modes. Which modes are actually observable then depends on the mechanisms, so far poorly understood, which limit the amplitudes. However, given also the often rapid rotation, it is likely that complex spectra may be produced. Such spectra are indeed observed: a recent example is shown in Fig. 2.

The theoretical interpretation of the observations, and hence the asteroseismic use of the data, is greatly complicated by uncertainty about the mode identification in the observed spectra and the need to include a possibly non-uniform rotational splitting (*e.g.* Goupil *et al.* 1993; Pérez Hernández, Claret & Belmonte 1995). If a sufficient number of modes could be identified, it might be possible to carry out inversion to determine localized aspects of the interiors of the δ Scuti stars. Goupil *et al.* (1996) found that a fairly well-resolved determination of the internal rotation profile of such a star could indeed be obtained from observed rotational splittings, assuming that modes of degree as high as $l = 6$ could be observed.

5.3. EC 14026 STARS

Information about late stages of stellar evolution are available in the EC 14026 stars. They are hot subdwarfs at the blue end of the horizontal branch, believed to be in the core helium burning phase. The discovery of their pulsations (Kilkenny *et al.* 1997) was made in parallel with the initial theoretical suggestion that the stars might be unstable, due to the operation of the radiative mechanism in the opacity bump caused by the iron-group elements (Charpinet *et al.* 1996). For a recent review on these stars, see Koen *et al.* (1998). Recent multi-site observations of the star PG1605+072 have found more than 50 frequencies, making this one of the richest known cases of oscillations of a distant star (Kilkenny *et al.* 1998).

A particularly interesting aspect of these stars is that their excitation may be related to element segregation in the stellar interior. It was pointed out by Charpinet *et al.* (1997) that instability at the observed stellar parameters requires enhancement of the iron abundance in the region of the opacity bump. This was shown to arise naturally from levitation due to selective radiative pressure. Thus the result constitutes one of the rare occasions where seismic information may be obtained directly from the observed presence of specific oscillations.

5.4. PULSATING WHITE DWARFS

Moving to even later stages of evolution, extensive data on oscillations have been obtained for several different classes of white dwarfs (for a review, see *e.g.* Winget 1991). Unlike most of the stars discussed so far, white dwarfs pulsate in high-order g modes: the typical periods are of order 5 – 20 minutes, much longer than the characteristic value of $t_{\rm dyn}$ of a few seconds. Thus the spectra are characterized by a nearly uniform spacing in period. Departures from this pattern may provide information about the compositional stratification of the outermost layers of the star (*e.g.* Bradley & Winget 1994).

As a specific example I consider the DB white dwarf GD 358 (Winget *et al.* 1994). It was observed in an extensive campaign of the Whole Earth Telescope (WET), over 11 days and utilizing 8 observatories spanning the globe. In the resulting spectrum more than 180 frequencies were identified. From analysis of the frequencies the mass of the star was determined as $M_* = (0.61 \pm 0.03)\, M_\odot$; the mass of the outer He envelope was found to be $(2 \pm 0.5) \times 10^{-6}\, M_*$, providing important constraints on the evolution preceding the white-dwarf phase. Furthermore, the analysis gave estimates of the luminosity and hence distance of the star, whereas frequency splittings allowed determination of the rotation rate of the star and suggested the presence of a weak magnetic field. Even though some of these inferences may require further verification, the results indicate the wealth of information obtainable from observations of compact-star oscillations. It should also be noted that in several cases frequency changes, resulting from evolutionary changes in structure, have been observed; an interesting case is the pre-white dwarf PG0122+200, where the evolution is dominated by cooling due to neutrino losses (O'Brien *et al.* 1998).

6. Conclusion

This brief overview of some of the current aspects of asteroseismology clearly illustrates the potential for investigating a wide range of stellar parameters and evolutionary phases. This provides an essential complement to the detailed solar investigations. For example, data on double-mode Cepheids and high-amplitude δ Scuti stars have indicated the need for substantial opacity increases at temperatures around 2×10^5 K; this could not have been inferred from solar data, since at such temperatures energy transport in the Sun is dominated by convection. Convective cores cannot be studied in the present Sun but may be reflected in frequencies observed for δ Scuti stars, and in solar-like oscillations in stars with $M \gtrsim 1.3\, M_\odot$. These and other examples highlight the importance of pushing for detailed and accurate observations of as broad a range of stars as possible, in addition to the continuing efforts on observing solar oscillations.

Major advances have in fact been made in recent years in observations of "classical" variable stars such as δ Scuti stars; the most striking examples are probably the large-scale efforts towards coordinated observation, as exemplified by the recent campaign on FG Vir (Breger *et al.* 1998). Although very demanding in terms of resources (both human and telescopic) such campaigns are essential to resolve the complex spectra of oscillation and hence get closer to identifying the modes of oscillation. They will no doubt be extended to other types of pulsators of fairly large amplitude (such as the β Cephei and SPB stars). In the slightly longer term networks of automated telescopes could perhaps be dedicated to such observations; it is also likely that the increasing sophistication of their equipment will allow amateur astronomers to play a significant rôle in these efforts.

Solar-like oscillations in other stars remain an elusive goal, where current observational techniques appear to be on the borderline of detection at the expected amplitudes. Although further ground-based effort is highly desirable, it is likely that incontrovertible evidence will be obtained only from observations from space. The French COROT mission (*e.g.* Catala *et al.* 1995; Baglin *et al.* 1998) has been approved for launch in 2002, whereas other missions are in the planning stage. It is to be hoped that a mission as ambitious as the STARS mission proposed to ESA (*e.g.* Fridlund *et al.* 1995) will eventually become feasible; this would allow detailed seismic investigation of solar-like stars in several open clusters and hence would provide tight constraints on models of stellar evolution.

Acknowledgements. I am grateful to H. Kjeldsen and M. J. Thompson for comments on earlier versions of the paper. This work was supported in part by the Danish National Research Foundation through the establishment of the Theoretical Astrophysics Center.

References

Audard, N., Provost, J. & Christensen-Dalsgaard, J., 1995. *Astron. Astrophys.*, **297**, 427
Baglin, A. and the COROT team, 1998. In *Proc. IAU Symp. 185: New eyes to see inside the Sun and stars*, eds Deubner, F.-L., Christensen-Dalsgaard, J., & Kurtz, D. W., Kluwer, Dordrecht, in the press.
Baker, N.H. & Gough, D.O., 1979. *Astrophys. J.*, **234**, 232
Balona, L.A., Krisciunas, K. & Cousins, A.W.J., 1994. *Mon. Not. R. astr. Soc.*, **271**, 905
Bogdan, T.J., Cattaneo, F. & Malagoli, A., 1993. *Astrophys. J.*, **407**, 316
Bradley, P.A. & Winget, D.E., 1994. *Astrophys. J.*, **430**, 850
Breger, M., Zima, W., Handler, G., *et al.*, 1998. *Astron. Astrophys.*, **331**, 271
Brown, T.M. & Gilliland, R.L., 1994. *Ann. Rev. Astron. Astrophys.*, **32**, 37
Brown, T.M., Christensen-Dalsgaard, J., Mihalas, B. & Gilliland, R.L., 1994. *Astrophys. J.*, **427**, 1013
Brown, T.M., Gilliland, R.L., Noyes, R.W. & Ramsey, L.W., 1991. *Astrophys. J.*, **368**, 599
Brown, T.M., Kennelly, E.J., Korzennik, S.G., Nisenson, P., Noyes, R.W. & Horner, S.D., 1997. *Astrophys. J.*, **475**, 322
Catala, C., Auvergne, M., Baglin, A., *et al.*, 1995. In: *Proc. Fourth SOHO Workshop: Helioseismology*, eds Hoeksema, J.T., Domingo, V., Fleck, B & Battrick, B., ESA SP-376, ESTEC, Noordwijk, p. 549
Charpinet, S., Fontaine, G., Brassard, P. & Dorman, B., 1996. *Astrophys. J.*, **471**, L103
Charpinet, S., Fontaine, G., Brassard, P., Chayer, P., Rogers, F., Iglesias, C.A & Dorman, B., 1997. *Astrophys. J.*, **483**, L123
Christensen-Dalsgaard, J., 1993. In *Proc. GONG 1992: Seismic investigation of the Sun and stars*, ed. Brown, T.M., ASP Conf. Ser., **42**, 347
Christensen-Dalsgaard, J., 1995. In *Frontiers of Astrophysics, Proceedings of the Rosseland Centenary Symposium*, eds Lilje, P.B. & Maltby, P., Institute of Theoretical Astrophysics, Oslo, p. 161
Christensen-Dalsgaard, J. & Däppen, W., 1992. *Astron. Astrophys. Rev.*, **4**, 267
Christensen-Dalsgaard, J. & Frandsen, S., 1983. *Solar Phys.*, **82**, 469
Christensen-Dalsgaard, J. & Petersen, J.O., 1995. *Astron. Astrophys.*, **299**, L17
Christensen-Dalsgaard, J., Bedding, T.R. & Kjeldsen, H., 1995. *Astrophys. J.*, **443**, L29
Christensen-Dalsgaard, J., Dilke, F.W.W. & Gough, D.O., 1974. *Mon. Not. R. astr. Soc.*, **169**, 429
Cox, A.N., Morgan, S.M., Rogers, F.J. & Iglesias, C.A., 1992. *Astrophys. J.*, **393**, 272
Cox, J.P. & Whitney, C., 1958. *Astrophys. J.*, **127**, 561
Dziembowski, W.A., 1997. In *Proc. IAU Symp. No. 181: "Sounding solar and stellar interiors"*, eds Provost, J. & Schmider, F.-X., Kluwer, Dordrecht, p. 317

Dziembowski, W.A. & Jerzykiewicz, M., 1996. *Astron. Astrophys.*, **306**, 436
Dziembowski, W.A. & Pamjatnykh, A.A., 1991. *Astron. Astrophys.*, **248**, L11
Dziembowski, W.A., Moskalik, P. & Pamyatnykh, A.A., 1993. *Mon. Not. R. astr. Soc.*, **265**, 588
Eddington, A.S., 1926. *The internal constitution of the stars*, Cambridge University Press, Cambridge.
Edmonds, P.D. & Cram, L.E., 1995. *Mon. Not. R. astr. Soc.*, **276**, 1295
Edmonds, P.D. & Gilliland, R.L., 1996. *Astrophys. J.*, **464**, L157
Frandsen, S., 1997. In *Proc. IAU Symp. No. 181: "Sounding solar and stellar interiors"*, eds Provost, J. & Schmider, F.-X., Kluwer, Dordrecht, p. 331
Fridlund, M., Gough, D.O., Jones, A., *et al.*, 1995. In *Proc. GONG'94: Helio- and Astero-seismology from Earth and Space*, eds Ulrich, R.K., Rhodes Jr, E.J. & Däppen, W., ASP Conf. Ser., **76**, 416
Gautschy, A. & Saio, H., 1995. *Ann. Rev. Astron. Astrophys.*, **33**, 75
Gautschy, A. & Saio, H., 1996. *Ann. Rev. Astron. Astrophys.*, **34**, 551
Gilliland, R.L., Brown, T.M., Kjeldsen, H., *et al.*, 1993. *Astron. J.*, **106**, 2441
Goldreich, P. & Keeley, D.A., 1977. *Astrophys. J.*, **212**, 243
Goupil, M.J., Dziembowski, W.A., Goode, P.R. & Michel, E., 1996. *Astron. Astrophys.*, **305**, 487
Goupil, M.J., Michel, E., Lebreton, Y. & Baglin, A., 1993. *Astron. Astrophys.*, **268**, 546
Guzik, J.A., 1998. In *Proc. IAU Symp. 185: New eyes to see inside the Sun and stars*, eds Deubner, F.-L., Christensen-Dalsgaard, J. & Kurtz, D.W., Kluwer, Dordrecht, in the press.
Houdek, G., Gough, D.O., Balmforth, N.J. & Christensen-Dalsgaard, J., 1998. *Astron. Astrophys.*, submitted
Kennelly, E.J., 1995. In *Proc. GONG'94: Helio- and Astero-seismology from Earth and Space*, eds Ulrich, R.K., Rhodes Jr, E.J. & Däppen, W., ASP Conf. Ser., **76**, 568
Kilkenny, D., Koen, C., O'Donoghue, D. & Stobie, R.S., 1997. *Mon. Not. R. astr. Soc.*, **285**, 640
Kilkenny, D., Koen, C., O'Donoghue, D., van Wyk, F., Larson, K.A., Shobbrook, R., Sullivan, D.J., Burleigh, M. & Dobbie, P.D., 1998. *Mon. Not. R. astr. Soc.*, submitted.
Kjeldsen, H. & Bedding, T.R., 1997. In *Proc. IAU Symp. No 189, Fundamental Stellar Properties: the Interaction between Observation and Theory*, eds Bedding, T.R., Booth, A.J. and Davis, J., Kluwer, Dordrecht, p. 279
Kjeldsen, H., Bedding, T.R., Viskum, M. & Frandsen, S., 1995. *Astron. J.*, **109**, 1313
Koen, C., O'Donoghue, D., Kilkenny, D. & Stobie, R.S., 1998. In *Proc. IAU Symp. 185: New eyes to see inside the Sun and stars*, eds Deubner, F.-L., Christensen-Dalsgaard, J. & Kurtz, D.W., Kluwer, Dordrecht, in the press.
Moskalik, P. & Dziembowski, W.A., 1992. *Astron. Astrophys.*, **256**, L5
Moskalik, P., Buchler, J.R. & Marom, A., 1992. *Astrophys. J.*, **385**, 685
Nordlund, Å. & Stein, R.F., 1998. In *Proc. IAU Symp. 185: New eyes to see inside the Sun and stars*, eds Deubner, F.-L., Christensen-Dalsgaard, J. & Kurtz, D.W., Kluwer, Dordrecht, in the press.
O'Brien, M.S., Vauclair, G., Kawaler, S.D., *et al.*, 1998. *Astrophys. J.*, **495**, 458
Pérez Hernández, F., Claret, A. & Belmonte, J.A., 1994. *Astron. Astrophys.*, **295**, 113
Pottasch, E.M., Butcher, H.R. & van Hoesel, F.H.J., 1992. *Astron. Astrophys.*, **264**, 138
Stein, R.F. & Nordlund, Å., 1991. In *Challenges to theories of the structure of moderate-mass stars, Lecture Notes in Physics*, vol. **388**, eds Gough, D.O. & Toomre, J., Springer, Heidelberg, p. 195
Waelkens, C., Aerts, C., Kestens, E., Grenon, M. & Eyer, L., 1998. *Astron. Astrophys.*, **330**, 215
Winget, D.E., 1991. In *White dwarfs, Proc. 7th European workshop on white dwarfs, Toulouse, France*, eds Vauclair, G. & Sion, E., Kluwer, Dordrecht, p. 129
Winget, D.E., Nather, R.E., Clemens, J.C., *et al.*, 1994. *Astrophys. J.*, **430**, 839
Zhevakin, S.A., 1953. *Astron. Zh.*, **30**, 161

SPACE ASTEROSEISMOLOGY ON LATER-TYPE STARS

IAN W. ROXBURGH & MICHAEL J. THOMPSON
Astronomy Unit, Queen Mary & Westfield College,
University of London, England

Helioseismology has been remarkably successful at revealing the internal structure and rotation of the Sun. We expect other stars with appreciable convective envelopes to exhibit similar, detectable oscillations. For example, if the Sun were at the distance of the Hyades, its low-degree 5-min oscillations should easily be detectable photometrically with a 1 m telescope in space. We focus on the possibility of using observational frequencies to probe the interiors of later-type stars, with masses around $0.8M_\odot$, which are important in determining globular cluster ages. The star's frequencies must be measured to a fraction of a μHz to make precise seismic diagnosis. This is not feasible for members of globular clusters, which are too far away and faint, but it would be possible for similar stars closer to the solar neighbourhood.

When observing stars other than the Sun, velocity measurements from spectroscopy are strongly constrained by photon fluxes. Ground-based photometric measurements of intensity fluctuations are dominated by atmospheric scintillation noise. But from space the photometric measurements can reach the photon noise limit, making feasible the detection of intensity fluctations associated with solar-type oscillations of distant stars. The photon noise must be low enough to give a signal-to-noise ratio (SNR) of at least a few. The photon noise $\sigma_{\rm amp}$ in the amplitude of the intensity fluctuation measurements varies as ${n_{\rm ph}}^{-1/2}$. Here $n_{\rm ph}$ is the number of photons received, which is proportional to the apparent luminosity of the star, the collecting area of the telescope, and the total duration T of the observations. Using estimates from [1],

$$\sigma_{\rm amp} = \left\{ 10^{0.4(11-m_v)} \left(\frac{d}{0.8\ {\rm metre}}\right)^2 \left(\frac{T}{1\ {\rm month}}\right)\right\}^{-1/2} {\rm ppm}\ ,$$

where d is the diameter of the telescope's effective aperture, and m_v is the star's apparent magnitude. Thus one can achieve 0.1 ppm in one month with

Astrophysics and Space Science is the original source of publication of his article. It is recommended that this article is cited as: *Astrophysics and Space Science* **261**: 13–15, 1999.

a 0.8m telescope for a $m_v = 6$ star, or 1 ppm if $m_v = 11$, both sufficiently below the expected mode amplitude of 5 ppm – the order of magnitude of the solar signal at a wavelength of 5500Å – to make oscillations detectable.

Having detected oscillations, we need to measure their frequencies. For a SNR ratio greater than a few, the error $\sigma(\nu)$ on the measured frequency ν varies with observation time T and intrinsic mode linewidth $\Delta\nu$ as roughly $\sigma(\nu) = (\Delta\nu/4\pi T)^{1/2}$ [5]. Thus for a width of 0.1 μHz, a precision of 0.3 μHz in the frequencies can be achieved in 1 month.

The solar-type modes are believed to be stochastically excited by turbulent convection. The energy E in a mode is therefore determined by a balance between the rate $\dot{E}$ at which energy is put into each mode, by the interaction with convection, and the rate η at which the mode is damped: $E \sim \dot{E}/\eta$. The mean velocity amplitudes of stochastically excited modes in stars between about 0.9 and 2.0 $M_\odot$ has been estimated in [3]. Kjeldsen & Bedding [4] deduce an empirical fit $v_{osc} \propto L/M$ to the theoretically predicted mode amplitudes in [2]. To relate this to amplitudes of the observable intensity fluctuations requires a knowledge of the mode physics, and in principle a detailed time-dependent radiative transfer calculation. In the absence of such a calculation, the amplitude of relative perturbations in intensity at a given wavelength (5500Å) have been estimated as related to the velocity amplitude by $(\delta L/L)_\lambda \propto v_{\rm osc}/T_{\rm eff}^2$ [4]; hence

$$\left(\frac{\delta L}{L}\right)_\lambda = \left(\frac{L}{L_\odot}\right)\left(\frac{M}{M_\odot}\right)^{-1}\left(\frac{T_{\rm eff}}{5770{\rm K}}\right)^{-2} \times 5\,{\rm ppm}\,.$$

Oscillations in ZAMS stars of mass smaller than 1 $M_\odot$ will have small amplitudes because of their relatively low luminosities, though this is partially offset by the inverse dependence on mass and effective temperature. The situation is easier in the case of evolved low-mass stars, however, because their evolution in the HR diagram during their hydrogen-burning lifetimes is towards higher luminosities.

We emphasise the large number of stars that could be monitored with a space-borne asteroseismology mission. A mission such as STARS could monitor 100-200 bright stars (i.e. brighter than magnitude 10) per field, and 5000-6000 stars could be monitored down to a reasonable faint-magnitude limit. If each field were observed for 1 month, then in a two-year mission this would amount to 150 000 stars.

The scientific pay-off of performing asteroseismology of later-type stars is considerable. Such stars have substantial convective envelopes and the oscillation measurements will test the models of the entropy in the convection zone as a function of effective temperature and surface gravity. This will allow us to calibrate stellar convection theory for use in all areas of stellar structure theory. Asteroseismology will also allow the calibration, via model comparisons, of aspects of the microphysics (e.g. opacities). The possibly competing effects of mixing and settling (anti-mixing) will also be constrained by measuring stellar frequencies, as they have been in the Sun. Another area where observing stars of different masses and ages may provide useful constraints on poorly modelled physics is the area of nuclear-driven instabilities in the early stages of stars: the role of the ^{3}He instability in these stars is not understood. Observable stars currently in the solar neighbourhood will reflect conditions in other parts of the galaxy where they were formed, and possibly parts through which they have passed. We have in mind particularly the high-velocity halo stars but more generally the orbits of stars from other parts of the galaxy may be traced with the Hipparcos catalogue data. Thus we will be able to use asteroseimic measurements of stars of different ages and chemical compositions to put observational constraints on models of the chemical evolution of the galaxy. By making observations of the later-type stars we shall also be testing the physics of stars which play a critical role in the determination of ages for globular clusters.

References

1. Badiali M., Catala F., Favata M., et al. (1996), *STARS Seismic Telescope for Astrophysical Research from Space: report of the Phase A study*, ESA D/SCI(96)4
2. Christensen-Dalsgaard J., Frandsen S. (1983), *Solar Phys.*, **82**, 469
3. Houdek G., Rogl J., Balmforth N.J., Christensen-Dalsgaard J. (1995), in *Proc. GONG'94: Helio- and Astero-seismology from Earth and Space*, eds Ulrich R.K., Rhodes, E.J., Däppen W., ASP Conf. Ser. **76**, 641
4. Kjeldsen H., Bedding T.R. (1995), *A&A*, **293**, 87
5. Libbrecht K.G. (1992), *ApJ*, **387**, 712

SOLAR ROTATION INVERSIONS FROM SPLITTINGS AND FROM A-COEFFICIENTS

F.P. PIJPERS
Theoretical Astrophysics Center
Aarhus University
Ny Munkegade
8000 Aarhus C
Denmark

1. Of modes, kernels and rotation

From helioseismic telescope networks such as the Global Oscillation Network Group (GONG) and also from the Solar Heliospheric Observatory satellite (SoHO), helioseismologists now have data on in excess of 20000 oscillation frequencies ν of the modes known collectively as the 5 minute oscillations. Depending on the order and degree of the modes these data are available either as individual frequencies of modes with known radial order n, spherical harmonic degree l, and order m, or as so-called 'a-coefficients' which are obtained by fitting polynomials to ridges of power in the (l, ν) diagram.

The internal rotation of the Sun affects the frequencies of the eigenmodes of oscillation because a wave propagating along with the solar rotation experiences a different Doppler shift than a wave propagating in the opposite direction. The mathematical relation between the frequency shift and the rotation rate is :

$$2\pi \frac{\nu_{nlm} - \nu_{nl\,-m}}{2m} = \int_0^1 \mathrm{d}r \int_{-1}^{1} \mathrm{d}\cos\theta \, K_{nlm}(r,\theta)\Omega(r,\theta)$$

where $\Omega(r,\theta)$ is the rotation rate as a function of radius and latitude (but only the part that is symmetric in northern and southern hemispheres). The K are known functions of radius and latitude related to the displacement in the solar oscillation, and n, l and m are three numbers fully characterizing

Astrophysics and Space Science is the original source of publication of his article. It is recommended that this article is cited as: *Astrophysics and Space Science* **261**: 17–18, 1999.

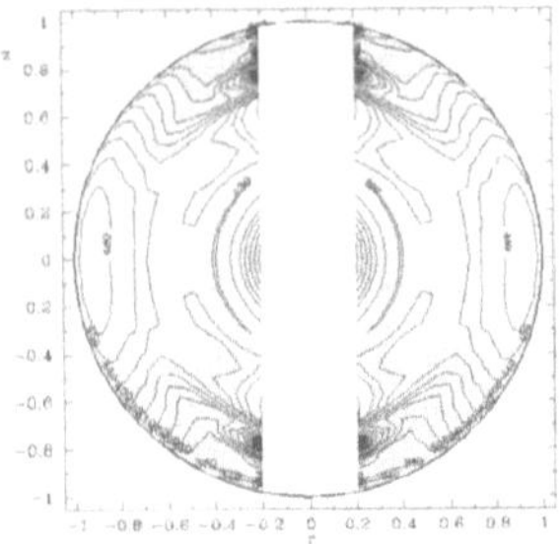

Figure 1. image of the internal rotation rate of the Sun (isorotation contours). Below the base of the convection zone the rotation rate appears to be essentially independent of radius and latitude. The surface regions near the pole rotate slower whereas the surface regions near the equator rotate faster than the radiative interior.

each oscillation : n is the number of radial nodes of the oscillation, l is the degree of a spherical harmonic, m is the order of a spherical harmonic.

Often the frequencies for each individual m cannot be determined with sufficient accuracy. Instead the $2l+1$ frequencies for a given n, l pair are then expressed as a fit to a number $j_{\max}$ known functions of m with $j_{\max} < 2l$. For these a-coefficients there is a similar relation with the solar internal rotation rate as shown above. Explicit expressions for the kernels can be found for instance in Pijpers, (1997).

2. 'Imaging' the solar internal rotation rate

Inverting the integral equation to obtain the rotation rate is a quite difficult mathematical problem. Unless one is careful the measurement errors in the data are combined in such a way that the error bar on the inferred rotation rate is very large. An inversion method which has proved very reliable in practice is known as Subtractive Optimally Localized Averages (SOLA), (cf. Pijpers and Thompson, (1994)). The SOLA method produces a set of coefficients c_i such that a function $\mathcal{K}$ can be constructed : $\mathcal{K}(r, \theta) \equiv \sum_{i \text{ over all } n,l,m} c_i K_i(r, \theta)$ which is reasonably well localized around a certain point (r_0, θ_0). In a sense it is the *point spread function* of the *seismic telescope* we are constructing with a computer algorithm. When the same coefficients are combined with the data we get a weighted average of the rotation rate, with the seismic point spread function as the weighting. In this way it is possible to image, with finite resolution and reasonable accuracy, the internal solar rotation (see fig. 1).

References

Pijpers, F.P. (1997), *A&A*, **326**, 1235
Pijpers, F.P., Thompson, M.J. (1994), *A&A*, **281**, 231

COROT: SEISMOLOGY OF STARS

IAN W. ROXBURGH FOR THE COROT TEAM
Astronomy Unit, Queen Mary and Westfield College,
University of London, London E1 4NS, UK.

COROT was approved as a CNES Petite Mission in December 1995 and successfully reviewed in June 1997, the PI IS Annie Baglin.. The mission is now scheduled for launch in early 2002. It will be launched on a Proteus platform into a polar inertial orbit with an altitude of 900 km with a mission lifetime of 2.5 to 3 years

COROT has two science goals: the primary goal is the study of the structure and dynamics of stars using the techniques of asteroseismology, the measurement of the frequencies, amplitudes and lifetimes of stellar oscillations, and the use of these data to diagnose the physical processes that govern the internal structure and evolution of stars. The second goal is the detection of earth-like planets around other stars by the detection of occultations as the planet passes across the stellar disc.

The instrument is a 25cm diameter, 115cm focal length telescope, with a single focal plane of four, large area, cooled CCDs. One set of detectors (2.8^o x1.4^o) is devoted to the asteroseismology experiment, the second set to both planet search and asteroseismology. The CCDs are based on the EEV CCD42-80, 4kx4k back illuminated device, operated in a transfer mode with a store shield fitted. This gives an image area of 2kx2k pixels per CCD, or 4kx2k for each focal plane area.

In the asteroseismology segment the stellar image (of selected targets) is spread over 300 pixels with a read out time of 1sec, in the planet segment the image is spread over 44 pixels with a read out time of 30secs. The seismology experiment should yield a sensitivity of 0.6 mmag for 5 days observation on a 6th magnitude star, and 2.5 mmag on a 9th magnitude star. 5 months observation should give a precision on the frequencies of 0.1μHz. In the planet segment it should be possible to detect an intensity decrease due to the transit of a planet across the stellar disc of 10^{-4} for stars brighter than 15th magnitude.

The mission has two programmes, The "Exploratory Programme" is devoted to asteroseismology (and possibly observations of Jupiter) in which

Astrophysics and Space Science is the original source of publication of his article. It is recommended that this article is cited as: *Astrophysics and Space Science* **261**: 19–20, 1999.

a selection (≈ 60) of stars brighter than 9th magnitude will be observed for intervals of order 20 days, giving broad coverage across the H-R diagram; and the "Central Programme" in which some 5 selected target fields will be observed for intervals of the order of 150 days each.

In the stellar segment observations will be restricted to the stars brighter than 9th mag but there will be no such limit on the planetary segment. The target fields will be selected prior to launch so as to maximise the scientific return from the asteroseismology programme, and should contain some stars of Type A, F, G, of magnitude 5 or 6.

References

Baglin A et al., 1997 COROT, Convection and Rotation, Phase A study, CNES, France.

ON THE DIAGNOSTIC PROPERTIES OF LOW DEGREE ACOUSTIC MODES

IAN W ROXBURGH AND SERGEI V VORONTSOV
Astronomy Unit, Queen Mary and Westfield College,
University of London, London E1 4NS, UK.

Standard asymptotic descriptions of stellar p-modes use $1/\omega$ as a small parameter; with either degree ℓ kept constant (e.g. Tassoul 1980), or $\tilde{w} = (\ell + 1/2)/\omega$, which specifies the position of the inner turning point (e.g. Vorontsov 1991). At low degree, these expansions are known to produce rather poor results, due to the strong effects of gravity perturbations, even when developed to higher order (Roxburgh and Vorontsov 1994a). Here, we employ an alternative asymptotic expansion, capable of providing much better accuracy at low degree—with $\tilde{w}$ as a small parameter instead of $1/\omega$.

The asymptotic eigenfrequency equation, up to terms of order $\tilde{w}^2$, is

$$F_0(\omega) + \tilde{w}^2 F_2(\omega) \simeq \frac{\pi}{\omega}\left[n + \ell/2 + 1/4 + \alpha(\omega)\right]$$

where n is radial order and $\alpha(\omega)$ is the surface phase shift, determined by the structure of the outermost layers. The two functions $F_0(\omega), F_2(\omega)$ are governed by the structure of the deep interior. When terms of order $\tilde{w}^2$ are neglected, the frequencies degenerate with respect to $n + \ell/2$; the function $F_0(\omega)$ is an effective acoustic radius. Small frequency separations are described by $F_2(\omega)$:

$$F_2(\omega_{\ell,n}) \simeq \frac{\pi}{2}\left(n + \frac{\ell}{2} + \frac{1}{4}\right)\frac{\omega_{\ell,n} - \omega_{\ell+2,n-1}}{2\ell + 3}$$

After $F_2(\omega)$ has been measured and approximated by linear regression $F_2(\omega) \simeq A + B\omega$, the surface term $\alpha(\omega)$ in its combination with $F_0(\omega)$ can be inferred as

$$\frac{\beta^*(\omega)}{\omega^2} = \frac{d}{d\omega}\left[\frac{F_0(\omega)}{\pi} - \frac{\alpha(\omega)}{\omega}\right] \simeq \frac{\bar{\omega}}{\omega_{\ell,n+1} - \omega_{\ell,n}} - n - \frac{\ell}{2} - \frac{3}{4} + \frac{(\ell + \frac{1}{2})^2}{\pi\bar{\omega}}(2A + B\bar{\omega})$$

at frequency $\bar{\omega} = (\omega_{\ell,n+1} + \omega_{\ell,n})/2$.

Astrophysics and Space Science is the original source of publication of his article. It is recommended that this article is cited as: *Astrophysics and Space Science* **261**: 21–22, 1999.

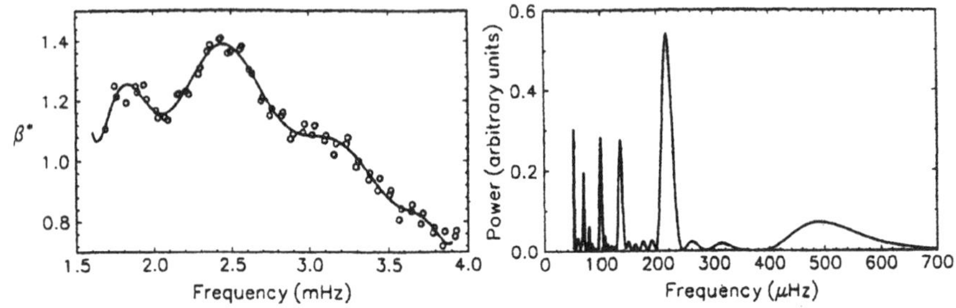

Figure 1. a) $\beta^*(\omega)$ inferred from modes of degree $\ell = 0$ to 3: the quasi periodicity is produced by the HeII ionisation zone. b) Fourier transform of the residuals: the peak at 220 μHz is produced by the base of the convective zone.

Figure 1a illustrates the result of an experiment where $\beta(\omega)$ was estimated using the frequencies of a standard solar model. Modes with ℓ from 0 to 3 were used in this analysis. The solid line shows a best fit polynomial of degree 10. The prominent quasi-periodic variation with frequency is produced by the HeII ionisation zone; this signature can be used to estimate the He abundance.

We then took a discrete Fourier transform of the residuals, convolving the residuals with $exp(i\omega/\omega_0)$. The result is shown in Figure 1b. The prominent peak at about 220μHz is produced by the rapid variation of the sound speed gradient at the base of the convective zone. At lower frequencies there is signal with ω_0 corresponding to the large separation , half this frequency and their combinations. The position of the peak produced by the base of the convective zone can be used to determine the (acoustic) depth of the convective zone. The strength of the signal is sensitive to possible effects of penetrative convection and magnetic fields (Roxburgh and Vorontsov 1994b).

References

Tassoul, M. 1980, *Astrophys. J. Supp* **43**, 469

Roxburgh, I. W. and Vorontsov, S. V. 1994a, *Mon. Not. R. astr. Soc.*, **268**, 143

Roxburgh, I. W. and Vorontsov, S. V. 1994b, *Mon. Not. R. astr. Soc.*, **268**, 880

Vorontsov, S. V. 1991, *Soviet Astron* **35**, 400

WHAT WE HAVE LEARNED FROM HELIOSEISMOLOGY

MICHAEL J. THOMPSON
Astronomy Unit, Queen Mary & Westfield College,
University of London, England

Abstract. The Sun's global oscillations, which are studied both in spatially unresolved ("Sun-as-a-star") and resolved observations of the solar disk, have enabled helioseismology to probe in detail the solar interior. I review first what is learned from the unresolved measurements, since this gives an idea of what we may in the not too distant future be able to learn about the interiors of other stars undergoing solar-type oscillations. I then look at the main results from resolved observations, which have begun to reveal the structure and dynamics of the interior of a star in exquisite detail.

1. Introduction

Measurements of the frequencies and other properties of global resonant oscillations of stars have tremendous diagnostic potential for understanding stellar interiors, through asteroseismology (see Christensen-Dalsgaard 1998, these proceedings). In the case of the Sun, for which we have very precisely measured frequencies of many thousands of such global modes, the seismology of the interior (helioseismology) is already well advanced. Because the Sun is very nearly spherical, the horizontal structure of each mode is well described by a spherical harmonic $Y_l^m(\theta, \phi)$ of degree l and azimuthal order m ($m = -l, \ldots, l$). Here θ is colatitude and ϕ is longitude. The mode is completely classified by l, m and a third quantum number n, the radial order: $|n|$ is roughly the number of nodes in e.g. the vertical displacement as a function of radius r. In a perfectly spherical star, the frequencies ν_{nlm} would be independent of m, so for each (n, l) multiplet there would be a $(2l+1)$-fold degeneracy and a single multiplet frequency ν_{nl}. Rotation and other departures from spherical symmetry raise this degeneracy. For the Sun, modes are observed with degrees l from 0 to several thousand. Frequencies of the largest-amplitude modes are in the approxi-

Astrophysics and Space Science is the original source of publication of his article. It is recommended that this article is cited as: *Astrophysics and Space Science* **261**: 23–34, 1999.

mate range 2–4mHz, though modes can be observed and measured at both lower and higher frequencies.

The horizontal wavenumber of the mode can be identified with L/r, where $L = l+1/2$. Hence low degrees correspond to low horizontal wavenumber, i.e. large-scale waves, and high degrees correspond to small-scale waves. It follows that for a distant star, where the surface cannot be resolved, it is the low-degree modes that can be observed, while the high-degree modes will be averaged by the observation over the stellar surface and hence be undetectable. Consequently, in this review I propose to take a look at helioseismology from a slightly unusual standpoint. In practice, of course, helioseismologists use all the data that are available to them and which are relevant for answering a given scientific question. Thus historically, the inferences about the solar interior have generally been drawn from all the available data. However, I shall start by considering what could be learned in helioseismology if we had only had the frequencies of the low-degree modes, which we might hope to detect also for more distant stars. I shall then fill in the picture by reviewing what has been learned from the higher-degree modes.

The global modes observed to date on the Sun are set up either by surface gravity waves (the f modes) or – the majority – by acoustic waves (the p modes). As an acoustic wave propagates down into the solar interior, it senses the increasing sound speed c (predominantly due to the sound speed increasing like the square root of temperature) and hence gets refracted. It undergoes total internal reflection at the lower turning point, the radial location of which is just a function of ν/L. Waves corresponding to low-degree modes penetrate deep into the star, whereas for the high-degree modes the waves are confined close to the surface. The frequencies of the modes are determined largely by conditions in the acoustic cavity where they have the character of the propagating waves. This dependence can then be inverted: the frequency of a mode can be used to constrain the properties of the Sun in the mode's acoustic cavity. Because different modes penetrate to different depths (and also have different sensitivities within their acoustic cavities, depending on the radial variation of their eigenfunctions), the observed frequencies of many modes can be used in helioseismology to build up a detailed image of the Sun's internal structure and dynamics.

2. Seismology of solar p modes of low degree

When the mode order n is large (as it is for the observed low-degree modes) or when l is large, the modes are reasonably described by an asymptotic description. Then the frequencies can be related to the underlying structure of the Sun by JWKB theory, so that

$$\int_{r_t}^{R_t} K\,\mathrm{d}r = \pi(n+\epsilon)\,. \tag{1}$$

Here K is the local vertical wavenumber, given by

$$K^2 = \frac{\omega^2 - \omega_c^2}{c^2} - L^2\left(1 - \frac{N^2}{\omega^2}\right), \tag{2}$$

(I neglect the perturbation to the gravitational potential, to a first approximation, and I neglect effects of rotation until §2.3); $\omega \equiv 2\pi\nu$ is the angular frequency of the mode; r_t and R_t are respectively the lower and upper turning points of the mode, where K^2 vanishes; the buoyancy frequency $N(r)$ and cut-off frequency $\omega_c(r)$ are characteristic frequencies of the local solar structure; and ϵ is a constant. For low-degree modes, equation (1) can be manipulated to give approximately

$$\nu_{nl} = \left(n + \frac{1}{2}L + \varepsilon\right)\nu_0 - \left(AL^2 - \delta\right)\nu_0^2/\nu_{nl} + \mathrm{O}(\nu_0^4/\nu_{nl}^3)\,, \tag{3}$$

(Gough 1986) where ε, ν_0, A and δ are constants that depend on the internal structure. In particular,

$$\nu_0 = \left(2\int_0^R c^{-1}\,\mathrm{d}r\right)^{-1}$$

is the inverse of the sound travel time from the surface to the centre and back again, and

$$A = \frac{1}{(2\pi)^2\nu_0}\left[\frac{c(R_t)}{R_t} - \int_{r_t}^{R_t}\frac{1}{r}\frac{\mathrm{d}c}{\mathrm{d}r}\,\mathrm{d}r\right]. \tag{4}$$

Equation (3) is essentially the formula of Tassoul (1980).

2.1. CORE STRUCTURE

Equation (3) is important for the interpretation of the low-degree spectrum of frequencies. It implies first that the frequency spacing between modes of the same degree that differ in n by 1 roughly $\nu_0 \approx 140\,\mu\mathrm{Hz}$: this is the so-called large separation, and is therefore an observational measure of the sound crossing time inside the Sun. Also, since the first term of (3) is dominant, it follows that there is a near degeneracy $\nu_{n\,l} \approx \nu_{n-1\,l+2}$: they differ by the so-called small separation

$$d_{nl} \equiv \nu_{n\,l} - \nu_{n-1\,l+2} \approx (4l+6)A \approx (1 + 2l/3)\times 10\,\mu\mathrm{Hz}\,.$$

The small separation depends sensitively on the conditions in the core. Mathematically this is because of the $1/r$ factor in the integrand of the second term in expression (4) for A. Physically, one can envisage that the principal difference in sensitivity to solar conditions between a (say) $l = 0$ mode and an $l = 2$ mode of almost the same frequency is that the former penetrates slightly more deeply into the core, and so the difference in frequency comes principally from that region between the two modes' respective lower turning points. By either line of reasoning one can also see why the small separation will vary with frequency (i.e. with n) and why d_{nl} for higher n (higher frequency and therefore deeper turning point) will be sensitive to conditions even deeper in the core than the corresponding small separation at smaller n. The small separation is often averaged over a range of n, in part to reduce the effect of measurement errors, but the variation with n may itself convey valuable information.

The small separation is able to discriminate between models with different core structure. Thus it has been used to rule out specific solar models which have been proposed in attempts to solve the solar neutrino problem: models containing WIMPs, which have values of the small separation which are smaller than observed, and models with partially mixed cores which have small separations which are bigger than observed (Faulkner et al. 1986, Däppen et al. 1986). Models with gravitational settling of helium produce small separations which are in better agreement with observation than models without settling (Christensen-Dalsgaard et al. 1993, Elsworth et al. 1995) which tends to confirm evidence from inversion of higher-degree data that gravitational settling does occur in the Sun and is needed in solar models.

The variation of d_{nl} with n can also be used to constrain the size of a convective core in the present Sun. A small convective core causes the small separations to peel away from the values for a model without a convective core, at sufficiently high n. The lack of any such discernible behaviour in the observations puts a limit on the radius of any such core in the present Sun of about $0.05R$, where R is the solar radius (Roxburgh 1998, these proceedings).

The difference in small separation between one model and another, or between the Sun and a model, can be understood also in terms of differential kernels. Starting from an variational principle obeyed by the eigenmodes of the star, one can relate the difference in frequencies between the Sun and a reference model to the differences in their structure. So for example, after suitable manipulations and assuming the equation of state to be known, one can find kernels $K^{nl}_{\rho,Y}(r)$, $K^{nl}_{Y,\rho}(r)$ relating relative differences in mode frequencies to relative differences in density ρ and differences in helium abundance Y:

$$\frac{\delta\omega_{nl}}{\omega_{nl}} = \int_0^R \left(K^{nl}_{\rho,Y}(r)\delta\ln\rho(r) + K^{nl}_{Y,\rho}(r)\delta Y(r)\right) \mathrm{d}r \,, \tag{5}$$

where I have neglected a possible contribution from the surface region. The difference in small separation d_{nl} between the Sun and the model can be expressed as a linear combination of the above frequency differences, and hence kernels can easily be written down relating differences in d_{nl} to differences in ρ and Y. Gough & Kosovichev (1993) have displayed such a kernel relating the difference in n-averaged d_{n0} value to the density differences between the Sun and a standard model. This shows that the difference in n-averaged d_{n0}, which can be deduced from observation, is predominantly a positive weighted average of the density difference, between about $0.1R$ and $0.3R$. (The modes in their average had frequencies in the range 2.5-3.8 mHz.) The advantage of formulating equations (5) is that one can now take linear combinations of these equations other than the combination corresponding to the small separation, and see whether one can get a linear combination of the kernels $K^{nl}_{\rho,Y}$ that is even better localized in the core. With the same dataset, and using an optimally localized averages technique (cf. Christensen-Dalsgaard et al. 1990), Gough & Kosovichev (1993) find that it is possible to produce a weighted average of the density that is indeed better localized to the core. In fact, they show that it is possible with a set of $l = 0, 1, 2$ frequencies in the range 2.5-3.8 mHz to resolve the desnity in $0.05 \lesssim r/R \lesssim 0.35$ with a resolution scale of better than $0.10R$. Thus it is possible to perform spatially localized inversions for the density stratification in the core, with low degree modes only.

2.2. NEAR-SURFACE PHYSICS

So far we have used the small but significant difference in penetration depth between different low-degree modes to probe the core. The other principal variation of a low-degree set of p modes that we can exploit is the range of frequencies. The local vertical wavenumber (given by eq. 2), and hence the spatial sensitivity of the mode, depends on the frequency. In fact the frequency dependence dominates the vertical wavenumber close to the surface, because the sound speed is relatively small there. Using eq. (2), eq. (1) can be rewritten to give Duvall's law

$$\int_{r_t}^R \left(1 - \frac{L^2c^2}{\omega^2 r^2}\right)^{1/2} c^{-1}\,\mathrm{d}r = \pi\frac{(n+\alpha(\omega))}{\omega} \,, \tag{6}$$

where the predominantly frequency-dependent near-surface behaviour has been absorbed into a phase function $\alpha(\omega)$. The importance of this representation of the data is that the left-hand side depends on the observables

ω and l only in the combination ω/L, so by seeking a function $\alpha(\omega)$ that gives the right-hand side this same dependence, both the integral function on the left of eq. (6) and the function α can be determined. With a sufficient range of degrees l, and hence of ω/L, the integral (which is of Abel type) can be inverted to infer the sound speed: this requires the intermediate-degree data that we consider in the next section. Here we concentrate on α, which contains information about the near-surface structure.

In fact, α is only determinable from the data to within an additive multiple of ω, since adding such a multiple would $\pi(n+\alpha)/\omega$ only by a constant. Hence it is sometimes preferred to work in terms of $\beta \equiv -\omega^2 \mathrm{d}(\alpha/\omega)/\mathrm{d}\omega$, which can be determined from the low-degree data as

$$\beta(\omega) = \frac{\omega - n(\mathrm{d}\omega/\mathrm{d}n) - L(\mathrm{d}\omega/\mathrm{d}L)}{\mathrm{d}\omega/\mathrm{d}n} .$$

The function $\beta(\omega)$ depends predominantly on the near-surface structure of the Sun. The uppermost layers contribute to $\beta(\omega)$ a component with a slowly-varying frequency dependence: this is a potential diagnostic of the superadiabatic layer and atmosphere, but there is sufficient uncertainty about the solar structure and the mode physics in the atmosphere that any calibration of these uncertainties from low-degree modes alone requires strong additional assumptions. Relatively deeper layers contribute components to β that have characteristic oscillatory dependence on frequency. In particular, the region of the second ionization of helium, located at about $0.98R$, contributes strongly to β. The signal depends on the location and strength of the "bump" in adiabatic exponent in Γ_1 due to the partial ionization of helium. This in turn depends on the helium abundance Y by mass, the equation of state (EOS) and the specific entropy.

Various investigations of the near-surface layers have been undertaken, using β or fairly related observational quantities (Vorontsov et al. 1991, 1992; Christensen-Dalsgaard & Pérez Hernández 1991; Pérez Hernández & Christensen-Dalsgaard 1994, 1998), though all but the last paper use the intermediate-degree data. Nonetheless they show the potential of β as a diagnostic of these layers. The principal results are that β is a fine discriminator of the equation of state, and that the helium abundance in the convective envelope is $Y = 0.25 \pm 0.01$. This value of Y is lower than the initial value of $Y \approx 0.28$ required in evolutionary solar models to match the present Sun. This is again indicative of gravitational settling of helium during the Sun's lifetime. As Pérez Hernández & Christensen-Dalsgaard (1998) show, frequency errors substantially below $0.5\,\mu$Hz are required if one is to impose significant constraints on the equation of state or helium abundance of solar-type stars using the low-degree modes alone: but such small frequencies are indeed attained by the current low-l solar measurements, and the COROT mission aims at determining stellar frequencies with errors as small as $0.1\,\mu$Hz (Catala et al. 1995).

Rather abrupt transitions in structure even deeper in the Sun contribute oscillatory signals to β, provided they are located sufficiently far from the centre that the low-l radial eigenfunctions are still essentially independent of degree. The base of the convection zone, where in solar models either the first or second derivative of sound speed has a discontinuity, contributes to β a signal with "period" given by the acoustic depth of the discontinuity beneath the surface (cf. e.g. Monteiro et al. 1998, these proceedings).

An alternative approach to determining the depth of convection zone and the helium abundance from low-degree frequencies, based on structural kernels as in eq. (5), is presented by Kosovichev (1993).

2.3. INTERNAL ROTATION FROM LOW-DEGREE MODES

I have so far neglected entirely the m-dependence of the frequencies. In a nonrotating star, the frequencies would be independent of m. The Sun's rotation raises this degeneracy and causes each frequency ω_{nlm} to differ from the frequency ω_{nl0} of an axisymmetric (i.e. $m = 0$) mode with the same n and l by an amount that is essentially m times a weighted average of the rotation rate Ω in the mode's acoustic cavity:

$$\omega_{nlm} - \omega_{nl0} = m\langle\Omega\rangle_{nl\,|m|}$$

(see e.g. Schou et al. 1994 for details of the kernel describing the weighted average). The so-called splitting between mode frequencies within a given (n, l) multiplet can therefore be used to make inferences about the rotation inside the Sun. Using the splittings of all the low-degree modes collectively, it is therefore possible to estimate a mean internal rotation rate. This would give a value of around $450 \pm 20\,$nHz for the mean internal rotational cyclic frequency (e.g. using the GOLF measurements of the rotational splitting, Lazrek et al. 1996). This rate is comparable to the surface rotation rate from Doppler measurements and surface tracers, which varies from about 460 nHz at the equator (a period of about 25 days) to 370 nHz at latitude 60° (about 30 days) and even slower rotation at higher latitudes. It thus rules out a very rapidly rotating solar interior. Likewise the relatively small variation in mean rotational splitting between $l = 1$ and $l = 2, 3$ can rule out models with a very fast or very slowly rotating core. But the measurement of the splitting at low degree is very difficult, and the uncertainties on the values of the splittings are still sufficiently large that making a spatially resolved inversion for the core rotation from low-degree data alone is not possible without unacceptably large error bars on the resulting inferences. This contrasts with the more optimistic scenario for structural inversion, the difference in part arising from the absence of any diagnostic potential from the $l = 0$ modes for the rotation problem. Nonetheless, getting averaged rotation rates in the Sun and other stars, and ruling out classes of models for the rotation, is immensely valuable, and in the case of the Sun this is supplemented by inversions with fine spatial resolution from the higher-degree data, to which we now turn.

3. Helioseismic results from intermediate and high degree modes

3.1. SOLAR INTERNAL STRUCTURE

The measurement of frequencies of degrees from $l = 0$ up to 200 or more enables us to resolve the solar internal structure and rotation in quite fine detail. In the radial direction, this is made possible largely by the variation of the lower turning point with ν/L. The principle is that shallowly penetrating modes (high l) enable us to determine conditions in the outer part of the star, and then modes penetrating successively more deeply differentially allow the structure and rotation in the deeper layers to be determined. There are basically two approaches. One is based on the asymptotic expression (6), or variants, and has been used to determine the sound speed $c(r)$ as a function of radius (e.g. Christensen-Dalsgaard et al. 1985). Various aspects of the physics can be constrained by knowing the sound speed. The sound speed is related to Γ_1, temperature T and mean molecular weight μ ($c^2 \propto \Gamma_1 T/\mu$) and so is sensitive to the element abundances (through Γ_1 and μ), the equation of state (thought Γ_1) and the physics of the heat transport (through $T(r)$) including opacities and convection. The other approach is via a differential inversion using numerically derived kernels, as in eq. (5) (e.g. Dziembowski et al. 1990). What has been learned from such inversions is that the convection zone is deeper than in some solar models current in the mid-seventies, with its base at around $r = 0.713R$ (Christensen-Dalsgaard et al. 1991). The sound-speed inversions of the mid-eighties pointed to a shortfall of the theoretical opacities just beneath the convection zone, which indeed has subsequently been rectified in modern opacity calculations. The sound-speed inversions beneath the convection zone also indicate that models with gravitational settling are in closer agreement with observation than models without settling. Together with the evidence from the envelope helium abundance and the small frequency separations, it therefore seems conclusive that settling does take place in a star like the Sun. There are still discrepancies between the inferred sound speed and that in the current models: though the differences are small (a fraction of one per cent) they are statistically very significant (Fig. 1). The most robust features are that the sound speed in the Sun is about 0.2% higher than that in the model in a shallow region just beneath

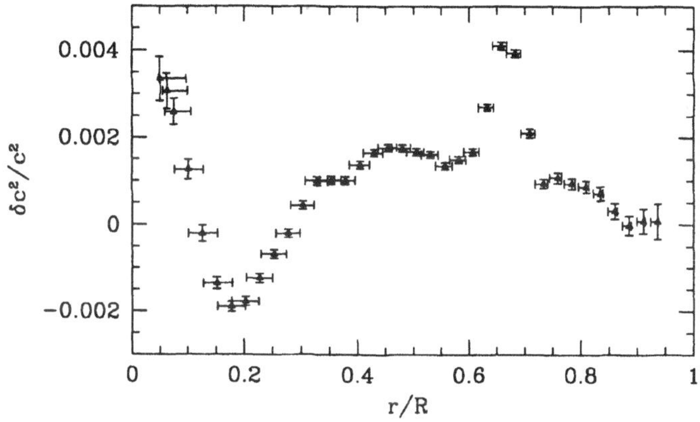

Figure 1. Inferred relative difference in sound speed squared between the Sun and a standard solar model with helium diffusion (Model S of Christensen-Dalsgaard et al. 1996), in the sense (Sun) - (model). The data are the 'Best Set' combininination (Basu et al. 1997) of data from BiSON and LOWL. The 1-σ standard errors and the radial resolution are indicated by the vertical and horizontal bars, respectively. (Adapted from Basu et al. 1997.)

the convection zone, and about 0.1% lower around $r = 0.2R$. The former could indicate that the settling of helium out of the convection has been slightly overestimated in the model, increasing μ locally in the model and hence reducing the model sound speed there: possibly the effect of settling is counterbalanced by some macroscopic mixing. The lower sound speed in the outer core of the Sun could also be an indication of mild mixing of helium-enriched material from the inner core. For a good review of the state of the subject, see Gough et al. (1996a, 1996b).

3.2. SOLAR INTERNAL ROTATION

Considerable progress has been made in the past ten years to reveal the Sun's internal rotation rate Ω as a function of radius and latitude. In the same way that the radial extent of the mode's cavity depends on ν/L, the latitudinal extent of the cavity depends on m/L. (Asymptotically, the modes are sensitive to a region betwee latitudes $\pm\cos^{-1}(m/L)$.) Thus the region averaged by $\langle\Omega\rangle_{nl|m|}$ varies in radial and latitudinal extent from one mode to another, and this is exploited to resolve the internal rotation in radius and latitude. Such inversions (e.g. Brown et al. 1989, Thompson et al. 1996), one of which is illustrated in Fig. 2, have revealed that the surface differential rotation as a function of latitude largely persists through the convection zone: the rotation rate there is not constant on cylinders aligned with the rotation axis, as was predicted by simulations in the early 1980s.

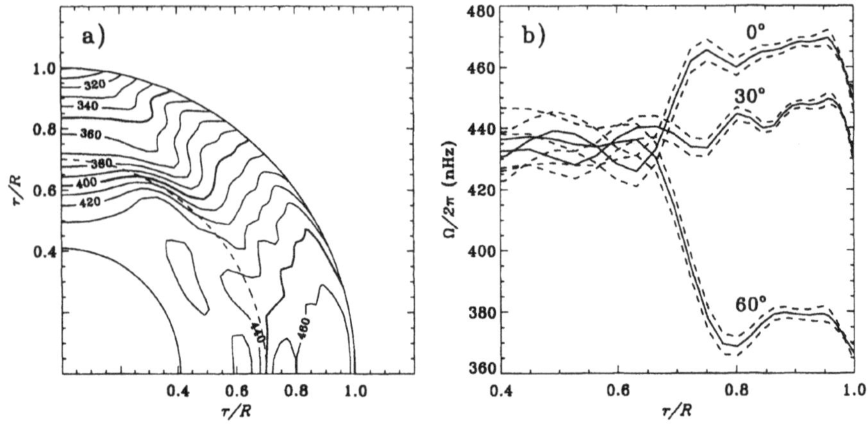

Figure 2. (a) Solar internal rotation rate in the region $r > 0.4R$ as inferred from GONG data by Thompson et al. (1996), using a 2D least-squares technique. (b) Cuts through the inferred rotation rate shown in panel (a), at latitudes 0°, 30° and 60°. This figure corrects an error in the near-surface representation of the cuts in Fig. 4b of Thompson et al. (1996), which was caused by using insufficient resolution in the graphical representation.

There is a transition to latitudinally independent rotation near the base of the convection zone, in a layer which has become known as the tachocline (Spiegel & Zahn 1992). This is quite sharp, indeed at or below the current resolution of the inversions (see Kosovichev 1996). Deeper still, present results would seem to be consistent with rigid-body rotation, though little can yet be said with any precision about the rotation of the core, because of the difficulty of measuring the splittings of the low-degree modes. A reasonably up-to-date review of the state of our knowledge of the internal rotation from helioseismology is provided by Thompson et al. (1996).

3.3. HIGH-RESOLUTION OBSERVATIONS

At high degree, say around $l = 200$ and above, it becomes impossible observationally to isolate modes of individual l values (e.g. Howe & Thompson 1998). Possibly by this stage the modes are no longer global in character. It may still be the case that analysing the wave field in terms of spherical harmonics is useful, but recently other, local analyses have been devised which promise to reveal new detail the near-surface structure and dynamics probed by the high wavenumber modes. Such developments have been made of practical relevance by the beautiful high-resolution observations from the MDI instrument on board the SOHO satellite (Scherrer et al. 1991). Particularly promising is time-distance helioseismology, which probes the sound travel time along rays by correlating the wave field at a point with the wave field in annuli of different radii centred upon it. Not only does this reveal the propagation speed (and hence the temperature stratification) as a function of depth, it can also reveal anisotropies in propagation speed due to e.g. horizontal and vertical flows. In this way it is possible to build up three-dimensional maps of subsurface inhomogeneities and flows. Initial results of applying this new technique to MDI data have been presented by Duvall et al. (1997).

4. Conclusion

Helioseismology has revealed the radial structure (in particular the sound speed) and rotation throughout much of the solar interior with fine spatial resolution. The helium abundance Y in the convective envelope has been measured to an accuracy of about ± 0.01. Various aspects of the microphysics in situ within a stellar interior, in particular the opacity and equation of state, have been constrained, providing fresh impetus to the theoretical calculation of these aspects of the physics. The rotation was rather poorly constrained by theory prior to helioseismology, and the theoretical expectations from numerical simulations of the interaction rotation and convection have been shown to have been wrong, presumably because the simulations were far from solar conditions in terms of Rayleigh number and hence far from simulating the highly turbulent solar envelope. Only now may such simulations on massively parallel computers be beginning to come close to reproducing the helioseismic observations (Clune et al. 1997). Also on a macroscopic scale, the helioseismic inversions hint at mixing both beneath the convection zone and in the core, though the extent of mixing in the core is constrained by the small frequency separation.

For more distant Sun-like stars, we cannot hope to resolve their interiors in such detail, since we only expect to detect low-degree oscillations. Nonetheless I have sought to show from our experience with helioseismology that the low-degree modes have considerable diagnostic potential, for core stratification, helium abundance and possibly other near-surface diagnostics, mean interior rotation rate, and locating convection-zone boundaries.

Progress in helioseismology is assured by the flow of exquisite data from MDI, VIRGO and GOLF on board SOHO, and from the various ground-based single-site and networked observations such as BiSON, IRIS, LOWL, GONG and TON. Even lower uncertainties on the frequencies, and the extension of the measured mode properties to lower frequencies and to higher degrees, will allow the existing tools for seismic inference, and new tools such as the local area analyses, to be pushed further both in terms of resolution and spatial coverage. Experience shows that what is revealed will present new challenges to our theoretical understanding of the physics of stellar interiors.

References

Basu S., Chaplin W.J., Christensen-Dalsgaard J., et al. (1997), *MNRAS* **291**, 243

Brown T.M., Christensen-Dalsgaard J., Dziembowski W.A., Goode P., Gough D.O., Morrow C.A. (1989), *ApJ*, **343**, 526

Catala C., Auvergne M., Baglin A., et al. (1995), in *Proc. Fourth SOHO Workshop: Helioseismology*, eds Hoeksema J.T., Domingo V., Fleck B., Battrick B., ESA SP-376, p. 549

Christensen-Dalsgaard J. (1998), these proceedings

Christensen-Dalsgaard J., Pérez Hernández F. (1991), in *Challenges to theories of the structure of moderate-mass stars*, eds Gough D.O., Toomre J., Lect. Notes Phys., **388**, 43

Christensen-Dalsgaard J., Gough D.O., Thompson M.J. (1991), *ApJ*, **378**, 413

Christensen-Dalsgaard J., Proffitt C.R., Thompson M.J. (1993), *ApJ*, **403**, L75

Christensen-Dalsgaard J., Schou J., Thompson M.J. (1990), *MNRAS*, **242**, 353

Christensen-Dalsgaard J., Duvall T.L., Gough D.O., Harvey J.W., Rhodes E.J. (1985), *Nature*, **315**, 378

Christensen-Dalsgaard J., Däppen W., Ajukov S.V., et al. (1996), *Science*, **272**, 1286

Clune T.L., Elliott J.R., Glatzmaier G.A., Miesch M., Toomre J., (1998), *J. Comp. Phys.*, submitted

Duvall T.L., Kosovichev A.G., Scherrer P.H., et al. (1997), *Solar Phys. 170*, **170**, 63

Dziembowski W.A., Pamyatnykh A.A., Sienkiewicz R. (1990), *MNRAS*, **244**, 542

Däppen W., Gilliland R.L., Christensen-Dalsgaard, J. (1986), *Nature*, **321**, 229

Elsworth Y., Howe R., Isaak G.R., McLeod C.P., Miller B.A., Wheeler S.J., New R. (1995), in *Proc. GONG'94: Helio- and Astero-seismology from Earth and Space*, eds Ulrich R.K., Rhodes E.J., Däppen W., ASP Conf. Ser., **76**, 51

Faulkner J., Gough D.O., Vahia M.N. (1986), *Nature*, **321**, 226

Gough D.O. (1986), in *Hydrodynamic and Magnetohydrodynamic Problems in the Sun and Stars*, ed Osaki Y., Dept. of Astronomy, Univ. Tokyo, p. 117

Gough D.O., Kosovichev A.G. (1993), in *Proc. GONG 1992: Seismic Investigation of the Sun and Stars*, ed Brown T.M., ASP Conf. Ser., **42**, 351

Gough D.O., Leibacher J.W., Scherrer P.H., Toomre J., (1996a), *Science*, **272**, 1281

Gough D.O., Kosovichev A.G., Toomre J., et al. (1996b), *Science*, **272**, 1296

Howe R., Thompson M.J. (1998), *A&A*, submitted

Kosovichev A.G. (1993), *MNRAS*, **265**, 1053

Kosovichev A.G. (1996), *ApJ Lett.*, **469**, L61

Lazrek M., Pantel A., Fossat E., et al. (1996), *Solar Phys.*, **166**, 1

Monteiro M.J.P.F.G., Christensen-Dalsgaard J., Thompson M.J. (1998), these proceedings

Pérez Hernández F., Christensen-Dalsgaard J. (1994), *MNRAS*, **269**, 475

Pérez Hernández F., Christensen-Dalsgaard J. (1998), *MNRAS*, in press

Roxburgh, I.W. (1998), these proceedings **[poster paper]**

Scherrer P.H., Hoeksema J.T., Bush R.I. (1991), *Adv. Space Res.*, **11**, 113

Schou J., Christensen-Dalsgaard J., Thompson M.J. (1994), *ApJ*, **433**, 389

Spiegel E.A., Zahn J.-P. (1992), *A&A*, **265**, 106

Tassoul M. (1980), *ApJ Suppl.*, **43**, 469

Thompson M.J., Toomre J., Anderson E.R., et al. (1996), *Science*, **272**, 1300

Vorontsov S.V., Baturin V.A., Pamyatnykh A.A. (1991), *Nature*, **349**, 49

Vorontsov S.V., Baturin V.A., Pamyatnykh A.A. (1992), *MNRAS*, **257**, 32

THE EFFECT OF SOLAR OSCILLATIONS ON SPACE GRAVITATIONAL WAVE EXPERIMENTS

G GIAMPIERI, A POLNAREV, I ROXBURGH, S VORONTSOV
Astronomy Unit, Queen Mary and Westfield College, University of London, London E1 4NS, UK.

The proposed space gravitational wave detectors LISA and OMEGA, typically consist of six spacecraft located in pairs at the corners of an equilateral triangle; the triangular configuration is kept stable by careful choice of the spacecraft orbits. The two spacecraft at each vertex are phase locked through the exchange of a reference signal. Each spacecraft sends a laser beam to one of the distant spacecraft where the signal is received and coherently transponded back; the interferometer readout is eventually obtained by interfering the incoming signal with the outgoing one. This Michelson-type interferometer will be sensitive to the passage of gravitational waves in the frequency range below 1 Hz, a range currently inaccessible on the ground due to seismic noise.

To analyse the effect of solar oscillations on a such spaceborne interferometer located at a distance of ~ 1 AU from the Sun we determine the external time dependent gravitational potential in the form

$$\psi = \sum \frac{GM_\odot}{R_\odot} J_{\ell mn} \left(\frac{R_\odot}{r}\right)^{\ell+1} Y_{\ell m} e^{i\omega_{\ell mn} t} \tag{1}$$

where $J_{\ell mn}$ is the multipole moment corresponding to an oscillation with cyclical frequence $\omega_{\ell mn}$. We determined the relative magnitude of the $J_{\ell n}$ for unit displacement of the solar surface for solar p and g modes by solving the oscillation equations for a standard solar model. Since the field of a multipole J_ℓ decreases as $r^{-(\ell+1)}$ only modes with $\ell = 2$ are of significance. (The external potential for modes with $\ell = 0$ is time independent and those with $\ell = 1$ are identically zero.) For given surface amplitude the most significant modes are the low order g modes although the low order p modes (n=1,4) are also significant and might well have higher amplitude.

At a distance of 1 AU the detector would be affected by the changing tidal Newtonian field and at higher frequencies by gravitational waves, generated by solar oscillations. The two effects correspond respectively to the

Astrophysics and Space Science is the original source of publication of his article. It is recommended that this article is cited as: *Astrophysics and Space Science* **261**: 35–36, 1999.

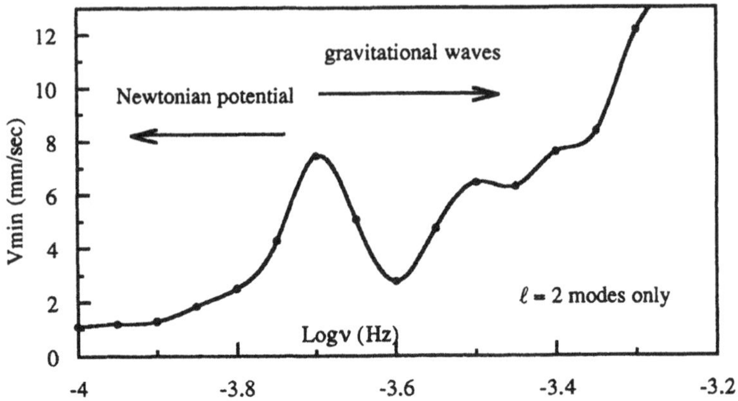

Figure 1. Upper limits on the surface velocity amplitude of solar oscillations ($\ell = 2$ modes) detectable by the propsed LISA gravitational wave detector.

detector being located in the near zone (for frequencies $\nu < \nu_w \sim 3 \times 10^{-4}$ Hz), where the main contribution to phase shift is due to Newtonian tidal perturbations, or in the wave zone ($\nu > \nu_w$), where the effect is dominated by gravitational waves. The phase shift as measured in the interferometer is related to solar oscillations by

$$\frac{\delta\phi}{\phi}(\nu) \equiv \psi(\nu) = \psi_0(\nu)\left(\frac{\nu}{\nu_w}\right)^{\alpha} \tag{2}$$

where $\alpha = -2$ in near zone ($\nu < \nu_w$) and $\alpha = +2$ in the wave zone ($\nu > \nu_w$), and

$$\psi_0(\nu) \simeq 2 \times 10^{-12} J_2(\nu) \frac{V(\nu)}{2\pi\nu R_\odot} \tag{3}$$

$J_2(\nu)$ is the Fourier component of the solar quadrupole moment, normalized to unit relative displacement of the surface, and $\delta R(f)/R_\odot = V(\nu)/2\pi\nu R_\odot$ is the Fourier component of the solar surface displacement expressed in terms of surface velocity $V(\nu)$. Using the expected sensitivity of LISA to stochastic signals we calculated the reponse of the detector to the $\ell = 2$ oscillations expressed in terms of the velocity amplitude of the modes on the solar surface. The results are shown in Figures 1. Modes with surface velocities of mm/sec to cm/sec should be detectable in such experiments.

References

LISA, 1994. Laser Interferometric Gravitational Wave Detector in Space, ESA Assessment study, ESA Sci (94) 9

OMEGA: Orbiting Medium Explorer for Gravitational Astrophysics, Midex Proposal (NASA/JPL, Pasadena).

HELIOSEISMIC CONSTRAINTS ON THE SOLAR CORE

IAN W ROXBURGH
Astronomy Unit, Queen Mary and Westfield College, University of London, London E1 4NS, UK.

We show that the observed frequencies of low ℓ oscillation modes are consistent with the sun having a small convective core occupying the inner 5% of the radius.

The frequencies of solar oscillations measured by both ground based networks and by experiments on SOHO are, to a first approximation, consistent with the theoretical predictions of a "standard solar model" (eg model S Christensen-Dalsgaard, 1996). The actual value of the frequencies are sensitive to the structure of the surface layers which are poorly understood, but the run of sound speed with radius $c(r)$ obtained by inversion of the data is in good agreement with the models save in the layers just below the convective zone and in the central core where the resolution is poor.

Only modes of low degree $\ell = 0, 1, 2, 3$ penetrate deep into the solar core before being refracted back to the surface. The question addressed here is how well constrained is the structure of the very central regions by observations of the frequencies of these modes? Of particular interest is whether or not the sun may have a small convective core that could perhaps cause mixing of chemically processed matter through overshooting and or waves. We approach this problem by accepting that the hydrostatic structure of the bulk of the sun (pressure and density) are as determined by a "standard" solar model, or alternatively as deduced from the measured oscillation frequencies, and modify the internal strucure of the very central core leaving the rest of the solar structure unchanged. Such a change will only affect modes of low degree.

We constructed a set of models with small convective cores of radii $r_c = x_c R_\odot$ which matched smoothly onto the rest of the solar model at some greater radii $r_f = x_f R_\odot$. In the core the Brunt Vaisaila frequency $N(r) = 0$. In the transition region it was taken as:

$$N^2(r) = N_0^2(r)\left(1 + a\,(x_f^2 + c\,x^2)\,(x_f^2 - x^2)^m\right), \quad x_c \le r/R_\odot \le x_f \qquad (1)$$

Astrophysics and Space Science is the original source of publication of his article. It is recommended that this article is cited as: *Astrophysics and Space Science* **261**: 37–38, 1999.

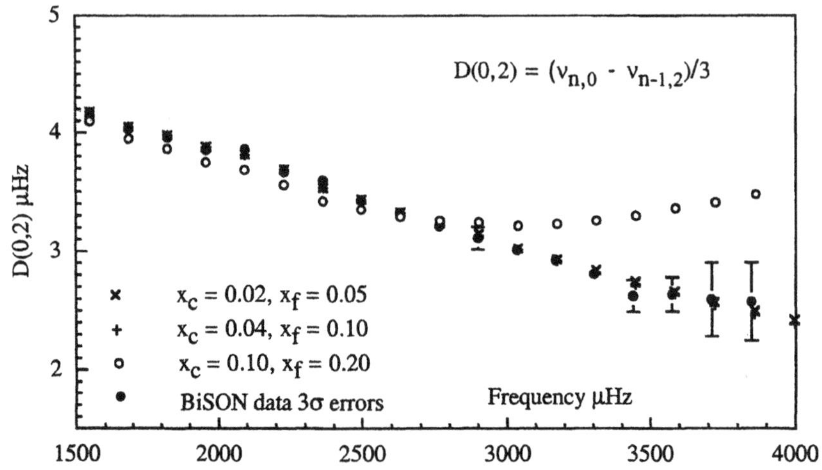

Figure 1. Small separations $D(0,2) = (\nu_{2,n} - \nu_{0,n-1})/3$ for solar models with small convective cores and the values observed by the BiSON network.

where $N_0(r)$ is the value in the original solar model. The parameters a, c, central density ρ_c and central pressure P_c, are determined such that the M_r, P, ρ are continuous at x_f and that the boundary of the core is at x_c. The continuity of derivatives at x_f is ensured by taking $m \geq 2$.

We then calculated the oscillation frequencies for these models and compared them with observational values obtained by the BiSON network (Elseworth et al 1996). Figure 1 compares the values for the small separations $D(0,2) = (\nu_{2,n} - \nu_{0,n-1})/3$ which are the most appropriate comparators since they subtract out the unquantified surface effects and give a measure of the structure of the central regions (as do D(1,3), D(0,1), D(1,0) not shown here). The values for a very small core ($x_c = 0.02, x_f = 0.05$) are indistinguishable from those for the standard solar model, the results for a large core ($x_c = 0.1$) are incompatible with the observations and show the periodic behaviour characteristic of convective cores (Roxburgh and Vorontsov 1994).

With the present precision on these measurements one cannot rule out the possibility that the sun has a small convective core.

References

Christensen-Dalsgaard, J., et al, 1996, *Science* ,**272**, 1286

Elseworth Y. et al,: 1996, *Private communication*

Roxburgh, I W, Vorontsov, S V, 1994, *Mon. Not. R. astr. Soc.*, **268**, 880

LINE PROFILES OF SOLAR P-MODES

IAN W ROXBURGH AND SERGEI V VORONTSOV
Astronomy Unit, Queen Mary and Westfield College,
University of London, London E1 4NS, UK.

We address the problem of the opposite asymmetry of low-frequency p-mode line profiles observed in intensity and velocity measurements, using a simple model to illustrate that this feature can be explained by including a contribution from the stochastic excitation velocity field to the non-resonant background in the doppler measurements.

The asymmetry of line profiles in the power spectrum of solar p modes was discovered in intensity observations by Duvall et al (1993); even larger asymmetry has been detected in doppler velocity measurements. The most striking feature is that the line asymmetry seen in intensity and velocity at the same frequencies (below 3 mHz) appears to be in opposite directions.

The dynamics of wave excitation is expected to be very non-linear and there is a strong evidence that it occurs in a shallow super-adiabatic region at the top of the solar convection zone, close to the observation level (within one pressure scale height from the photospheric level, see e.g. Kumar 1994 and references therein). In such a situation it is natural to expect that individual excitation events contribute to velocity measurements not only through the generation of acoustic waves, but also through the kinematic velocity producing a coherent non-resonant background in the observational power spectra, which is different in intensity and velocity measurements.

Our analysis is based on the simple model of acoustic power spectra described in (Roxburgh and Vorontsov 1995). We extend this theoretical description slightly, allowing for an instantaneous point-like source—an excitation which is described by δ−function in both space and time. We make a simple order of magnitude esimate of the source function for monopole and dipole excitations, and also estimate the direct contribution to the doppler velocity signal from the kinetic velocity of the convective eddy (Roxburgh and Vorontsov 1997). We define the total complex amplitude $A(\omega)$ of the doppler velocity measurements as

$$A(\omega) = k_{mon} A_{mon}(\omega) + k_{dip} A_{dip}(\omega) + k_{kin} A_{kin}(\omega), \qquad (1)$$

Astrophysics and Space Science is the original source of publication of his article. It is recommended that this article is cited as: *Astrophysics and Space Science* **261**: 39–40, 1999.

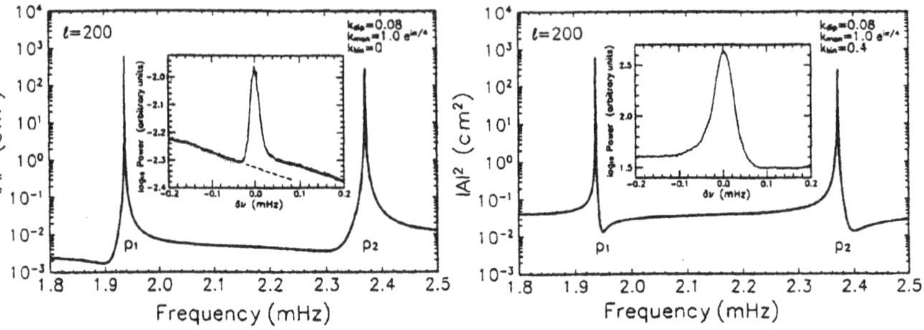

Figure 1. a) Artificial power spectrum for a composite excitation source. b) The same as a) but with an additional contribution from the kinematic effect. The insert in a) shows the line profiles observed in intensity and in b) the line profiles in velocity (Duvall et al, 1993).

with three weighting coefficients k_{mon}, k_{dip} and k_{kin}. The first two coefficients are the efficiency factors of monopole and dipole excitation; k_{kin} is a factor of "visibility" of the kinematic velocity field. When the kinematic effect is neglected, the line profiles seen in intensity and in doppler measurements are similar. We expect the intensity data to be less sensitive to the kinematic velocity field, and simulate the results of the intensity measurements simply by using the Eq.(1) with k_{kin} set to zero. We also expect the efficiency factors to have nearly the same order of magnitude, if our hypothesis is correct.

Figure 1a shows the power spectrum produced by a composite excitation source. The line profiles depend on the relative magnitude and phase of the even and odd components of the excitation; the observational asymmetry of intensity measurements (smaller amplitude at the lower-frequency wing of the line) can be easily reproduced. The results obtained with the same composite excitation source but with the kinematic effect taken into account are shown in Figure 1b. The line asymmetry is changed by the coherent contribution to the acoustic background of the kinematic-velocity signal.

References

Duvall, T. L., Jr., Jefferies, S. M., Harvey, J. W., Osaki, Y. and Pomerantz, M. A. 1993, *Astrophys. J.* **410**, 829

Kumar, P. 1994, *Astrophys. J.* **428**, 827

Roxburgh, I. W. and Vorontsov, S. V. 1995, *Mon. Not. R. astr. Soc.* **272**, 850

Roxburgh, I. W. and Vorontsov, S. V. 1997, *Mon. Not. R. astr. Soc.*, **293** L23

DETECTION OF THE LOWER BOUNDARY OF STELLAR CONVECTIVE ENVELOPES FROM SEISMIC DATA

MÁRIO J.P.F.G. MONTEIRO,
Departamento de Matemática Aplicada, Faculdade de Ciências, and Centro de Astrofísica, Universidade do Porto, Portugal
JØRGEN CHRISTENSEN-DALSGAARD,
Teoretisk Astrofysik Center, and
Institut for Fysik og Astronomi, Aarhus Universitet, Denmark
AND
MICHAEL J. THOMPSON
Astronomy Unit, Queen Mary & Westfield College, University of London, England

A sharp feature in a star's internal structure gives rise to a characteristic periodic signal in the frequencies of its global p-modes. Such a sharp feature occurs at the borders of a convective region in the star's interior. Several groups have used this characteristic signal to study the base of the solar convective envelope ([1], [2], and references therein).

In this work, we apply the analysis of [1] and [2] to the low degree modes that one might detect on distant stars, to show what kind of diagnostic the oscillations might provide for convective envelopes of low-mass stars. We regard the effect of the sharp feature on the frequencies as a perturbation $\delta\omega{=}A(\omega)\cos[2(\omega\tau_{\rm d}{+}\phi)]$, added to the frequencies of a fictitious star with the sharp feature smoothed out. The amplitude $A(\omega)$ depends on the properties of the star in the vicinity of the base of the convective envelope, in particular on whether or not overshooting is present: see [3] for a detailed analysis. The quantity $\tau_{\rm d}$ is the acoustical depth (i.e., the sound travel time) of the edge of the convection zone measured from the surface of the star. The effective surface phase ϕ of the eigenfunctions varies with frequency; thus when we measure the 'period' of $\delta\omega$ we measure not $\tau_{\rm d}$ alone but $\bar{\tau}_{\rm d} \equiv \tau_{\rm d} + a_\phi$, where $a_\phi = {\rm d}\phi/{\rm d}\omega$.

We consider models of ZAMS stars of mass $0.85 M_\odot$ to $1.3 M_\odot$. The models were calculated as described by [4]. From the p-mode frequencies of these models we have determined the amplitude and $\bar{\tau}_{\rm d}$ using the least-

Astrophysics and Space Science is the original source of publication of his article. It is recommended that this article is cited as: *Astrophysics and Space Science* **261**: 41–42, 1999.

squares method described by [1]. The modes used have degrees l between 0 and 4. These modes might in principle be observed for distant stars.

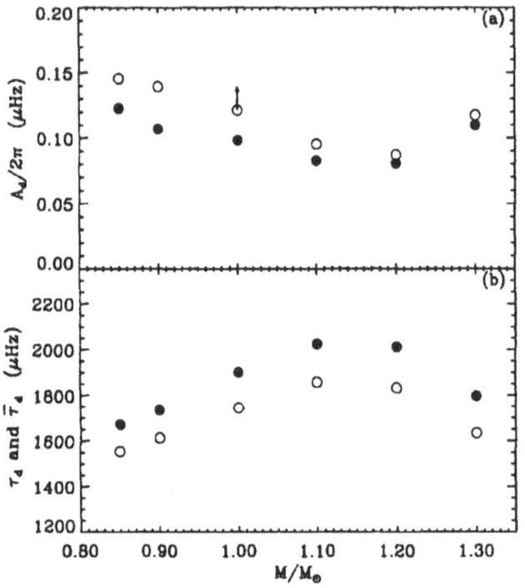

Fig. 1: (a) Expected dependence on stellar mass of the amplitude of the periodic signal in the frequencies arising from the edge of a convective envelope; open circles correspond to predicted values (see [3] for details), while filled circles are the values found by fitting the signal in the frequencies. Shown as an arrow is the shift of the values if we consider an overshoot layer of depth about 7% of the local pressure scale height (this is the upper limit found for the Sun - see [2]). (b) Extent, in terms of acoustic depth, of the convective envelope versus stellar mass. Open circles are the values τ_d calculated from the models, while filled circles are the values $\bar{\tau}_d$ extracted from the signal in the frequencies of oscillation.

To facilitate the comparison between different stars, we consider the amplitude, $A_d = A(\tilde{\omega})$, evaluated at a fiducial frequency $\tilde{\omega}$ (as in [3]). Comparison of the predicted and fitted amplitudes shows a reasonable agreement for all masses (Fig. 1a). The fit of the signal in the frequencies also provides an estimate of the acoustic depth of the convective envelope. The values found are shown in Fig. 1b, together with the actual acoustic depths in the models. As discussed above, we expect these two to differ by an amount a_ϕ associated with the reflection of the modes at the surface. From the models, a_ϕ is about 100 sec; and it appears ([1]) that in practice it is fairly insensitive to the physics used to model the upper layers of a star. Thus it may be accounted for, allowing an actual measurement of the depth of the outer convective region in these solar-type stars.

Figure 1a also illustrates that if the models incorporate convective overshooting (as simulated in [2]), the amplitudes of the signal become even larger and easier to detect. Moreover, given that the mass of the star can also be constrained from oscillation data, the greater than expected amplitude could then be used to deduce the extent of overshooting beyond the convectively unstable regions in these stars.

References

1. Monteiro M.J., Christensen-Dalsgaard J., Thompson M.J. (1994), *A&A* **283**, 247
2. Christensen-Dalsgaard J., Monteiro M.J., Thompson M.J. (1995), *MNRAS* **276**, 283
3. Monteiro M.J., Christensen-Dalsgaard J., Thompson M.J. (1998), in *Proc. IAU Symp 185*, (eds) F. Deubner, J. Christensen-Dalsgaard, & D. Kurtz, in press
4. Christensen-Dalsgaard J. (1982), *MNRAS* **199**, 735

CONVECTIVE OVERSHOOTING AND MIXING IN STELLAR EVOLUTION

IAN W. ROXBURGH
Astronomy Unit, Queen Mary and Westfield College, University of London, London E1 4NS, UK.

1. Introduction

Stars are a major constituent of the Universe. They provide the basic measurements of masses, distances and ages. Stars are the sites of nuclear reactions and the origin of the elements; they determine the chemical and hence luminosity evolution of galaxies. The Sun is the driving force in the evolution of the Solar System and the Earth's environment. Understanding stellar structure and evolution is at the core of our quest to understand the structure, history and evolution of the Universe.

The internal structure of stars is governed by the balance between pressure and self gravity, the distribution of the chemical elements, the transport of energy by radiation and convection, and the liberation of energy by nuclear reactions. The evolution of stars is primarily determined by the changing composition due to the nuclear burning of elements in the central parts of the star, and the redistribution of the products of these reactions by mixing processes. The dominant mixing process is convection: it governs the extent of the mixed cores in moderate and large mass main sequence stars and their subsequent evolution; it mixes (or dredges up) nuclear processed material into the envelopes of giants, affecting the composition of material ejected into the interstellar medium and thereby affecting the chemical and luminosity evolution of galaxies.

In simple models of stars convection is assumed to mix material on a short time scale only in the regions that are convectively unstable. But this is unreasonable, motion driven in the stable regions will penetrate into the surrounding stable layers affecting the internal structure of stars and enlarging the chemically mixed region, which in turn affects the subsequent evolution. Understanding convective penetration is central to understanding stellar evolution.

Astrophysics and Space Science is the original source of publication of his article. It is recommended that this article is cited as: *Astrophysics and Space Science* **261**: 43–50, 1999.

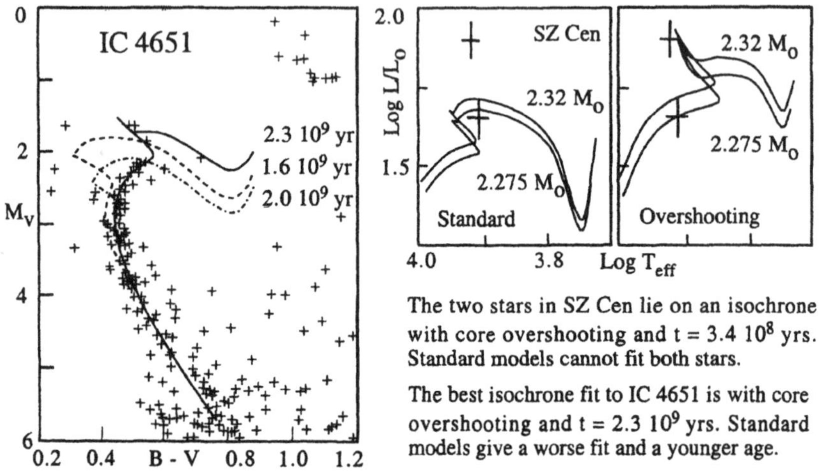

Figure 1. Isochrone fit to IC 4651 and SZ Cen with and without core overshooting (Adapted from Dowler and VandenBerg 1995).

2. Isochrone fitting to stellar clusters and binary systems

Since the evolution of a star is affected by the extent of its convective core so too will be the shape of the isochrone (or locus at constant age) of a group of stars in the H-R diagram. Overshooting can therefore be investigated by seeking the best fit to the observed H-R diagram of individual clusters, the width of the main sequence band, and to binary stars

In studies by Maeder and Mermillod (1992) and Meynet et al. (1993), the convective core was assumed to be extended by αH_p ($H_p = P/|dP/dr|$). These authors found that the best fit to the isochrones of a large number of clusters indicated that $\alpha \approx 0.2 - 0.3H$. In a recent study using HIPPARCOS data, Perryman et al (1998) found a similar degree of overshooting gave the best fit the observed H-R diagram of the Hyades. Dowler and VandenBerg (1995) found that cluster isochrones could be well fitted if core overshooting was as given by the integral constraint I derived some years ago (Roxburgh 1978, 89), and that the same model of core overshooting improved the fit of evolutionary models with the observations of binary stars with well determined properties (see Figure 1).

The inclusion of overshooting in general leads to a larger estimated age for the cluster. The age is essentially given by the turn-off point of the cluster, that is by stars that have just exhausted hydrogen in their cores. Overshooting increases the mass of the core and hence the available supply of hydrogen, thereby increasing the main sequence life time.

3. Non-local models of convective overshooting

Convection penetrating into the surrounding stable layers carries a negative energy flux, requiring an enhanced radiative flux, hence a steeper (and therefore less stable) temperature gradient, which in turn enhances the penetration. This feed back has been incorprated in models by several authors [Shaviv and Salpeter (1973), Maeder (1975), Roxburgh (1978, 1985), Schmitt et al. (1984), Zahn (1991)]. In the classical mixing length model the velocity, v, and temperature excess, δT, of an eddy are determined by

$$v\frac{dv}{dz} = g\frac{\delta T}{T}, \quad \frac{d}{dz}\left(\frac{\delta T}{T}\right) = \Delta\nabla\frac{1}{H_p}, \quad \Delta\nabla = \nabla - \nabla_{ad}, \quad \nabla = \frac{dLogT}{dLogP} \tag{1}$$

where $F_{rad} = \chi(\nabla_{ad} + \Delta\nabla)$ with $\chi = 4acT^4/(3\kappa\rho H_p)$. An eddy starting some distance inside the unstable layer arrives at the boundary $z = 0$, where $\Delta\nabla = 0$, with velocity v_0. Following Zahn (1991), we expand $F_r = F(1 + z/H_\chi + ...)$, where $H_\chi = |\chi'/\chi|$, use the standard expression for the convective flux $F_c = c_p\rho T v\delta T/T = F - F_r$, multiply equation (1) for v by v and integrate, we obtain

$$v^3 = v_0^3 - \frac{3}{5}\frac{F}{\rho}\frac{z^2}{H_\chi H_p.} = 0 \ \text{ when } \ z = \left(\frac{5\rho v_0^3}{3F}\right)^{1/2}(H_\chi H_p)^{1/2} \tag{2}$$

giving a penetration distance of the order of $0.5H_p$. This result, whilst crude, shows that inclusion of the feedback on the thermal structure can result in a substantial penetration distance.

However since the convective mixing time is small compared with the thermal relaxation time, the equilibrium structure of a convecting region is likely to be globally, rather than locally, determined.

4. The Integral Constraint on convective overshooting

One attempt at a global model is the integral constraint I proposed some years ago (Roxburgh 1978, 89, 92). On using the equation of continuity $\partial\rho/\partial t + div(\rho\mathbf{v}) = 0$, the thermal energy equation ($dS = dQ/T$) can be expressed as:

$$\frac{\partial}{\partial t}(\rho S) + div(\rho S\mathbf{v}) = \frac{\Phi + \epsilon\rho - div\mathbf{F}_r}{T} \tag{3}$$

where S is the entropy per unit mass, ρ the density, $\mathbf{v}$ the velocity, $\mathbf{F_r}$ the radiative flux, ϵ the energy generation per unit mass and Φ the viscous dissipation per unit volume. If the motions are statistically stationary, then on integrating this equation over a volume V containing the convective core and overshoot region, so that outside V, $\mathbf{v} = 0$ and $\mathbf{F}_r = \mathbf{F}$ gives

$$\int_V (\mathbf{F}_r - \mathbf{F}).\nabla\left(\frac{1}{T}\right) dV = \int_V \frac{\Phi}{T} dV > 0 \tag{4}$$

where **F** is the total flux and the inequality follows from the fact that the viscous dissipation Φ is positive definite. If viscous dissipation is neglected (as is done in the simple non-local models described above), then since $F_r < F$ inside the unstable layer there must be a penetration region where $F_r > F$ whose extent can be estimated from equation (4). Simple stellar models using this condition gave an enhancement of the core mass of the order of 50% (Roxburgh 1978) and it was this condition (with Φ neglected) that was used in the work of Dowler and VandenBerg described in section 2 and Figure 1. In resolved 2-D numerical siulations Roxburgh and Simmons (1993) found that at small values of the Prandtl number viscous dissipation was unimportant and that the mean properties of the solution were adequately described by the integral condition with $\Phi = 0$. One should however be cautious over extrapolating this result to stellar conditions.

5. Numerical simulation of convection

There are several groups undertaking numerical simulation of convective overshooting in the sun and stars, mostly for plane parallel geometries rather than for convective cores (Nordlund and Stein 1996, Hulbert et al. 1994, Muthsam et al. 1995, Singh, Roxburgh and Chan 1994, 95, 98, Roxburgh and Simmons 1993, Roxburgh 1998). Whilst such simulations can advance our understanding of convection it is important to remember that they do not simulate the real astrophysical conditions. These calculations fall into two categories: laminar calculations in conditions of very high viscosity, "convection in treacle", and large eddy simulations with some sub-grid scale modelling of the turbulent flow. Sub-grid modelling is an area that requires much detailed study; Canuto (1996) has recently claimed that many schemes in current use are inconsistent as they do not satisfy Galilean invariance - this criticism (fortunately!) does not apply to the Smagorinski scheme used by myself and colleagues Kwing Chan and Harinder Singh.

3-D simulations by Nordlund and Stein (1996) of turbulent convection in the solar envelope include detailed modelling of radiative losses at the upper boundary. Their results on overshooting at the base of the layer suggest a rather smooth transition between the convective region and the radiative interior, the sharp transition predicted by simple overshoot models being smoothed out by averaging over the effect of many plumes penetrating to different depths. Similar results were obtained in resolved 2D and 3D simulations (Roxburgh and Simmons, 1993, Roxburgh 1998) and using a Smagorinski sub grid model for both entropy and momentum transport.

These calculations are for parameter regimes far removed from stellar conditions or use some some form of (unjustifiable) sub-grid modelling. One should be cautious about extrapolating these results to stellar conditions.

6. Helioseismology of the base of the solar convective zone

Helioseismology provides an invaluable tool for probing the internal structure of the sun and in particular for probing the structure at the base of the solar convective zone. Discontinuities in the derivatives of the sound speed, or rather changes on a scale short compared to the wavelength, produce a phase shift in the acoustic waves $\alpha(\nu)$ of the form (Roxburgh and Vorontsov 1994)

$$\frac{\alpha(\nu)}{\nu} = A \sin\left(2\omega\tau_b + \psi\right), \; A = \left[\left(\frac{A_f}{\nu^3}\right)^2 f^2(\epsilon) + \left(\frac{A_h}{\nu^2}\right)^2 h^2(\epsilon)\right]^{1/2}, \quad (5)$$

where $f(\epsilon), g(\epsilon)$ are functions of the overshoot distance $d = \epsilon H_p$, and of the local opacity derivatives. $\alpha(\nu)$ is quasi periodic with period $1/(2\tau_b)$ where τ_b is the acoustic depth of the interface. The term A_f comes from the discontinuity in the second derivative of the sound speed and A_h from the discontinuity in the first derivative. This provides a diagnostic tool with which to probe the structure of the region at the base of the zone. Using a second order asymptotic analysis Roxburgh and Vorontsov (1994) placed an upper limit on the adiabatic penetration distance of less that $0.25H_p$; using a variational approach Monteiro et al (1994) placed an even tighter limit of 0.1 on ϵ. However for ϵ small, A is only weakly dependent on ϵ, so it is difficult to determine ϵ if it is less than ≈ 0.15.

An new high resolution inversion technique, based on the Born approximation analysis of solar oscillation modes, has recently been developed by our group in London (Roxburgh and Vorontsov 1994,1996, Marchenkov, Roxburgh and Vorontsov, 1998 see these proceedings). Applying this to the GONG frequencies we found that below the convective zone the gradient dc^2/dr is less than that of a standard solar model, indicating that the layer is less stably stratified, whilst at greater depths the reverse is true. These features may be interpreted as evidence for convective penetration below the solar convective zone since a decreasing (negative) contribution to energy transport from mild overshooting - decaying with depth - has the effect of smoothing the sharp change in the gradient of dc^2/dr. The steep gradient in dc^2/dr below the convective zone in standard models is in part due to the gravitational settling of Helium, this suggests that this settling is offset by mixing from convective penetration until some greater depth, where the gradient in Helium is then responsible for the excess of dc^2/dr in the inversion over that in the model. An alternative possibility is that the overshooting is stopped by interaction with an internal magnetic field (which may be responsible for the uniform rotation of the solar interior).

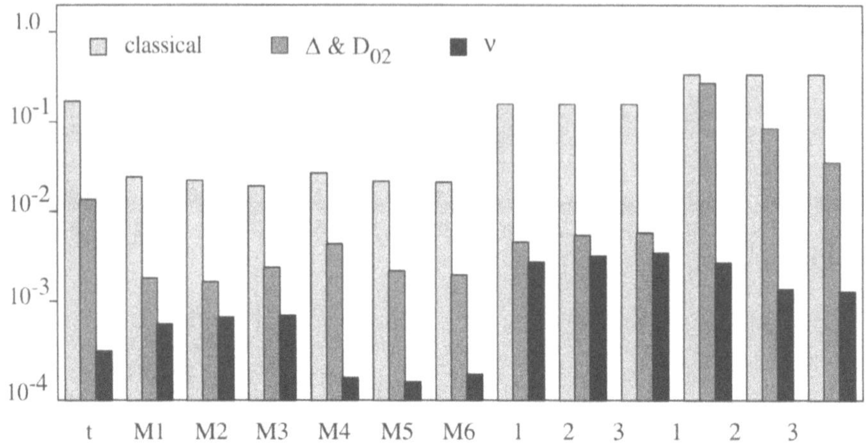

Figure 2. Error estimates on the determination of the age t, stellar masses M_i, the increase in core mass from overshooting $\beta_i(M_i)$, and the mixing length parameters $\alpha_i(M_i)$, for a group of 6 stars: a) without oscillation data; b) with the large and small separations Δ, D_{02}, c) with a set of frequencies ν (Audard and Roxburgh 1998).

7. Asteroseismology and convection

A major advance in our understanding of convective overshooting should be achieved through asteroseismology, especially through high precision space observations such as those planned for the French satellite mission COROT (cf. Baglin et al. 1997). Advances should also be achieved through the development and application of diagnostic and inversion techniques to coordinated ground based observations of (for example) δ Scuti and β Cephei stars. Any region of sharp change in acoustic properties, such as the boundary of a convective core, produces a periodic signature in the oscillation frequencies giving a method of determining the acoustic radius of the core. Given a relatively small set of frequencies one can apply new inversion techniques, using only low ℓ modes, that separate off the unknown surface layers and enable us to probe the internal structure. With a sufficiently rich of modes set one can even use standard targeted kernel inversion techniques (Roxburgh et al 1997). One can also of course use model fitting - that is finding a (not necessarily unique!) model that gives a good fit to the observed frequencies. In the author's opinion this is a dangerous procedure, the actual value of the frequencies are sensitive to the structure of the surface layers and surface layer models are subject to many uncertainties - not least the details of strongly superadiabatic convection.

For a group of stars in a cluster, or in a binary system, model fitting offers many more potentialities since the stars may be assumed to have the same age and initial composition, and in the case of binaries possibly good constraints on the masses. By parametrising the unknowns in models of stellar evolution, we can seek to determine these parameters by a simultaneous fit to the classical observables (M_V, T_{eff}, X, Y, Z) and the measured oscillation frequencies. Figure 2 shows the results of a study by Audard and Roxburgh (1998) of a group of 6 stars in a model cluster similar to the Hyades. One should get detailed estimates of age, convective core overshooting and the entropy of stellar convective envelopes from measurements of the large and small separations, and better constraints if one had more detailed knowledge of the frequencies. (The Hyades can just, and I trust will, be observed by COROT.)

8. Conclusions

Convection and convective overshooting is one of the most important and least understood processes that determine the structure and evolution of the sun and stars. There is evidence from observations of clusters and binary systems that penetration above convective cores is significant. A major advance can be expected from seismic observations of other stars, from the ground and from space, providing data that can be used to diagnose the properties of convective core overshooting and the efficiency of surface layer convection. Numerical simulations, whilst restricted in the range of parameters they can study, can also be used to address problems in this field and to guide theoretical developments. A major effort however needs to be directed towards understanding sub-grid scale modelling and eddy viscosity.

References

Audard N, Roxburgh I W, 1998, in Sounding Solar and Stellar Interiors, Poster Volume, eds J Provost and F-X Schmider, p213, Nice, France.

Baglin A et al., 1997 COROT, Convection and Rotation, Phase A study, CNES, France.

Canuto V, 1996, Paper presented at meeting on *Windows on the Sun*, Bombay, 1996.

Christensen-Dalsgaaard J, et al., 1996, *Science*, **272**, 1286.

Dowler P, 1995, Thesis, University of Vancouver, Canada.

Dowler P., VandenBergh D., 1995, (private communication).

Hulbert N E, Toomre J, Massagauer J, Zahn J-P, 1994, *Astrophys. J.*, **421**, 245

Marchenkov K I, Roxburgh I W, Vorontsov S V, 1998 *Mon. Not. R. astr. Soc.*, (sub)

Maeder A, 1975, *Astron. Astrophys.*, **40**, 303

Maeder A, Mermilliod J-C, 1992, *Astron. Astrophys.***98**, 477.

Meynet G, Mermilliod J-C, Maeder A, 1993, *Astron. Astrophys.*, **98**, 477.

Monteiro M, Christensen-Dalsgaard J, Thompson M, 1994, *Astron. Astrophys.*, **283**, 247.

Muthsam H, Gob W, Kupka F, Leibich W, Zochlin J, 1995, *Astron. Astrophys.*, **293**, 127

Nordlund Å, Stein R F, 1996, in *Stellar Evolution: What Should be Done? 32nd Liege Colloquium.* Eds Noels A, Fraipont-Car D, Gabriel M, Grevesse, N, Demarque, 75.
Perrymann M et al. 1998, *Astron. Astrophys.*, in press
Roxburgh I W, 1978, *Astron. Astrophys.* **65**, 281
Roxburgh I W, 1985, *Solar Physics*, **100**, 21-51
Roxburgh I W, 1989, *Astron. Astrophys.* **211**, 361
Roxburgh I W, 1998, to be published
Roxburgh, I W, et al., 1998. in Sounding Solar and Stellar Interiors, Poster Volume, eds J Provost and F-X Schmider p245, Nice, France.
Roxburgh I W, Simmons J, 1993, *Astron. Astrophys.*,, **277**, 93
Roxburgh I W, Vorontsov S V, 1994a, *Mon. Not. R. astr. Soc.*, **268**, 880.
Roxburgh I W, Vorontsov S V, 1994b, *Mon. Not. R. astr. Soc.*, **267**, 297.
Roxburgh I W, Vorontsov S V, 1996, *Mon. Not. R. astr. Soc.*, **278**, 940.
Schmitt J H M M, Rosner R, Bohm H U, 1984,*Astrophys. J.* **282**, 316.
Shaviv G and Salpeter E, 1973, *Astrophys. J.*, **184**, 191
Singh H P, Roxburgh I W, Chan K L, 1994, *Astron. Astrophys.*, **281**, L73,
Singh H P, Roxburgh I W, Chan K L, 1995, *Astron. Astrophys.*, **295**, 703
Singh H P, Roxburgh I W, Chan K L, 1998, *Astron. Astrophys.*, in press
Zahn J-P, 1991, *Astron. Astrophys.* **252**, 179

NON LINEAR INVERSION FOR THE HYDROSTATIC STRUCTURE OF THE SOLAR INTERIOR

K I MARCHENKOV, I W ROXBURGH AND S V VORONTSOV
Astronomy Unit, Queen Mary and Westfield College, University of London, London E1 4NS, UK.

The inversion technique which we use is based on the "quasi-asymptotic" description of solar p-modes developed by Roxburgh & Vorontsov (1996) which incorporates a Born approximation to permit a detailed treatment of the regions of rapid variation of seismic parameters with depth and strong influence of gravity perturbations.

The eigenfrequency equation is represented as second-order inhomogeneous Airy equation with a frequency dependent "acoustic potential", which is solved by a Born-type perturbation analysis. Matching the solutions in the interior with "exact", non-asymptotic solutions in the surface layers, where asymptotic analysis becomes locally invalid, leads to the eigenfrequency equation

$$\int_{r_1}^{R} sdr \simeq \frac{\pi}{\omega}\left(n + \frac{5}{72}\frac{1}{\pi^2 n} + \alpha^{int} + \alpha^{out}\right), \quad s^2 = \frac{1}{c^2} - \frac{\tilde{w}^2}{r^2}, \quad \tilde{w} = \frac{\ell + 1/2}{\omega}.$$

The outer layers contribute to the eigenfrequency equation with a surface phase shift α^{out}, similar to that of the standard asymptotic analysis. α^{int} is the frequency and degree dependent internal phase shift, given in the first order by an integral over the interior of the acoustic potential weighted with the square of an Airy function. r_1 denotes the position of the turning point $s^2(r_1) = 0$, c the sound speed, ω the cyclical frequency.

To construct an inversion technique based on this analysis we proceed as follows. First we infer the leading-order approximation to $F(\tilde{w}) = \int_{r_1}^{R} sdr$ by setting $\alpha^{int} = 0$. In this stage we separate the input data $(\pi n/\omega)$ on the right-hand side of equation (1) into three parts: a) the leading-order asymptotic term $F(\tilde{w})$, b) the frequency dependent surface phase shift for low-degree modes $\alpha_0^{out}(\omega)$, c) the first-order correction to the surface phase shift for intermediate and high degree modes $\tilde{w}^2\alpha_2^{out}(\omega)$. Next, we invert $F(\tilde{w})$ using the standard Abel integral transform to get a first approximation for the sound-speed profile. We then use the equation of hydrostatic

Astrophysics and Space Science is the original source of publication of his article. It is recommended that this article is cited as: *Astrophysics and Space Science* **261**: 51–52, 1999.

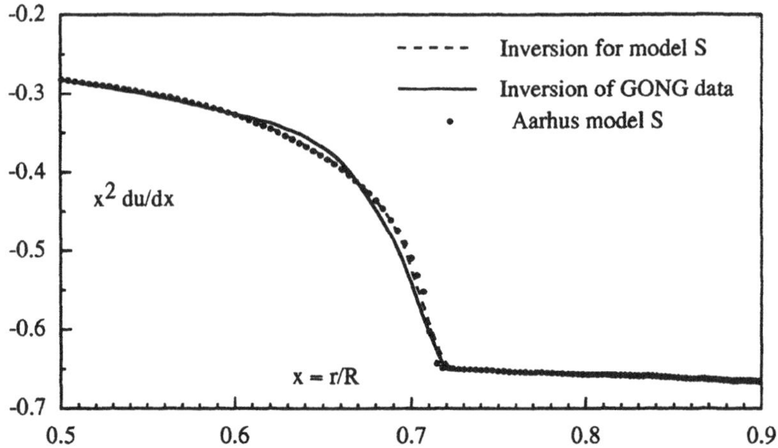

Figure 1. Variation of $r^2\, dc^2/dr$ in the neighbourhood of the base of the solar convective zone as found by applyiing the Born inversion technique to frequencies obtained by GONG, and using the frequencies of Aarhus model S.

support to determine the Brunt-Vaisala frequency profile and calculate the internal acoustic Born potential and corresponding Born phase shift for each mode used in the analysis. With the surface phase shift determined in the first step and with the internal phase shift determined in the previous step we infer an improved value of $F(\tilde{w})$ from equation (1) and an improved estimate of the sound speed-profile. We repeat the above steps until there is no change in the sound speed profile between successive iterations. The procedure is stable and converges in few iterations. We used two data sets: the GONG set with 1803 modes with frequencies in the range $1 \leq \omega/(2\pi) \leq 4.5$ mHz. and the SOI/MDI 144 days data set with 1802 modes. We also used artificial data from the Aarhus standard Solar Model S Christensen-Dalsgaard *et al.*, 1996) In Figure 1 we present the results for the dimensionless sound speed gradient, $r^2\, dc^2/dr$, which is well suited to investigate the structure beneath the convective zone. The inversions show a slightly shallower gradient than the model immediately below the zone, perhaps a signature of some mixing and/or convective penetration.

References

Christensen-Dalsgaard, J, et al.: 1996, *Science*, **272**, 1286.

Marchenkov, K I, Roxburgh, I W, Vorontsov, S V : 1998, *Mon. Not. R. astr. Soc.*(sub)

Roxburgh, I W, Vorontsov, S V : 1996, *Mon. Not. R. astr. Soc.*, **278**, 940.

A PARAMETRIC STUDY OF 3-D SIMULATION OF PENETRATIVE CONVECTION

HARINDER P SINGH
Department of Physics, Sri Venkateswara College, University of Delhi, India.

IAN W ROXBURGH
Astronomy Unit, Queen Mary and Westfield College, University of London, London E1 4NS, UK.

AND

KWING L CHAN
Department of Mathematics, Hong Kong University of Science & Technology, Hong Kong, China.

We report on recent results from our ongoing programme of three dimensional numerical simulations of compressible convection to study the penetration of convection from the unstable layers into the surrounding stable layers (Singh, Roxburgh & Chan 1994,95,96,98). We adopt a simple 3 layer configuration of an unstable layer surrounded by two stable layers, and model the small scale motions using the Smagorinski sub-grid scale closure procedure. The goal of the investigation reported here was examine whether the numerical simulations satisfy the scaling relationship $\Delta \propto V_{zo}^{3/2}$ suggested by simple non-local eddy and plume models of convective penetration (Schmitt et al. 1984, Zahn 1991, Roxburgh 1997). Δ is the penetration distance and V_{zo} the vertical velocity at the bottom of the unstable convective layer.

The computational domain consists of a rectangular box of depth $7.5H_p$ (H_p the pressure scale height) with the central unstable region occupying $5.4H_p$ and the lower stable $1.2H_p$ or 40% of the box. The input flux is prescribed at the bottom boundary and the horizontal boundaries are periodic. In the standard configuration the aspect ratio of the box is 1.5. Four models with different values of the input flux F_b were computed and the used to test whether they statisfied the equation

Astrophysics and Space Science is the original source of publication of his article. It is recommended that this article is cited as: *Astrophysics and Space Science* **261**: 53–54, 1999.

TABLE 1. Relationship between penetration distance and rms vertical velocity

Model No.	Aspect Ratio	$(F_b(1), F_b(2))$	LHS Eqn.(1)	RHS Eqn.(1)
1	1.5	(0.1875,0.125)	1.18	1.24
2	1.5	(0.1875,0.0625)	1.72	1.78
3	1.5	(0.1875,0.03125)	2.46	2.66
4	1.5	(0.125,0.0625)	1.46	1.43
5	1.5	(0.125,0.03125)	2.08	2.14
6	1.5	(0.0625,0.03125)	1.43	1.49
7	1.5	(0.125,0.0625)	1.40	1.41
8	2.0	(0.125,0.0625)	1.26	1.41
9	3.0	(0.125,0.0625)	1.44	1.54

$$\frac{\Delta_d(F_b(1))}{\Delta_d(F_b(2))} = \frac{V_{zo}^{3/2}(F_b(1))}{V_{zo}^{3/2}(F_b(2))} \tag{1}$$

We obtained six values for the left and the right hand sides of above equation from our four cases. The results are given in Table 1 (models 1-6). The relationship $\Delta_d \propto V_{zo}^{3/2}$ is a reasonable representation of the results.

We then changed the location of the interface between the unstable and the lower to the mid point of the box giving a lower stable layer of $2H_p$. Two cases were studied with different values of the input flux F_b. The result (model 7) satisfies the scaling relationship between the penetration distance and the vertical velocity. We then increased the aspect ratio of the box from 1.5 to 2.0 (model 8) and 3 (model 9). The resulting scaling is shown in Table 1. These results are in broad but not exact agreement with the relationship $\Delta \propto V_{zo}^{3/2}$. This may be due to the coarse resolution in the two horizontal directions (35×35) which were not changed for the larger aspect ratios. Calculations with improved horizontal resolution for larger aspect ratios are currently under way.

References

Roxburgh, I. W. 1997. In *Solar Convection Oscillations and their Relationship*, eds Pijpers F, Christensen-Dalsgaard J and Rosenthal C, Kluwer, Dordecht, pp 213-50.

Schmitt, J. H. M. M., Rosner, R., Bohm, H. U. 1984, *Astrophys. J.*, **282**, 316

Singh, H. P., Roxburgh, I. W., Chan, K. L. 1994, *Astron. Astrophys.*, **281**, L73.
Astron. Astrophys., **295**, 703, 1995, *Bull. Astr. Soc. India* **24**, 281, 1996
Astron. Astrophys., 1998 (in press).

Zahn, J. P. 1991, *Astron. Astrophys.*, **252**, 179

STEADY CONVECTION IN DEEP COMPRESSIBLE LAYERS

IAN W ROXBURGH
Astronomy Unit, Queen Mary, University of London
Mile End Road, London E1 4NS, UK

The standard mixing-length model of convection traditionally used in astrophysics is based on an anology with the kinetic theory of gases; convective eddies travelling a distance ℓ before mixing with the background. The mixing length ℓ is assummed to be of the order of the local scale height of density or pressure, on the grounds that if, at any one level, upward moving material occupies half the area, then on travelling a scale height it would have expanded to fill all the available area (cf Biermann (1935). Simulations of steady convection do not find eddies to be of the order of the local scale height, but rather a single cell extending over several scale heights from top to bottom of the layer (cf Graham 1975, Chan and Wolf 1982, Hulburt et al 1994, Roxburgh and Simmons 1993). Simulations of turbulent convection with some form of sub-grid scale closure give plumes or downdrafts extending over several scale heights (cf Nordlund and Stein 1996, Singh, Roxburgh and Chan 1995). We here present a simple explanation of these results.

Consider a plane parallel convectively unstable layer surrounded by stable layers. Let the total energy Flux be $\mathbf{F}$. In the stable layers all the energy is carried by radiation $\mathbf{F}_r = \mathbf{F}$. In the unstable layer some energy is carried by convection so $\mathbf{F}_r < \mathbf{F}$. at the bottom boundary of the unstable convective zone $\mathbf{F}_r = \mathbf{F}$, it then decreases reaching a minimum at some point in the unstable layer before increasing again until, at the top boundary of the unstable layer, again $\mathbf{F}_r = \mathbf{F}$. Therefore the divergence of the radiative flux $div\mathbf{F}_r = d\mathbf{F}_r/dz$ is negative in the lower part of the convective zone and positive in the upper part.

The thermal equilibrium of the convective layer is governed by the thermal energy equation (the 2nd law of thermodynamics $T\,dS = dQ$) which can be written in the form

$$\frac{d}{dt}(\rho S) = \frac{\partial}{\partial t}(\rho S) + div(\rho S\mathbf{v}) = \frac{\Phi - div\mathbf{F}_r}{T}, \tag{1}$$

Astrophysics and Space Science is the original source of publication of his article. It is recommended that this article is cited as: *Astrophysics and Space Science* **261**: 55–56, 1999.

where $S = c_v \ell n(T/\rho^{2/3})$, is the entropy per unit mass, ρ the density, $\mathbf{v}$ the velocity, $\mathbf{F_r}$ the radiative flux and Φ the viscous dissipation per unit volume.

Now consider steady convection with closed stream tubes and integrate equation (1) over the volume V inside the stream-tube. By Gauss's theorem the integral of $div(\rho S\mathbf{v})$ over the volume equals the integral of $(\rho S\mathbf{v})$ over the surface and, since there is no flow across the surface of a stream tube, this integral is identically zero.

$$\int_V \frac{d\rho S}{dt}\, dV = \int_V div(\rho S\mathbf{v})\, dV = \int_\Sigma \rho S\mathbf{v}.d\Sigma = \int_V \frac{(\Phi - div\mathbf{F}_r)}{T}\, dV = 0 \quad (2)$$

Since the total production of entropy in a closed stream tube is zero the tube must have regions of positive entropy production which are balanced by regions of negative entropy production. Since $\Phi > 0$ then, with $div\mathbf{F_r} < 0$, the rate of production of entropy per unit volume in the bottom part of the unstable layer $(\Phi - div\mathbf{F}_r)/T > 0$. The region of negative entropy production is in the top part of the layer where $div\mathbf{F_r} > 0$. The stream tube must therefore span these two regions, the core (or eye) of the convective eddy being at the level where $(\Phi - div\mathbf{F}_r) = 0$. This result is independent of the depth of the layer or of the variation in density and pressure across the convective region. This argument is not rigorous, horizontal temperature variations and penetrative counter cells being neglected (cf Roxburgh 1989).

The result brings out an important point that is ignored in both local and non-local mixing-length models of astrophysical convection. The turn over time of convection is small compared to the thermal relaxation time so convection is a global phenomenon, the whole layer adjusting to a global equilibrium. In steady convection single eddies spanning the layer, in turbulent convection plumes or downdrafts extending over several scale heights. We would therefore strongly caution against using the mixing length model in situations where the details of the convective flows are important.

References

Beirmann L., 1935, *Astron. Nach.*, **257**, *270*

Chan K.L., Wolf C.L., 1982, J. Comp. Phys., **47**, *109*

Graham E., 1975, J. Fluid. Mech., **70**, *689*

Hulbert N. E., Toomre J., Massagauer J., Zahn J-P., 1994, Astrophys. J., **421** *245*

Nordlund Å.,Stein R F, 1996, Stellar Evolution: What Should be Done? eds Noels A, Fraipont-Car D, Gabriel M, Grevesse, N, Demarque P. pp 75-94

Roxburgh I.W., 1989, Astron. Astrophys., **211**, *361*

Roxburgh I.W., Simmons J., 1993, Astron. Astrophys., **277**, *93*

Singh H.P., Roxburgh I.W., Chan K. L., 1995, Astron. Astrophys., **295**, *703*

HELIOSEISMIC CONSTRAINTS ON SOLAR STRUCTURE AND THE SOLAR NEUTRINO PROBLEM

IAN W ROXBURGH
Astronomy Unit, Queen Mary and Westfield College, University of London, London E1 4NS, UK.

One of the major outstanding problems of solar (or particle) physics is that the measured flux of neutinos on earth (cf Kirsten 1996, 1997) is substantially less than that predicted by a standard solar model (eg Model S, Christensen-Dalsgaard 1996). See rows 1 and 2 in Table 1. The flux as measured by the GALLEX and SAGE Gallium experiments, $F(Ga)$, is just compatible with the energy being produced by nuclear burning of Hydrogen but not with the (^{3}He,^{4}He) branching ratio predicted by standard solar models. The flux of $B + Be$ neutrinos measured by the Homestake $C\ell$ experiment, $F(C\ell)$, is less than the flux of B neutrinos alone as measured by the Kamiokande and Super-Kamiokande experiments, $F(B)$. If the errors are substantially underestimated this inconsistency is removed. If the solar model is correct then the solution lies with particle physics: the MSW effect, vacuum oscillations, magnetic moment of the neutrino. But is the solar model correct?

The argument that the solar model is correct is based on the agreement between the internal structure of a "standard solar model" and that deduced from the measured solar oscillations. But since these oscillations are adiabatic over the vast bulk of the sun they depend only on the acoustic structure, that is pressure $P(r)$, density $\rho(r)$ and adiabatic exponent $\Gamma_1(r)$. They do **not** separately constrain the temperature $T(r)$ and hydrogen abundance $X(r)$, which are related through the equation of state, but depend on the details of energy transport and any mixing in the solar interior, an area of much greater uncertainty (cf Roxburgh 1997). We therefore take the acoustic structure (P, ρ, Γ_1) as given, and look for that $X(r)$ which minimises the neutrino fluxes. We also impose global thermal equilibrium (for which we have no direct evidence!) so $L_{nuc} = L_{\odot}$

If we impose the (reasonable) constraint that $T(r), X(r)$ decreases outwards, then the minimum neutrino flux is given by $X(r)$ constant in the energy generating core, increasing to the surface value outside this region.

Astrophysics and Space Science is the original source of publication of his article. It is recommended that this article is cited as: *Astrophysics and Space Science* **261**: 57–58, 1999.

We used standard reaction rates and an approximate equation of state - these accurately reproduced the results of the standard solar model. The minor species (including ^3He) were taken to be in local equilibrium since, given $X(r)$, it makes no difference whether we take local global or local equilibrium (Roxburgh 1996b). The resulting minimum had $X = 0.53$, the individual fluxes are given in row 3 of Table 1 (see Roxburgh 1996a).

We then explored the consequences of scaling the p-p cross section by f_1 and the (Be,e):(Be,p) branching ratio by f_7, which corresponds to increasing or decreasing one or other of these reaction rates. The results are also shown in Table 1 (rows 4 - 7).

The mixed core lowers the total neutrino flux but not to values in agreement with observation. Closer agreement is obtained if the p-p cross section is increased by $\sim$ 8%, the (Be,e) capture rate is drastically reduced, and the central hydrogen abundance increased. Since all the models studied here have the same P, ρ, Γ_1 as the standard model, they necessarily satisfy the constraints from helioseismology.

TABLE 1. Neutrino Fluxes: Observations and model prediction

Model	f_1	f_7	X_c	$F(C\ell)$	$F(Ga)$	$F(B)$
Observed				2.6±0.2	70±8	2.8±0.4
Standard	1.000	1.00	0.338	8.3	132	6.1
Mixed 1	1.000	1.00	0.530	2.52	107	1.23
Mixed 2	1.050	1.00	0.657	0.7	87	0.10
Mixed 3	1.075	1.00	0.692	0.5	86	0.05
Mixed 4	1.050	0.10	0.657	1.7	90	1.0
Mixed 5	1.075	0.02	0.692	3.0	92	2.4

References

Christensen-Dalsgaard, J., et al, 1996, *Science* **272**, 1286

Kirsten T., 1996 in *New Trends in Solar-Neutrino Physics* eds V Berezinsky and G Fiorentini , INFN, Laboratori Nazionali del Gran Sasso, pp 1-20

Kirsten T, 1997. In *Proceedings 4th International Solar Neutrino Conference,* ed W Hampel, , Neumann Druck, Heidelberg, pp 138-150.

Roxburgh, I. W., 1996a. In *Windows on the Sun,* Bombay, October 1995, Eds Chitre, S.M. and Antia, H.M. *Bulletin Astron Soc India,* **24**, 89-102.

Roxburgh I W 1996b, In *New Trends in Solar-Neutrino Physics* eds V Berezinsky and G Fiorentini, INFN, Laboratori Nazionali del Gran Sasso, pp 269-274.

Roxburgh I W 1997, In *Solar Convection, Oscillations and their Relationship,* eds F J Pipers and J Christensen-Dalsgaard, and C. Rosenthal, Kluwer, Dordecht, pp 23-50.

STELLAR ROTATION AND ACTIVITY

The Story of Solar-Type Stars

D. R. SODERBLOM
Space Telescope Science Institute
Baltimore MD, USA

Abstract. This review is limited to solar-type stars (late-F to early-K main sequence stars) and to recent developments in the study of the evolution of angular momentum in those stars. Observations of rotation in young clusters are discussed, together with the models that have been put forth to account for what is seen. One key question is whether or not the convective envelopes of solar-type stars decouple from the radiative cores when the stars near the Zero-Age Main Sequence. That question cannot yet be answered, but forthcoming observations are likely to address the issue. Another significant open question is the degree to which any one cluster of stars is typical of all stars at that age, and that too is near to being resolved as we reach deeper into the Galaxy at high spectroscopic resolution. Finally, some general properties of activity in solar-type stars are presented.

1. What is a Solar-Type Star?

What makes a star "solar-type" is the presence of a convection zone (CZ) in the surface layers, for the CZ lies at the heart of the many fascinating phenomena that we associate with the Sun. Most of those phenomena are related to "activity" in one form or another – sunspots, flares, the corona, etc. – and activity is an observable manifestation of the Sun's magnetic field. The magnetic field exists because of the dynamo mechanism, and the dynamo arises because of the complex circulation patterns created by the interaction of convection with rotation, especially differential rotation. This simplified description is the current paradigm for studying and understanding the behavior of stars like the Sun.

So the essence of "solar-typeness" is the intimate connection between rotation and activity. That connection, and the nature of rotation and activity separately, have been the subjects of entire topical symposia, and

Astrophysics and Space Science is the original source of publication of his article. It is recommended that this article is cited as: *Astrophysics and Space Science* **261**: 59–68, 1999.

cannot be recapitulated here. Instead I will dwell on a few recent developments. My goals are to use the Sun to understand the physics of other stars, and to use stars to place the Sun in context. In this way we can hope to know the past behavior of the Sun, and to understand the mechanisms that account for that behavior.

2. Recent Work on Rotation

Probably the most important observations of solar-type stars in recent years were made in 1981. A Dutch thesis student, Floor van Leeuwen, was surveying the Pleiades cluster and found periodic variations in the brightness of some of the K dwarfs (Alphenaar & van Leeuwen 1981). That in itself was not surprising, for the Pleiades is young (100 Myr old), leading us to expect its solar-type stars to be highly active and highly spotted. But the periods observed were as low as 0.25 day, corresponding to a rotation rate of 100 times the Sun's.

Spectroscopy soon confirmed that these variations were indeed due to rotation (Soderblom *et al.* 1983) and that a significant fraction of Pleiades K and M dwarfs spin rapidly (meaning 20 times the solar rate or more) (Stauffer *et al.* 1984). Soon similar stars were found in the α Persei cluster (Stauffer *et al.* 1985), which, at an age of 50 Myr, was also Zero-Age Main Sequence (ZAMS). This confirmed that the Pleiades stars weren't freaks, and that many ZAMS stars go through a stage where they are Ultra-Fast Rotators (UFRs).

Less clear was whether or not *all* stars go through a UFR stage. If so, the high proportion (about 80%) of slow rotators in the Pleiades meant that the transition from UFR to slow rotator had to take place within whatever spread in age exists within the cluster. Observations of stars forming regions indicate that the age spread within a cluster is at most a few million years. Going from rotating at 100 times the solar rate to, say, 10 times the solar rate in 5 Myr or less means losing angular momentum at an extremely high rate, higher that we believe solar-type stars can realistically achieve.

Two obvious alternatives exist to solve this problem. First, perhaps only a fraction of stars ever get enough angular momentum to go through a UFR stage at all. Recent studies of the distributions of rotation in star-forming regions (e.g., Choi & Herbst 1996) support that view. The second possibility is that only the surface of the star spins down within those few millions of years because the radiative core decouples from the CZ. This decoupling may occur, of course, whether or not the UFR phenomenon has its origins in the initial distribution of angular momentum.

The best way to understand this UFR phenomenon is to observe stars in open clusters, for there we have reasonably large samples of stars of the same age, with different clusters providing different ages. For example, the situation for solar-type stars in clusters was reviewed and illustrated by Soderblom *et al.* (1993b). To summarize:

1. In the Pleiades, membership is well-determined, and the observed sample should be bias-free. Also, the age and metallicity are well-determined, and observations are available for a large sample (Soderblom *et al.* 1993a).
2. The distribution of rotation versus color for the Pleiades is flat (i.e., independent of color or mass). The spread at any one color is about a factor of 20, too much to assign a characteristic rotation rate.
3. About 20% of the solar-type stars in the Pleiades are UFRs, and the distribution of rotation is bimodal, with a lack of stars with intermediate rotation rates.
4. In the Pleiades, the median rotation rate at 1 $M_\odot$ is about 10 km s^{-1}.
5. Membership for the Hyades is also well-determined, and it is a metal-rich cluster. Rotation periods are known for many stars (Radick *et al.* 1987), removing the ambiguity of aspect.
6. In the Hyades, the median rotation rate at 1 $M_\odot$ is about 7 km s^{-1}, only slightly lower than in the Pleiades despite the Hyades being much older (at about 700 Myr).
7. There is very little spread in rotation rates in the Hyades; only about 10% at any one mass.
8. These two facts indicate that in evolving from the Pleiades to the age of the Hyades, solar-type stars undergo an enormous convergence in their rotation rates, even though the median value changes little.
9. For α Persei, membership is more difficult to determine and there is probably contamination of the samples by field stars.
10. Nevertheless, about 50% of the α Persei stars observed are UFRs, and the highest rotation rates seen are about twice what is seen in the Pleiades.
11. Other ZAMS clusters (e.g., IC 2602 and IC 2391) also have UFRs.

3. Interpreting Observations of Rotation

Several possibilities suggest themselves, and each should have observable consequences that we can test. At present we have data for so few clusters that several mutually exclusive possibilities can be considered (e.g., Li & Collier Cameron 1993, Collier Cameron & Li 1994), but that is changing.

3.1. CORE-ENVELOPE DECOUPLING?

Many have noted that these observations can make sense if we allow the core and envelope of a solar-type star to decouple from one another as the star nears the ZAMS. Depending on how strongly the two zones couple, the angular momentum (AM) reservoir of the core will later replenish the envelope, perpetuating the envelope's spin for a time. Eventually the two achieve solid-body rotation, and eventually the stars in a cluster converge to a common rotation rate.

In this picture, the AM loss rate for UFRs is independent of rotation rate above some threshold. This produces rapid convergence in the surface rates of the stars in a cluster and the bimodal distribution (a star lingers in the high-surface-rotation phase, then suddenly drops to slow rotation). This hypothesis also implies that the UFRs have their origin in the high-AM tail of the initial distribution of AM among stars. In other words, all solar-type stars do *not* go through a UFR phase.

One piece of corroborating evidence for core-envelope decoupling is that the ratio of the moments-of-inertia of the core to the envelope change dramatically in going down the main sequence from 1.4 to 0.7 $M_{\odot}$ (the range of mass surveyed in the Pleiades observations). This ratio determines how much AM remains in the core to replenish the envelope, and can naturally account for the strong mass-dependence seen in rotation in evolving from the Pleiades to the Hyades.

3.2. ELEMENTS OF A MODEL OF ANGULAR MOMENTUM EVOLUTION

To construct a model we start with these assumptions:

- The distribution of rotation seen in the Pleiades has its origins in the pre-main sequence (PMS) phase of evolution, and so it must be explained in terms of the initial distribution of AM and PMS phenomena and processes.
- In the PMS phase, circumstellar disks can regulate stellar rotation among the classical T Tauri stars because of magnetic interactions between the disk and the stellar magnetic field. Those disks dissipate after $\sim$ 10 Myr, allowing weak-0lined T Tauri stars to spin freely.
- It is permitted, but not required, for the convective envelope of a star to spin at a different rate than the radiative core.
- Stellar magnetic fields, together with an ionized wind (as we see on the Sun), lead to AM loss, and the AM loss rate is higher when the rotation rate is higher.
- Observations suggest that the dynamo saturates at about 10 times the solar rotation rate.

3.3. PRE-MAIN SEQUENCE ROTATION AND ACCRETION DISKS

We know that classical T Tauri stars (CTTs) rotate at relatively slow rates, with a median period of 8.3 days; also there are few short-period systems. On the other hand, the weak-lined T Tauri stars (WTTs) have a median rotation period of 2.6 days, and there are few *long*-period systems. What distinguishes CTTs from WTTs is the presence of an accretion disk in the former; this keeps the rotation rate nearly constant for an extended time, preventing spin-up as the star contracts. The WTTs have no such regulating mechanism and so spin up freely. Infrared observations of T Tauris support this because they indicate that the disk lifetimes are 1 to 10 Myr.

3.4. PARAMETERS OF A MODEL

1. **The time scale for internal AM transfer.** This could also be termed the coupling constant between the core and envelope. If the time scale is short, then solid-body rotation takes place, and if it's long, then core-envelope decoupling occurs.
2. **The disk lifetime.** If the accretion disk lasts a long time, the PMS star remains slowing rotating all the way to the ZAMS, but if the disk dies early on, very rapid rotation results. There may be, of course, a distribution of disk lifetimes within an ensemble of stars.
3. **The AM loss relation** for PMS and main sequence stars.
4. **The initial distribution of AM** within an ensemble of stars.
5. **The critical rotation rate** at which saturation of the magnetic field occurs.

Using some different assumptions and different variables, a number of recent papers have presented models. Keppens *et al.* (1995) use a long coupling time (i.e., core-envelope decoupling). Krishnamurthi *et al.* (1997) undertook an extensive study that varies many of these parameters to test the sensitivity of the results. Bouvier *et al.* (1997) used a short coupling time (solid body rotation) with disk locking (a constant rotation rate as long as the disk is present) together with magnetic braking via a wind. Finally, Collier Cameron and co-workers (Li & Collier Cameron 1993, Collier Cameron & Li 1994) examined models with both decoupling and solid-body rotation.

4. Some Results

The goals in modeling rotational velocities and their distributions are:

1. To account for how T Tauri stars evolve to have a broad range of rotation rates on their surfaces when they reach the ZAMS.

2. To account for how the rotation of stars converges to nearly uniform rotation by the age of the Hyades.
3. To account for the large proportion of slow rotators seen in the Pleiades.

The UFRs and their evolution are very sensitive to the assumed form of the AM loss law, but the slow rotators reflect our assumptions about AM transport within stars.

4.1. THE INTERNAL COUPLING TIME SCALE

Krishnamurthi *et al.* (1997) conclude that the slow rotators determine the time scale for internal AM transport, and that the observations suggest that this time scale is comparable to or greater than the ages of the young open clusters. Bouvier *et al.* (1997) assume near-solid-body rotation, and show that this can account for distributions of rotational velocities that look like those seen in the clusters. This apparent contradiction with Krishnamurthi *et al.* results from trading off disk locking against internal AM transfer.

4.2. THE SATURATION THRESHOLD

Let $\Omega_{\rm crit}$ be the critical rotation rate above which the magnetic field saturates and the AM loss rate is a constant. If $\Omega_{\rm crit}$ is the same for all masses, a given set of initial conditions will always make the high-mass stars rotate faster than the lower-mass stars. However, the observations of cluster stars indicate that rapid rotation disappears first among the higher-mass stars, which contradicts this. Therefore a mass-dependent $\Omega_{\rm crit}$ is needed. This is consistent with dynamo models, but it could also arise from, say, the coverage of the magnetic field over the surface of a star.

If $\Omega_{\rm crit}$ is too low and the disk lifetime is also low, the range of rotation rates seen remains very high up to and beyond the age of the Hyades, which is unacceptable. A high value of $\Omega_{\rm crit}$ suppresses rapid rotation, preventing stars from achieving rates as high as $100\Omega_{\rm Sun}$, as seen in the Pleiades and α Persei clusters. This statement is independent of the coupling time scale. It turns out that $\Omega_{\rm crit}$ must be $\sim 10\Omega_{\rm Sun}$ to keep convergence from happening until the age of the Hyades. In this picture

$$d\Omega/dt \sim \Omega^3 \quad \text{for} \quad \Omega < \Omega_{\rm crit}, \text{ and}$$

$$d\Omega/dt \sim \Omega \quad \text{for} \quad \Omega \geq \Omega_{\rm crit}.$$

4.3. COMMENTS ON MODELS

These models are largely successful at reproducing the trends seen in rotation rates among young solar-type stars, and they use reasonable assumptions. However, the models are mostly phenomenological, not physical, and we have little genuine understanding of the mechanisms and processes that are involved.

The models also tend to have more free parameters than can be fitted well by the observations, and key assumptions remain untested. Observers have managed to keep one step ahead of the modelers for these reasons and because nature continues to present us with a bewildering variety.

4.4. KEY FUTURE OBSERVATIONS OF ROTATION

Open clusters remain the best laboratories for studying the evolution of AM in solar-type stars. The broad range of rotation seen in ZAMS stars means that large, coeval samples are needed to discern the evolutionary trends and dependence on mass among the inevitable innate star-to-star differences. The following observations could help to further elucidate the situation:

1. **A large sample at 10 to 20 Myr age, a key phase in PMS evolution.** Post-T Tauri stars (PTTs) have remained extraordinarily elusive and controversial, but the cluster NGC 2264 in Monoceros fits the bill nicely. High-quality photometry of prospective members has recently become available, and recent spectra have been obtained at high spectroscopic resolution.
2. **Redundant observations where possible.** We continue to have to assume that any one cluster is completely representative of the behavior of all stars at that age. Data are now available for M34 and for NGC 6475 which suggest that this assumption is not true.
3. **Massive high-quality photometry of a large and unbiased sample of solar-type stars.** A proposed space mission (Kepler) would obtain continuous coverage on $\sim 5,000$ G dwarfs, with enough sensitivity to detect variations of 1 part in 100,000. The primary goal is to detect planets as they transit stellar disks, but the resultant database would provide complete and rich information on many aspects of the behavior of stars like the Sun. It should be possible to detect rotation in virtually all the stars observed, and perhaps differential rotation as well. Maybe even spot coverage, cyclic variations, ...

5. Activity in Solar-Type Stars

5.1. MEASURING ACTIVITY

There are many ways of measuring and studying activity in late-type stars. Two of the most productive are by getting x-ray images of stars, and by getting spectra of the H and K lines of Ca II. X-rays arise in stellar coronae, where the temperatures exceed 10 MK. They must be observed from space, and telescope time can be difficult to obtain, but the key benefit of using x-rays is that they are very high contrast. In other words, if a star is detected in x-rays, you can be confident it is active. This makes x-rays a very effective means of finding members in young clusters, because the traditional means (such as proper motion studies with radial velocity follow-ups) are slow and painstaking, especially for the low-mass stars, which blend into the field because they are so faint.

However, x-rays introduce an significant bias in surveying for cluster members because only the most active stars are likely to be detected. This can prevent us from finding low-activity stars which may be *bona fide* cluster members and which may be revealing of the true distribution of AM and activity.

The H and K lines of Ca II are the most-often-used indicator of stellar activity, simply because it is possible to get a useful measure with a short exposure from the ground at modest spectroscopic resolution ($R \sim 2,000$).

5.2. SURVEYS OF ACTIVITY

Because of the intimate relation between rotation and activity, observations of activity alone can tell us much about the evolution of AM in stars. The first significant survey of activity in solar-type stars was done by Vaughan & Preston (1980) at Mount Wilson. In this context we will take "solar-type" to mean a star with a $(B-V)$ value between 0.50 and 1.00, and lying within 1 magnitude of the ZAMS. In this range Vaughan & Preston surveyed about 200 stars. One important result was the "Vaughan-Preston gap," which is a lack of stars of intermediate activity levels. This gap suggests either that intermediate age stars are missing from the solar neighborhood, or that the activity-age relation has a kink that produces the appearance of a gap.

The Vaughan-Preston survey was used by Soderblom (1985) to infer the distribution of rotation among solar neighborhood solar-type stars. Rotation is generally too low in stars less massive than the Sun to be measured reliably, and rotation periods are available for just a few stars. At present using the rotation-activity relation of Noyes *et al.* (1984) to transform HK emission into an estimate of rotation remains the only way to study the rotation of a large, volume-limited sample.

More recently, Henry *et al.* (1996) observed about 800 southern G dwarfs out to about 50 pc. They confirmed the existence of a gap in their much larger sample, and they illustrated two interesting groups in the sample: the very active stars and the very inactive stars. Many of the very active stars turn out to be RS CVn objects, but a few have lithium abundances and parallaxes that indicate that they are PMS stars, even though they are far from any star-forming regions. The very inactive stars may be older solar-type stars that are going through a phase analogous to the Sun's Maunder minimum.

5.3. ACTIVITY AS AN AGE INDICATOR

Activity is useful as a rough age indicator for solar-type stars, and for many purposes a rough age is good enough. Also, G dwarfs are relatively bright yet numerous; lower-mass stars get fainter faster than they increase in numbers, while high-mass stars drop in frequency faster than they get brighter. That means that solar-type stars are well-suited to studying some Galactic processes, such as "disk heating" (the increase in Galactic scale height with age). Solar-type stars also make ideal targets for planetary searches, not just because they resemble the Sun in a comfortable way, but also because their intrinsic properties make detection easier. Among those properties are low rotation rates among the older stars, and very low photometric variability, also among the older stars. Both properties have their roots in the rotation-activity connection and the inevitable decline of rotation with age that characterizes solar-type stars.

6. Questions

The following are issues that remain open at this time. Some are being addressed with models and observations being worked on right now, but others are more speculative.

1. What is the distribution of rotation in stars just before they reach the ZAMS? During this time the stars undergo huge changes in their moments of inertia because of overall shrinkage and the development of the radiative core. The changes in surface rotation should be dramatic if we can find the right sample to observe.
2. How does rotation evolve in stars older than the Hyades? Is there still a reservoir of AM in the core that continues to replenish the surface, or are Hyades stars solid-body rotators?
3. Do cores and envelopes really decouple? If they do, what are the details and what are the time scales?
4. We see latitudinal differential rotation on the Sun and we believe it is fundamental to the dynamo mechanism. How does latitudinal differential rotation depend on overall rotation rate, age, etc.?
5. How does mass loss depend on rotation rate, age, etc., and how does it affect angular momentum loss?
6. A recent paper (Ribes & Nesme-Ribes 1993) presents a fascinating analysis of notebooks from the 17th century Observatoire de Paris. They claim that the apparent solar rotation rate has changed over

the past three centuries. Is this really true? If so, and given that the solar rotation appears to be independent of depth (as indicated by helioseismology), how is it possible for AM to hide?

7. When we look at a cluster of stars, how representative is that cluster? Does metallicity, for example, affect the distributions of rotation that we see?

Acknowledgements

I would like to thank the following for contributing materials and for ongoing collaborations: Burton F. Jones (Lick Obs.), John Stauffer (CfA), David Latham and coworkers (CfA), Todd Henry (CfA), Lionel Siess and Jeremy King (STScI), Sallie Baliunas and Robert Donahue (CfA), and Marc Pinsonneault and Donald Terndrup (Ohio State). I would also like to thank Teresa Lago for the opportunity to come to Porto for this special occasion.

References

Alphenaar, P., & van Leeuwen, F. 1981, Inf. Bull. Var. Stars, No. 1957

Bouvier, J., Forestini, M., & Allain, S. 1997, *Astron. Astrophys.*, **326**, 1023–1043

Choi, P. I., & Herbst, W. 1996, *Astron. J.*, **111**, 283–298

Collier Cameron, A., & Li, J. 1994, *Mon. Not. Royal Astron. Soc.*, **269**, 1099–1111

Henry, T. J., Soderblom, D. R., Donahue, R. A., & Baliunas, S. L. 1996, *Astron. J.*, **111**, 439–465

Keppens, R., MacGregor, K. B., & Charbonneau, P. 1995, *Astron. Astrophys.*, **294**, 469–487

Krishnamurthi, A., *et al.* 1997, *Astrophys. J.*, **480**, 303–323

Li, J., & Collier Cameron, A. 1993, *Mon. Not. Royal Astron. Soc.*, **261**, 766–782

Noyes, R. W., Hartmann, L. W., Baliunas, S. L., Duncan, D. K., & Vaughan, A. H. 1984, *Astrophys. J.*, **279**, 763–777

Radick, R. R., Thompson, D. T., Lockwood, G. W., Duncan, D. K., & Baggett, W. E. 1987, *Astrophys. J.*, **321**, 459–472

Ribes, J. C., & Nesme-Ribes, E. 1993, *Astron. Astrophys.*, **276**, 549–563

Soderblom, D. R. 1985, *Astron. J.*, **90**, 2103–2115

Soderblom, D. R., Jones, B. F., & Walker, M. F. 1983, *Astrophys. J.*, **274**, L37–L41

Soderblom, D. R., Stauffer, J. R., Hudon, J. D., & Jones, B. F. 1993a, *Astrophys. J. Supp. Ser.*, **85**, 315–346

Soderblom, D. R., Stauffer, J. R., MacGregor, K. B., & Jones, B. F. 1993b, *Astrophys. J.*, **409**, 624–634

Stauffer, J. R., Hartmann, L. W., Burnham, J. N., & Jones, B. F. 1985, *Astrophys. J.*, **289**, 247–261

Stauffer, J. R., Hartmann, L. W., Soderblom, D. R., & Burnham, J. N. 1984, *Astrophys. J.*, **280**, 202–212

Vaughan, A. H., & Preston, G. W. 1980, *Pub. Astron. Soc. Pac.*, **92**, 385–391

ROTATION OF T TAURI STARS REVISITED

L.J.R.MACHADO
Centro de Astrofísica da Universidade do Porto
Rua do Campo Alegre, 823, 4150 Porto - Portugal

AND

M.T.V.T. LAGO AND J.J.G. LIMA
Centro de Astrofísica da Universidade do Porto
Departamento de Matemática Aplicada, FCUP

Abstract. This paper reviews the rotation of T Tauri stars. It summarizes the results of a complete bibliographical survey of all photometric and spectroscopic data available in connection with rotation studies of these stars. It is shown that T Tauri stars are slow rotators. More than half of the stars in the sample have been found to have $v \sin i < 25$ km/s.

Furthermore, the histogram for the photometric periods of rotation is clearly bimodal, with a peak at approximately 3 days and another at around 8 days.

1. Introduction

T Tauri stars (TTS) are low-mass (up to 3 $M_\odot$) pre-main-sequence stars of spectral types G to M (Herbig, 1983). They usually show H_α in emission and Li λ6707 in absorption, an indication of their youth. TTS can be separated in two classes: Classical TTS (CTTS) and Weak-line TTS (WTTS). CTTS have equivalent width of H_α ($W(H_\alpha)$) larger than 10Å and an infrared excess, WTTS have $W(H_\alpha) < 10$Å and a spectral energy distribution similar to that of a main-sequence star.

TTS were initially thought to rotate fast since being very young they would still be contracting towards the main-sequence. However the measurements of ($v \sin i$) made by Vogel and Kuhi (1981) showed for the first time that they are in fact slow rotators.

2. Rotational velocities and photometric periods

The rotational velocities can be measured either by the Fourier transform (Gray, 1992) or the cross-correlation methods (Benz and Mayor, 1981).

Combining the measurements of $v \sin i$ available in the literature we have built the histogram shown in Fig. 1 (left). The precision of the results

Astrophysics and Space Science is the original source of publication of his article. It is recommended that this article is cited as: *Astrophysics and Space Science* **261**: 69–70, 1999.

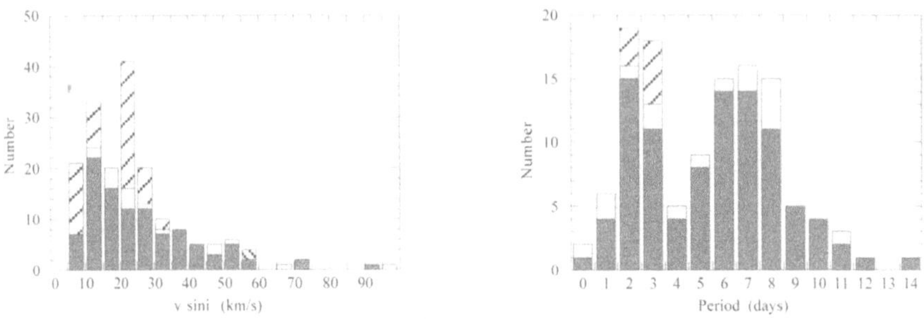

Figure 1. **left:** Histogram of the projected rotational velocity of TTS. **right:** Histogram of the rotational period of TTS. Dark grey are secure measurements of the quantity, light grey are uncertain measurements and the oblique stripes are upper limits.

has been taken into account. Around 62% of the 185 stars in this sample have $v \sin i < 25$km/s confirming the results of Vogel and Kuhi.

Some TTS show periodic changes in various photometric bands which are assumed to be associated with the presence of stellar spots. A periodogram is computed to determine the period.

The histogram shown in Fig. 1 (right) combines all available measurements of the rotational period. The distribution is clearly bimodal, with peaks around 3 and 8 days.

3. Discussion

Since TTS are slow rotators they must have lost angular momentum at some stage before this phase. An efficient way to loose angular momentum is through a magnetized stellar wind. However, to explain the slowest rotators another mechanism may in some cases be needed. A magnetic linkage at a very early phase between the protostar and a circumstellar disk is a possibility.

The bimodality in the histogram of the photometric period can be explained by the separation in the two classes CTTS and WTTS. On average CTTS rotate slower than WTTS. This could be due either to the presence of strong magnetic winds or a circumstellar disk.

Acknowledgments

L.J.R.M. acknowledges grant BD/5585/95 from PRAXIS XXI

References

Benz, W. and Mayor, M. (1981), *A&A*, **93**, p.235
Gray, D.F. (1992), in *The Observations and Analysis of Stellar Photospheres*
Herbig, G.H. (1983), in *Birth and Infancy of Stars*, p.535
Vogel, S.N. and Kuhi, L.V. (1981), *ApJ*, **245**, p.960

STELLAR TOMOGRAPHY

Differential rotation and magnetic polarity patterns on rapidly rotating stars

A. COLLIER CAMERON
University of St Andrews
School of Physics and Astronomy, North Haugh, St Andrews, Fife, SCOTLAND KY16 9SS

Abstract. I review recent progress in the field of stellar surface imaging, with particular reference to advanced methods for mapping surface-brightness inhomogeneities and the surface vector magnetic field on magnetically active late-type stars. New signal enhancement techniques, utilising profile information from hundreds or thousands of photospheric lines simultaneously, allow images to be derived for stars several magnitudes fainter than was previously possible. For brighter stars, the same techniques make it possible to map features as small as two or three degrees in extent on the stellar surface. This opens up whole new areas of research, such as the ability to use starspot tracking to study surface differential rotation patterns on single and binary stars, and to follow the secular evolution of the magnetic field itself.

1. Introduction

The problem of determining surface magnetic field distributions in active late-type stars has recently attracted a good deal of theoretical and observational attention. Schüssler et al (1996) have produced thin flux-tube models of flux emergence in G dwarfs, building on earlier work by Choudhuri (1987). The models trace the emergence of axisymmetric thin flux tubes models unstable to buoyant eruption from the base of the stellar convective zone. In models with solar (26-day) rotation periods, Schüssler et al find that buoyancy forces dominate over Coriolis terms, allowing flux tubes to rise radially from the generation region, giving rise to active regions only at low to intermediate latitudes. In models with more rapid (0.5-day) rotation rates, however, Coriolis forces dominate and the flux tubes are forced

Astrophysics and Space Science is the original source of publication of his article. It is recommended that this article is cited as: *Astrophysics and Space Science* **261**: 71–80, 1999.

to rise nearly parallel to rotation axis. These models predict that the flux tubes should emerge mainly at intermediate latitudes, but are not expected to produce polar caps or low-latitude spots.

The rapidly rotating K0 dwarf AB Doradus is an excellent object in which to test such theories. Its 12.4-hour rotation period is particularly well suited to surface imaging studies, allowing much of its surface to be mapped from a single telescope during the course of one night. The results presented in this paper have been secured as part of a long-term programme of Zeeman-Doppler imaging (ZDI) aimed at eliciting the geometry of the dynamo-generated magnetic field and to study the way in which surface differential rotation may drive changes in the global structure of the overlying corona, as traced by the slingshot-prominence system (Cameron & Robinson 1989a,b).

2. Doppler imaging: General principles

The main requirement for successful Doppler imaging is that the rotational Doppler effect should be the dominant line broadening mechanism for the majority of photospheric absorption lines. Each surface element on the star contributes a limb-darkened and Doppler-shifted copy of the local specific intensity profile. Integration over the visible disk at any instant then gives the rotationally-broadened spectrum. A dark spot on the stellar surface can then be regarded as a region where there is a deficit in the local spectrum. This has the effect of depressing the continuum slightly at all wavelengths, and leaving bright but narrow "bumps" in the photospheric absorption-line profiles. For the case of solid-body rotation, the displacement of the bump from line centre is directly proportional to the projected distance of the spot from the stellar rotation axis. A sequence of spectra (Fig. 1) then shows these bumps drifting across the profile at rates dictated by the latitudes of the spots in which they originate.

Given a map of the spots on the stellar surface, and a physical model for the specific intensity spectrum of radiation emerging from the spotted and unspotted regions on the stellar surface, Vogt & Penrod (1983) showed that it is relatively straightforward to predict the form of the observed spectral-line profiles at any given rotation phase. The stellar image is treated as a vector of dimension M, containing either local temperatures or spot filling factors, on a grid of pixels with pre-determined locations and surface areas. The observations can also be treated as a vector of dimension N, containing the spectral fluxes observed at each wavelength in a succession of spectra spanning the rotation cycle. The functional dependence of the local specific intensity on the local image pixel value and foreshortening angle at each observed rotation phase is encoded in the elements of a

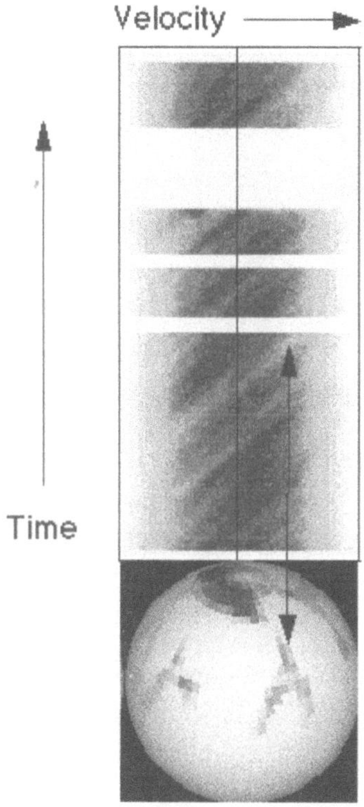

Figure 1. Schematic illustration of the basic principles of Doppler imaging for spotted stars. The spectral signature of the light "missing" from the spotted regions contains a Doppler-shifted absorption line, leaving bright bumps in the observed spectrum which migrate across the stellar disk at a rate that depends on the spot's stellar latitude.

$M \times N$ transformation matrix. The forward transformation of the image values into a synthetic data set is effected by computing the product of the transformation matrix with the image vector.

Vogt et al (1987) refined the technique further by applying the Maximum Entropy Method of Skilling & Bryan (1984) to the problem of recovering an image that was unique in the sense of determining the "simplest" image capable of yielding a satisfactory fit to the data. Several other groups developed Doppler imaging codes within the next few years, and achieved broadly similar results. The misfit between the resulting synthetic data and the observations is usually quantified by χ^2 or an equivalent statistic. A regularising function such as the image entropy (see, e.g. Brown et al 1991, Cameron 1992, Piskunov & Rice 1993) or the Tikhonov smoothness criterion (Rice et al 1989, Piskunov et al 1990) may also be computed, if the data are sparse or noisy enough for the problem to be ill-conditioned.

The image is then improved iteratively using a search algorithm based on the image-space gradients of both the fitting statistic and the regularising function. Other approaches to the iterative improvement of the image include Kürster's (1993) CLEAN-like algorithm, and Berdyugina's (1998) Occamian approach, which is essentially an adaptation of singular-value decomposition. In all these methods, the reconstruction is halted once a satisfactory fit to the data has been achieved (at an extremum of the regularising function, if appropriate).

3. Zeeman-Doppler imaging

It is possible to derive vector magnetograms of rapidly rotating stars using high-resolution circular spectropolarimetry. The development of the most successful instrument of this kind, the Semel Polarimeter, was described in detail by Donati et al. (1997). It consists of a polarimeter (mounted at the Cassegrain focus of the 3.9-m Anglo-Australian Telescope), a double-fibre link to the UCL Echelle Spectrograph (UCLES) and a Bowen-Walraven image slicer device. The polarimeter sends light analyzed for left- and right-circularly polarized light down each of the two fibres. Each Stokes V observation consists of a sequence of four observations, in which the polarizations in the two fibres are switched via rotation of the quarter-wave plate of the polarimeter.

For magnetic imaging, the model we use is that of Donati & Brown (1997), which assumes that the intrinsic profile is gaussian and constant in shape over the stellar surface. We further assume that this local profile is a Zeeman triplet and obeys the weak-field approximation. Although very simple, this model is found to give a good description of Stokes V profiles averaged over more than 1,500 individual spectral features (see below). The quantities we reconstruct are the radial, meridional and azimuthal components of the magnetic field vector in spherical coordinates, weighted by potential surface inhomogeneities in brightness, magnetic filling factor and central depth of the intrinsic profile

4. Least-squares deconvolution

Extreme rotation has the effect of making the photospheric absorption lines extremely broad and shallow. The amplitudes of the bumps produced by starspots are reduced accordingly, and in spectra with low signal-to-noise (S/N) ratios, faint spot features may be deeply buried in noise. In Zeeman Doppler imaging, the amplitude of the Stokes V signature in a typical photospheric line is no more than about 1 part in 10^3 of the local continuum level. Donati et al (1997) overcame these problems for both conventional and Zeeman-Doppler imaging problem through the use of a technique known as *least-squares deconvolution (LSD)*, which extracts averaged profile information from the thousands of photospheric lines that are recorded in a typical echelle spectrum. The observed spectrum r is treated as the convolution f of a "mean" line profile z with a set of weighted delta functions at the wavelengths of a comprehensive list of spectral lines. This operation can be written in matrix form as

$$\boldsymbol{f} = \boldsymbol{A}.\boldsymbol{z} \tag{1}$$

The line weights, incorporated in the convolution matrix A, are proportional to the central depths of the lines as computed from a Kurucz model atmosphere for the appropriate spectral type. The elements α_{jk} of the matrix A are computed by summing over all spectral lines i, the fractional contribution to data pixel j of the kth element of the profile when the centre of the deconvolved profile is shifted to the wavelength λ_i of each line in turn.

The mean profile is then recovered from the observed spectrum and the convolution matrix by the method of least squares. This entails solving the matrix equation

$$\boldsymbol{A}^T.\boldsymbol{V}.\boldsymbol{A}.\boldsymbol{z} = \boldsymbol{A}^T.\boldsymbol{V}.\boldsymbol{r}, \tag{2}$$

where the inverse variances σ_j^2 associated with the N elements r_j of the residual spectrum r are incorporated via the diagonal matrix

$$\boldsymbol{V} = \mathrm{Diag}[1/\sigma_1^2, \ldots, 1/\sigma_N^2]. \tag{3}$$

Since the square matrix $A^T.V.A$ is symmetric and positive-definite, the least-squares problem can be solved using efficient methods such as Cholesky decomposition (Press et al 1992).

In terms of signal improvement, this method is analogous to aligning and co-adding (with appropriate weighting factors for line strength and local continuum signal strength on the recorded frame) the profiles of the 2000 or so images of the 1500-odd individual photospheric absorption lines recorded on each echellogram. The deconvolution procedure compensates for local line blends, and so has the advantage over cross-correlation methods that outside the region occupied by the residual stellar profile, the deconvolved spectrum is flat. The formal errors on the M points of the deconvolved profile are obtained in the usual way from the diagonal elements of the $M \times M$ inverse matrix $C = [A^T.V.A]^{-1}$.

5. Images of AB Dor in 1995/6

Donati & Collier Cameron (1997) and Donati et al (1998) used the Semel Polarimeter to secure Stoves V and I spectra of the rapidly rotating southern K0 dwarf star AB Doradus in 1995 December and 1996 December. The

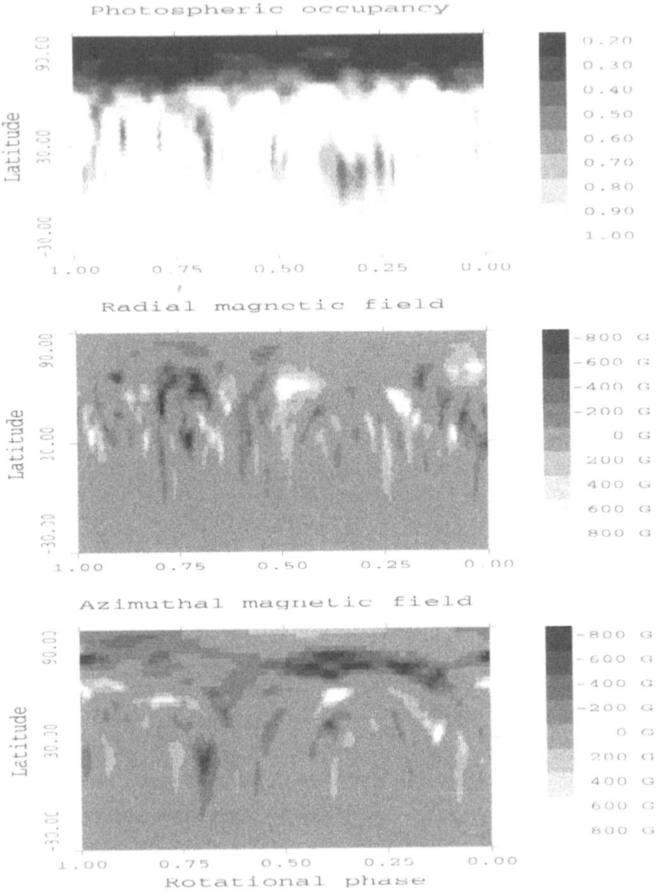

Figure 2. Maximum entropy brightness (upper row) and magnetic (lower rows) rectangular maps of AB Dor on 1996 Dec. 23 to 25. The reconstructed meridional field component is very small and is therefore not shown here. Note that rotational phase runs backwards (i.e. from right to left) on all maps.

resulting surface-brightness images in Figs. 2 through 4 were reconstructed using the two-component model of Cameron (1992). The image is defined in terms of a quantity f_j in the range $0 < f_j < 1$, representing the fraction of pixel j occupied by clean photosphere. The maximum entropy reconstruction code we use is that of Brown et al. (1991), which implements Skilling & Bryan's (1984) algorithm for maximum entropy optimisation problems. LSD profiles of the K0V star Gl 176.3 and the M4V object Gl 367 are used as template profiles describing the spectral contribution of the photosphere and spot respectively (see Donati & Cameron 1997 for details).

The radial and azimuthal components of the stellar surface magnetic

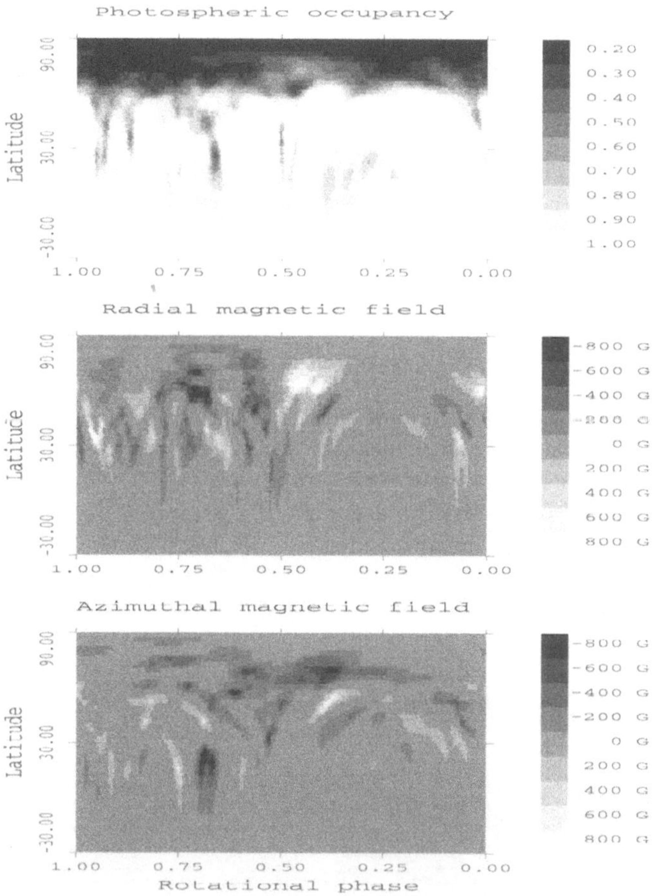

Figure 3. Same as Fig. 2 but for 1996 Dec. 27. The big gap in phase coverage between phase 0.06 and 0.42 is responsible for the poor definition of all image features located within the corresponding longitude range.

field derived from the Stokes V data are shown in the lower panels of Figs. 2 through 4. Note in particular the repeatability of fine structure in the left-hand halves of these maps, where good phase coverage was obtained throughout the run.

6. Differential rotation from starspot tracking

The deconvolved profiles, based on nearly 2000 line images, have a S/N ratio that is higher by a factor of 30 than any of the relatively few un-blended photospheric lines found in the spectral region used. This allowed us for the first time to reconstruct surface images without having to resort to binning

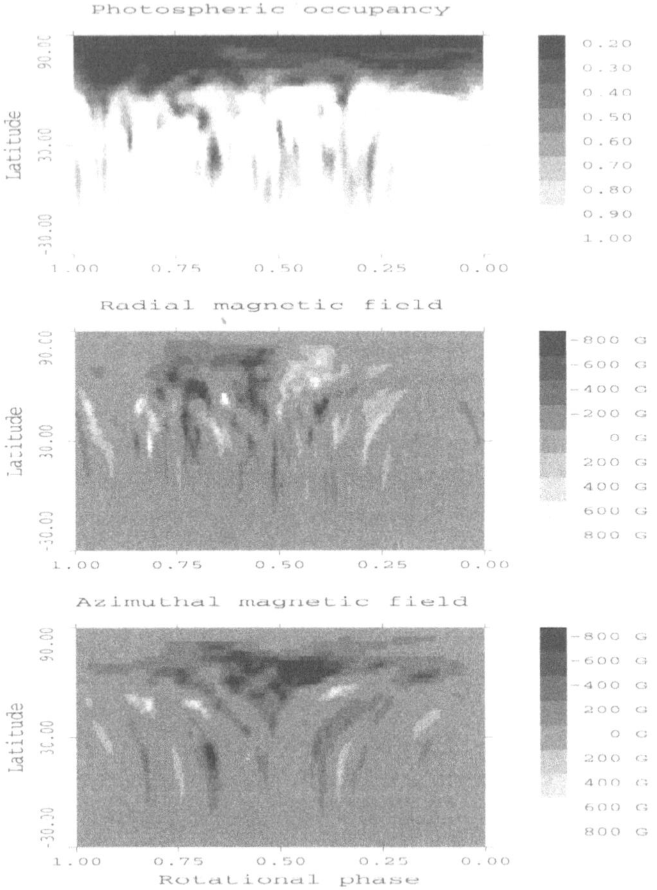

Figure 4. Same as Fig. 2 but for 1996 Dec. 29. The big gap in phase coverage between phase −0.05 (i.e. 0.95) and 0.30 is rseponsible for the poor definition of all image features located within the corresponding longitude range.

in either wavelength or time. The exquisite detail seen in Figs. 2 through 4 allowed us to use individual spots as tracers of the surface differential rotation pattern from the equator to the perimeter of the dark polar spot, at stellar latitude 70° or so. A latitude-by-latitude cross-correlation (Fig. 5) of images obtained up to 8 nights apart shows clearly that the star's equator rotates faster than its poles. Surprisingly, in spite of the fact that AB Dor rotates some 50 times faster than the Sun, we find that the length of time taken for the equator to pull one full turn ahead of the polar regions is about 120 days, very similar to the corresponding figure for the Sun.

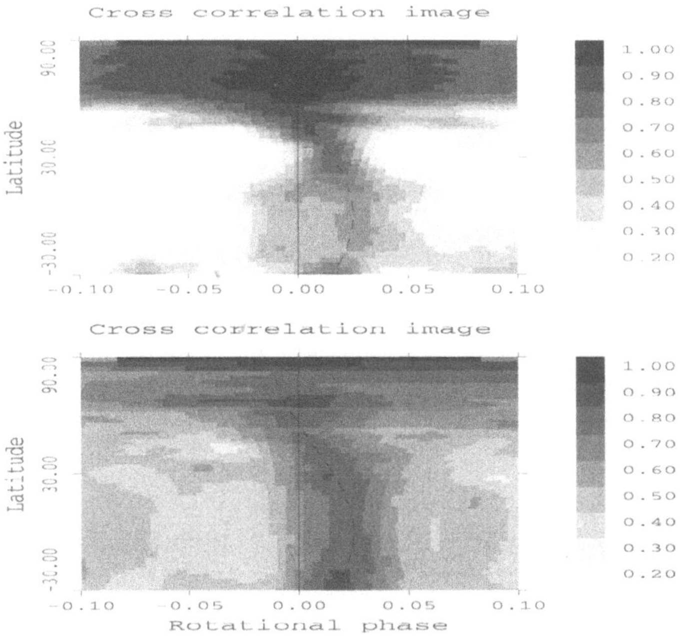

Figure 5. Cross-correlation images obtained by cross-correlating each latitude belt of Fig. 2 and Fig. 4 on phase interval [0.55,0.95], for brightness (upper panel) and magnetic field (lower panel). Note that the magnetic cross-correlation image is the geometrical mean of both radial field and azimuthal field cross-correlation maps. The dashed line in each panel depicts the surface differential rotation curve derived by Donati & Cameron (1997).

7. Magnetic topology

All magnetic regions detected in 1995 December by Donati & Cameron (1997) were located above a latitude of about 30°. The magnetic images clearly share the differential rotation found in the surface-brightness images. The higher quality data obtained in 1996 December reveal almost a full ring of negative azimuthal field encircling the visible stellar rotational pole (see Fig. 2), very similar to what is found on HR 1099, another very active though much more evolved object (Donati et al. 1992; Donati 1998). We also detect in the new images a low-latitude feature of negative azimuthal field. It strongly suggests in particular that the underlying large scale toroidal field structure undergoes at least two polarity switches in the upper hemisphere of AB Dor.

8. Summary and future prospects

The use of least-squares deconvolution to determine high-quality broadening functions for rapidly rotating stars has led to a significant improvement in our ability to discern fine detail on the surfaces of rapidly rotating stars. In particular, it now allows us to use starspots as tracers of the surface differential rotation patterns on stars other than the Sun. New maps of AB Dor show a solar-like differential rotation pattern and rate of surface shear. The distribution of starspot activity, and the presence of belts of azimuthal field, suggest that current models of buoyant eruption from a deeply-buried dynamo may require revision. Further ZDI and differential rotation observations will guide theoretical efforts to understand the generation and emergence of magnetic fields in these rapidly rotating analogues of the young Sun. Although at the time of writing differential rotation has been measured for only one star, AB Dor, further observations are scheduled in the next few months to secure differential rotation measurements for a further six or seven stars with half-day rotation periods and spectral types ranging from early G to early M.

References

Berdyugina S. V., Ilyin I., Tuominen I., 1998, in Donahue R. A., Bookbinder J. A., eds, The Tenth Cambridge Workshop on Cool Stars, Stellar Systems, and the Sun. ASP, San Francisco, p. CD
Brown S. F., Donati J.-F., Rees D. E., Semel M., 1991, A&A, 250, 463
Choudhuri A. R., Gilman P. A., 1987, ApJ, 316, 788
Collier Cameron A., Robinson R. D., 1989a, MNRAS, 236, 57
Collier Cameron A., Robinson R. D., 1989b, MNRAS, 238, 657
Collier Cameron A., 1992, in Byrne P. B., Mullan D. J., eds, Surface Inhomogeneities on Late-type Stars. Springer-Verlag, Berlin, p. 33
Donati J.-F., Brown S. F., 1997, A&A, In press
Donati J.-F., Collier Cameron A., 1997, MNRAS, 291, 1
Donati J.-F., Semel M., Carter B., Rees D. E., Collier Cameron A., 1997, MNRAS, 291, 658
Donati J.-F., Collier Cameron A., Hussain G. A. J., Semel M., 1998, MNRAS, In press
Donati J.-F., Semel M., Rees D. E., 1992, A&A, 265, 669
Donati J.-F., 1998, A&A, In press
Kürster M., 1993, A&A, 274, 851
Piskunov N. E., Rice J. B., 1993, PASP, 105, 1415
Piskunov N. E., Tuominen I., Vilhu O., 1990, A&A, 230, 363
Press W. H., Flannery B. P., Teukolsky S. A., Vetterling W. T., 1992, Numerical Recipes: The Art of Scientific Computing (2nd edition). Cambridge University Press, Cambridge
Rice J. B., Wehlau W. H., Khokhlova V. L., 1989, A&A, 208, 179
Schüssler M., Caligari P., Ferriz-Mas A., Solanki S. K., Stix M., 1996, A&A, In press
Skilling J., Bryan R. K., 1984, MNRAS, 211, 111
Vogt S. S., Penrod G. D., 1983, PASP, 95, 565
Vogt S. S., Penrod G. D., Hatzes A. P., 1987, ApJ, 321, 496

OUR NEW VIEW OF THE SOLAR CORONA FROM YOHKOH AND SOHO

C.E. PARNELL
Department of Mathematics
University of St. Andrews
The Mathematics Institute
North Haugh
St. Andrews
FIFE KY16 9SS
Scotland

Abstract. The Yohkoh satellite has now been orbiting the Earth for about 6 years and during this time it has revealed a number of new features in the solar corona. These include the discovery of X-ray jets and active region transient brightenings, as well as observations of hard X-ray sources above and at the feet of soft X-ray flare loops. SOHO, the newest solar space mission, is not orbiting the Earth, but is in fact orbiting the Sun and has been at the Earth's L1 point for about 2 years. During this time it too has identified some new and interesting characteristics of both the solar corona, the photosphere and the solar interior. For example, studies of high resolution MDI/SOHO magnetograms indicate that the magnetic carpet may play an important role in the heating of the coronal. Also polar plume studies from EIT/SOHO data suggest that they may be a possible source for the fast solar wind.

Together these two missions have dominated solar research throughout the 90s and are expected to continue to do so during the rise from solar minimum to the next solar maximum.

1. Introduction

The first recorded observations of the solar corona go back as far as 2000 B.C. These were made during eclipses, which were the only way of viewing the solar corona up until the invention of the coronagraph by Lyot in 1930.

Astrophysics and Space Science is the original source of publication of his article. It is recommended that this article is cited as: *Astrophysics and Space Science* **261**: 81–90, 1999.

This is because the Sun's surface, the photosphere, is incredibly bright, about a million times brighter than the solar corona and so it completely obscures the fainter emission from other regions. Lyot's coronagraph consisted of a telescope with an occulting disc that eliminated this glare and meant that white light images of the corona could be taken daily. Since then coronagraphs have been used not just at ground based observatories, but also on satellites. Indeed, the LASCO instrument on SOHO includes three coronagraphs.

Coronagraphs are, however, only able to observe the limb of the corona and cannot observe the corona above the disk of the Sun. Instead, to observe the coronal disk another instrument, in most cases a soft X-ray telescope, must be used. The corona is at a temperature of several million kelvin and is up to 200 times hotter than the lower atmospheres, so it is the only atmospheric layer of the Sun that emits thermally in X-rays. Unfortunately, however, soft X-rays are absorbed in the Earth's atmosphere, so X-ray telescopes can only be used from space based observatories. Both Skylab and SMM missions had X-ray telescopes that produced superb data, from which new phenomena, for instance X-ray bright points, were discovered. They also led to advances in the understanding of solar flares, as well as hinting at the amazing complex structure of the corona. The instruments on Yohkoh and SOHO produce much more detailed data with high spatial and temporal resolution giving stunning images of the corona that reveal yet more of the coronas secrets.

Before I discuss some of these wonderful results let me first say that to fully understand how the corona is heated or any phenomena is created it is essential to get as much information as possible, which, of course, involves observing in many different wavelengths over short periods with high cadences, as well as over long periods with lower cadence. Only then can the complete picture be drawn up and great strides be made in both theory and observations.

In the following sections I will start by talking about the magnetic structure at the Sun's surface, the magnetic carpet, as observed by MDI on board SOHO before moving on to discuss the solar corona, in particular the results from Yohkoh that seem to indicate magnetic reconnection. Next I will talk about solar flares in the light of results from Yohkoh. A discussion of coronal mass ejections then follows looking at some of the incredible images taken by LASCO on board SOHO. Finally, I will briefly mention the solar wind, in particular its possible origin, with the conclusions presented in the last section to round up the results.

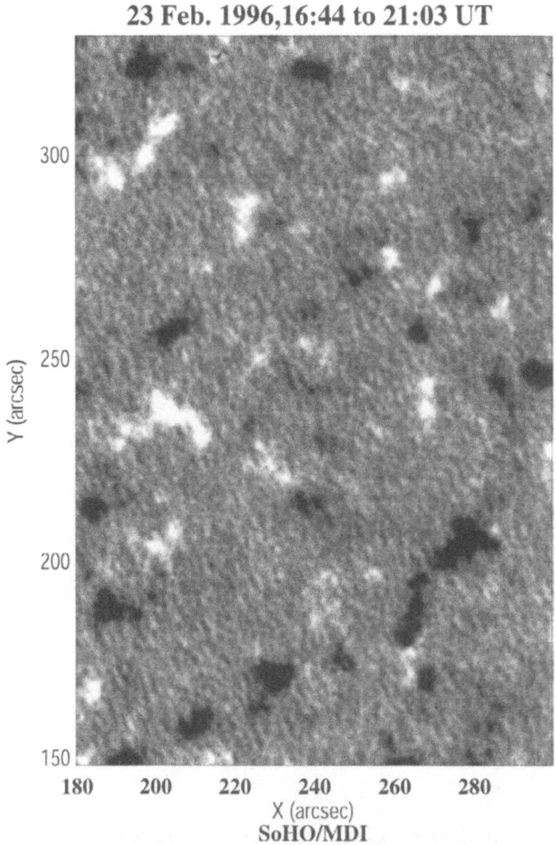

Figure 1. High resolution magnetogram image taken by MDI on SOHO showing the magnetic carpet (Courtesy of SOHO/MDI consortium. SOHO is a project of international cooperation between ESA and NASA.)

2. The Magnetic Carpet

The Michelson Doppler Imager (MDI) on board SOHO is a versatile instrument which not only produces data about the solar interior, but can also produce high resolution magnetograms (Figure 1). As SOHO was launched in 1995 the magnetogram images it has produced to date have been taken during solar minimum so are mostly of the quiet (less active) Sun. A study of high resolution magnetograms by Dr. Alan Title and his team at Lockheed (Palo Alto, USA) taken of the regions below the quietest, most inactive corona reveals that the surface of the Sun is far from inactive. Indeed there are literally hundreds of small magnetic fragments continuously popping up in pairs through the Sun's surface presumably indicating the emergence of small loops of flux pushed up from the convection zone. These small pairs

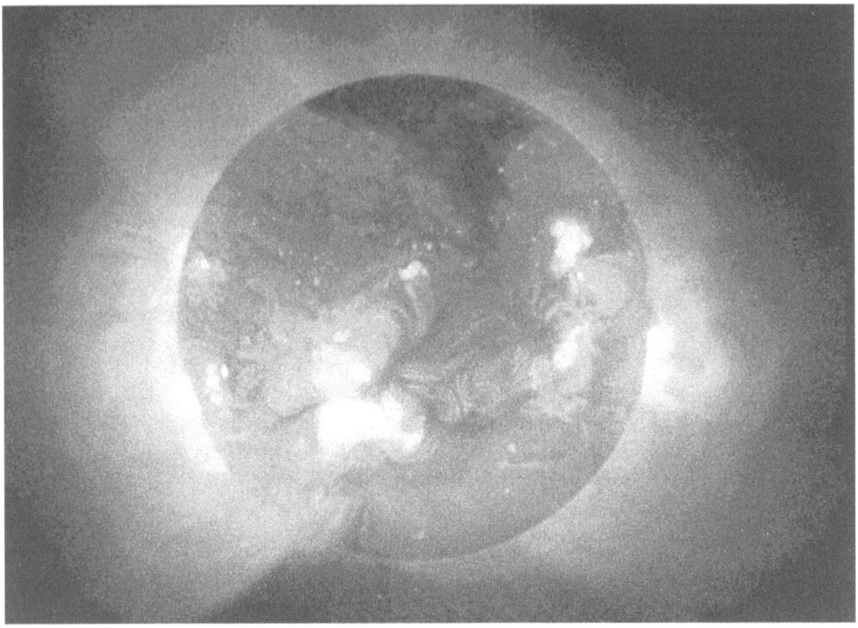

Figure 2. Full disc image of the solar corona taken by the SXT on Yohkoh (The soft X-ray telescope was prepared by the Lockheed Palo Alto Research Laboratory, the National Astronomical Observatory of Japan, and the University of Tokyo with the support of NASA and ISAS).

of fragments diverge until they either merge or cancel with surrounding invariably larger fragments. These larger fragments grow or shrink in size through their interaction with other flux until they eventually fragment due to changes in the surface flow patterns or merely their large size. It has been estimated by Schrijver et al. (1997) that the total magnetic flux on the quiet Sun is completely replaced every 40 hours through emergence of new flux and cancellation. This continual movement of flux and the incessant change of the magnetic field structures in the photosphere through fragmentation, merging, cancellation and emergence must affect the magnetic field structures of the corona above. In the following chapter some of the coronal signatures of these movements are discussed.

3. The Solar Corona

It is clear when looking at any high resolution soft X-ray image of the corona that there is a three-fold structure (Figure 2). The most notable of these structures are the many bright *loops*: small compact loops, long arcades and single isolated loops often dominate images especially those taken near

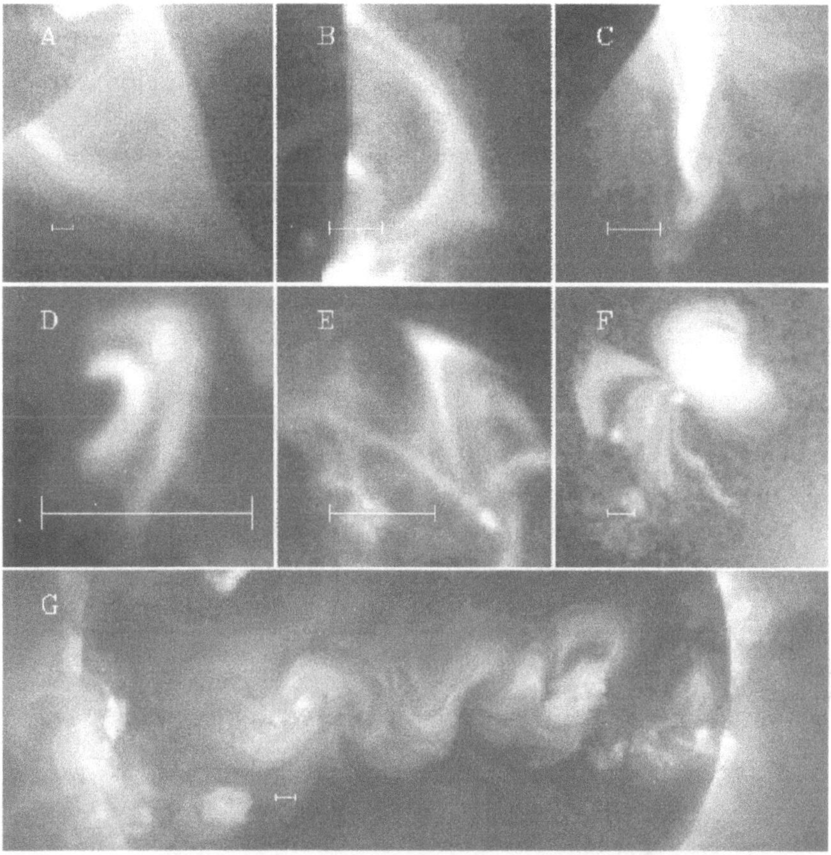

Figure 3. A collection of soft X-ray images showing various different types of loop structures present in the solar corona (The soft X-ray telescope was prepared by the Lockheed Palo Alto Research Laboratory, the National Astronomical Observatory of Japan, and the University of Tokyo with the support of NASA and ISAS).

solar maximum (Figure 3). In many images, however, the absence of bright regions may also be very striking. Dark regions, most commonly around the poles are open field regions known as *coronal holes.* In coronal holes and around the closed field regions (loops) it may be possible to identify small bright dots known as *X-ray bright points* that are the third structure that make up the corona. These bright points are the coronal counterparts to many of the cancelling or emerging bipolar fragments in the photosphere (Webb et al. 1993 and references therein, Priest et al. 1994).

When two opposite polarity magnetic fragments converge and then mutually lose flux (cancel) they must reconnect. The reconnection site is invariably high up in the corona above pairs of fragments where it causes hot dense plasma to be injected along newly reconnected field lines. These brighten and form the observed X-ray bright points (Parnell et al. 1994).

It is possible that further evidence for magnetic reconnection has been observed from soft X-ray images in the form of hot bright loop structures that look like X-points and bright hot cusp-like loop structures as one would expect near Y-points.

Yohkoh images have also revealed two new types of phenomena. Both are impulsive small scale features. The first, active-regions transient brightenings (ARTBs) appear in active regions and last for at most 15 mins (Shimizu et al. 1992, 1994). They have energies of 10^{25}-10^{27} ergs. The second, X-ray jets, are bright elongated features that extend out from small active regions and X-ray bright points either singly or in pairs (Shibata et al. 1992, 1994, and Shimojo et al. 1996). They have lifetimes of minutes, though in some cases can last up to 11 hours, and energies of 10^{25}-10^{28} ergs. It has been suggested (Parnell 1994 and Shibata 1996) that both of these features are caused by magnetic reconnection, but further studies need to be carried out before this can be put on a firmer footing.

4. Solar Flares

Arguably the Sun's most impressive phenomena, solar flares are large-scale dynamic eruptions that release energies of up to 10^{32} ergs and last from a few minutes to a few hours (Figure 4). They have signatures not only in soft X-rays, but also in hard X-rays and Hα. For many years solar physicists have suspected that solar flares are a result of reconnection in the corona. Several researchers (Tsuneta 1996 and Shibata 1996) now firmly believe that the Yohkoh satellite has really observed the key features of reconnection during flares. In one flare a hard X-ray source has been observed high in the corona above the bright soft X-ray loops (Masuda et al., 1994). This may correspond to the reconnection site. Cusp-shaped loops resembling newly reconnected field lines are often observed in flares and these also seem to point to an X-type reconnection site above. Theories suggest that downward out flow from the reconnection site causes standing slow shocks which would then heat the downward plasma (Forbes and Malherbe, 1991). Accelerated electrons are believed to stream down from these shocks along the newly reconnected field lines and impacting on the dense photospheric 'wall' below, giving rise to Bremsstrahlung emission which is observed in the form of hard X-ray sources at the feet of the loops and causes evaporation of plasma which fills the new loops making them visible in soft X-rays (Tsuneta, 1996). Furthermore, shrinkage of loops and the gradual widening of the loop foot points also may indicate reconnection (Forbes and Acton 1996).

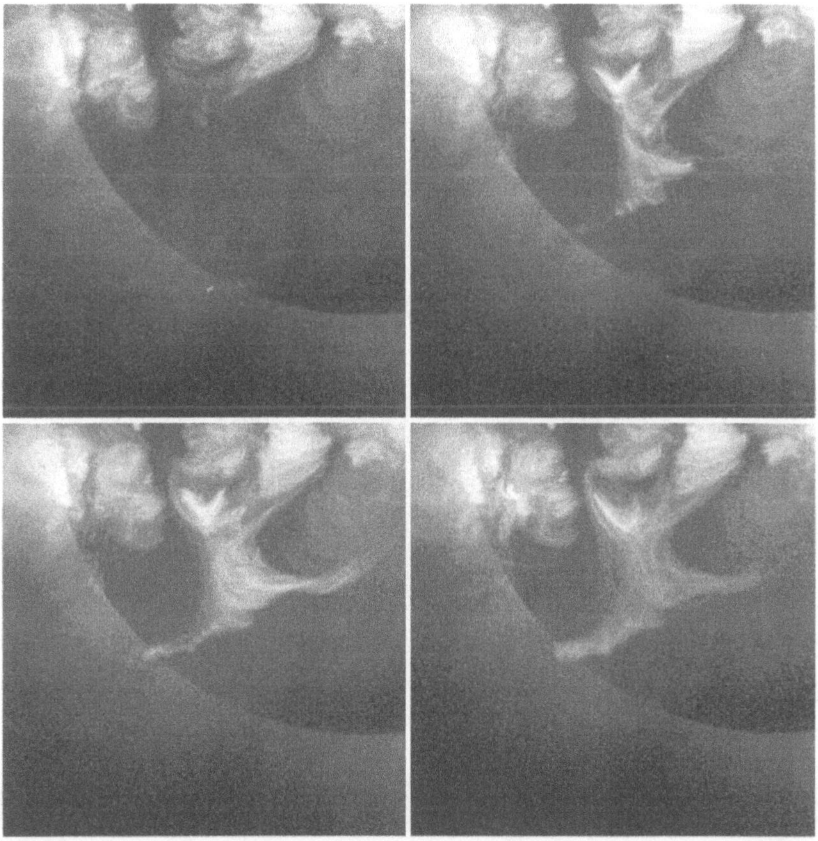

Figure 4. A time series of images taken by SXT on Yohkoh showing the reconstruction of the coronal magnetic field during a flare (The soft X-ray telescope was prepared by the Lockheed Palo Alto Research Laboratory, the National Astronomical Observatory of Japan, and the University of Tokyo with the support of NASA and ISAS).

5. Coronal Mass Ejections

The instrument on SOHO which has to date attracted more attention than any other is the Large Angle and Spectrometric COronagraph (LASCO). This is in fact a triple coronagraph capable of producing images with three different fields of view: C1 from 1.1 to 3 $R_\odot$, C2 from 1.5 to 6 $R_\odot$ and C3 from 3 to 30 $R_\odot$. The extensive range of coronagraph images produced allows researchers to follow any large scale ejection out to many solar radii. One of the main features tracked by LASCO has been coronal mass ejection (CMEs) (Figure 5). CMEs are huge (10^{13} kg) mass expulsions that are thrown out of the Sun's atmosphere at speeds of between 100 km s^{-1} to 1200 km s^{-1}. They have energies of about 10^{32} ergs and can have a devastating

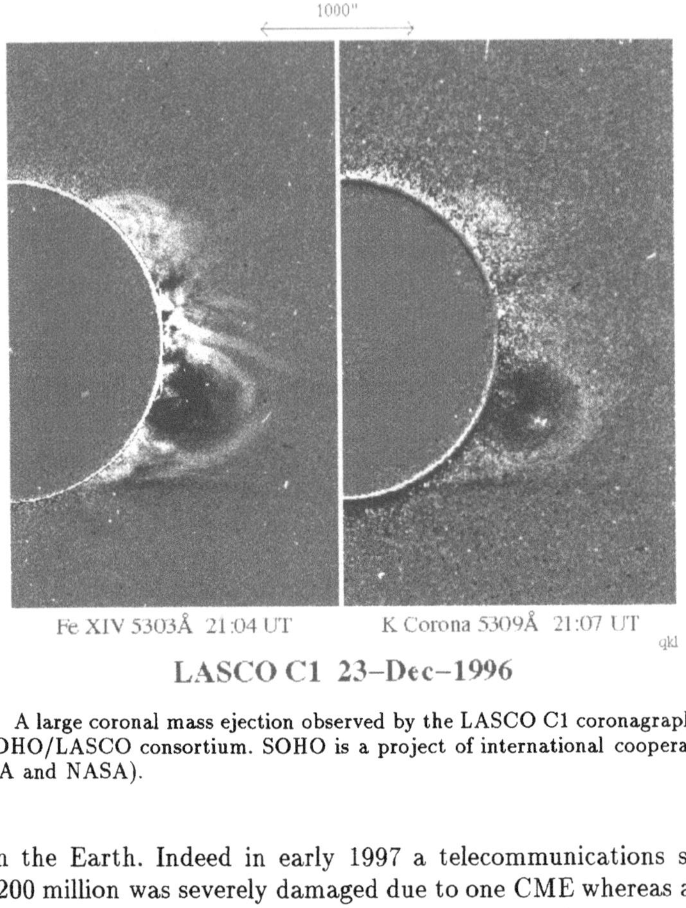

Figure 5. A large coronal mass ejection observed by the LASCO C1 coronagraph (Courtesy of SOHO/LASCO consortium. SOHO is a project of international cooperation between ESA and NASA).

effect on the Earth. Indeed in early 1997 a telecommunications satellite worth $200 million was severely damaged due to one CME whereas another earlier CME caused a major blackout in the US and Canada as power surges blew much of North Americas power distribution grid.

LASCO observes CMEs on a daily basis, though mostly they are not Earth directed, and has provided valuable data on what the actual cause of these huge ejections may be.

6. The Solar Wind

SOHO instruments such as SWAN are providing excellent data about the solar wind (Bertaux, these proceedings). Other coronal instruments, for instance the Extreme Ultraviolet Imager (EIT) on SOHO, also contribute to solar wind studies. For many years it has been known that polar winds are much faster than equatorial winds; this is probably due to the open field nature of coronal holes. Estimates of the solar wind at 2 solar radii above the poles suggest the wind is travelling at around 200 km s^{-1}. New

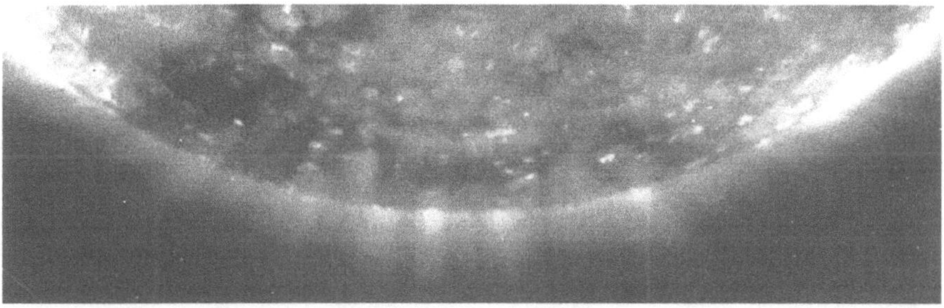

Figure 6. A close up of a coronal hole showing polar plumes taken by EIT in Fe IX/X (Courtesy of SOHO/EIT consortium. SOHO is a project of international cooperation between ESA and NASA).

research using EIT by Deforest et al. (1997) show that polar plumes may be the source to this fast wind. They are bright (dense) elongated structures emanating from coronal holes, which last for several hours (Figure 6). They have lengths of up to 15-20 solar radii and have variations that propagate outwards at speeds of 300 km s^{-1}.

7. Conclusions

This last decade has seen many advances in solar physics, especially from an observational point of view. We now have achieved images of both high spatial and temporal resolution which have enabled us to discover new small-scale dynamic features such as ARTBs and X-ray jets, as well as furthering our knowledge of the details of large-scale phenomena such as solar flares and CMEs. These observations have helped advance our understanding of coronal heating, in particular X-ray bright points, and the origin of the solar wind. Furthermore, the ability to take continuous 24 hour observations of the Sun has led to real steps being taken in determining the nature the solar interior (Thompson, these proceedings).

Solar Physics is in an extremely lively state on both the observational and theoretical fronts. Yohkoh and SOHO will soon be joined by the Transition Region and Coronal Explorer (TRACE) and will hopefully continue to feed the solar physicists with much useful data for many years to come.

For more information and the latest updates on the Yohkoh and SOHO missions the following web addresses may be of interest:

http://www.space.lockheed.com/SXT/homepage.html

http://sohowww.nascom.nasa.gov/

Acknowledgements

I would like to thank the organising committee for inviting me to share some of the wonderful findings from Yohkoh and SOHO at this meeting.

References

Deforest, C.E., Gurman, J.B. and Moses, J.D. (1997) Time Variability of Polar Plumes as Observed with SOHO/EIT and SOHO/MDI, *Bull. American Astron. Soc.*, **SPD 29**, pp. 803D

Forbes, T.G. and Acton, L.W. (1996) Reconnection and Field Line Shrinkage in Solar Flares, *Astrophysical Journal*, **459**, pp. 330

Forbes, T.G. and Malherbe, J.M. (1991) A Numerical Simulation of Magnetic Reconnection and Radiative Cooling in Line-tied Current Sheets, *Solar Physics*, **135**, pp. 361–391

Masuda, S., Kosugi, T., Hara, H., Tsuneta, S. and Ogawara, Y. (1994) A Loop-top Hard X-ray Source in a Compact Solar Flare as Evidence for Magnetic Reconnection, *Nature*, **371**, pp. 495

Parnell, C.E. (1994) Modelling of Dynamic Coronal Heating, *Proc. of the Third SOHO Workshop - Solar Dynamic Phenomena and Solar Wind Consequences (ESA SP-373)*, pp. 149–158

Parnell, C.E., Priest, E.R. and Golub, L. (1994) The Three-dimensional Structure of X-ray Bright Points, *Solar Physics*, **151**, pp. 57–74

Priest, E.R., Parnell, C.E. and Martin, S.F. (1994) A Converging Flux Model of an X-ray Bright Point and an Associated Cancelling Magnetic Feature, *Astrophysical Journal*, **427**, pp. 459–474

Schrijver, C.J., Title, A.M., van Ballegooijen, A.A., Hagenaar, H.J., and Shine, R.A. (1997) Sustaining the Quiet Photospheric Network: The Balance of Flux Emergence, Fragmentation, Merging and Cancellation, *Astrophysical Journal*, **487**, pp. 424

Shibata, K. (1996) New Observational Facts about Solar Flares from Yohkoh Studies - Evidence of Magnetic Reconnection and a Unified Model of Flares, *Adv. Space Res.*, **17**, pp. 9–18

Shibata, K., Ishido, Y., Acton, L.W., Strong, K.T., Hirayama, T., Uchida, Y., McAllister, A.H., Matsumoto, R., Tsuneta, S., Shimuzu, T., Hara, H., Sakuri, T., Ichimoto, K., Nishino, Y. and Ogawara, Y. (1992) Observations of X-ray Jets with the Yohkoh Soft X-ray Telescope, *Publ. Astron. Soc. Japan*, **44**, pp. L173–L179

Shibata, K., Nitta, N., Strong, K.T., Matsumoto, R., Yokoyama, T., Hirayama, T., Hudson, H. and Ogawara. Y. (1994) A Gigantic Coronal Jet Ejected from a Compact Active Region in a Coronal Hole, *Astrophysical Journal*, **431**, pp. L51-L53

Shimojo, M., Hashimoto, S., Shibata, K., Hirayama, T., Hudson, H.S. and Acton, L.W. (1996) Statistical Study of Solar X-ray Jets Observed with the Yohkoh Soft X-ray Telescope, *Publ. Astron. Soc. Japan*, **48**, pp. 123–136

Shimuzu, T., Tsuneta, S., Acton, L.W., Lemen, J.R. and Uchida, Y. (1992) Transient Brightenings in Active Regions Observed by the Soft X-ray Telescope, *Publ. Astron. Soc. Japan*, **44**, pp. L147–L153

Shimuzu, T., Tsuneta, S., Acton, L.W., Lemen, J.R., Ogawara, Y. and Uchida, Y. (1994) Morphology of Active Region Transient Brightenings with the Yohkoh Soft X-ray Telescope, *Astrophysical Journal*, **422**, pp. 906–911

Tsuneta, S. (1996) Magnetic Reconnection: Open Issues, *Magnetic Reconnection in the Solar Atmosphere ASP Conference Series*, **111**, pp. 409–418

Webb, D.F., Martin, S.F., Moses, D., and Harvey, J.W. (1993) The Correspondence Between X-ray Bright Points and Evolving Magnetic Features in the Quiet Sun, *Solar Physics*, **144**, pp. 15

EUV SPECTRAL VARIABILITY AND NON-EQUILIBRIUM IONISATION IN THE 'QUIET' SUN

D.H. BROOKS AND H.P. SUMMERS
Dept. Physics & Applied Physics
University of Strathclyde, Glasgow G4 0NG, Scotland

R.A. HARRISON AND J. LANG
Rutherford Appleton Laboratory
Oxon. OX11 OQX, England

AND

A.C. LANZAFAME
Istituto di Astronomia
Universita di Catania, I-95125 Catania, Italy

Abstract. Recent spectroscopic observations by the Solar and Heliospheric Observatory (SOHO) have revealed the dynamic nature of even the 'quiet' Sun. Spectral variability data clearly show that dynamics in the solar upper atmosphere take place on timescales shorter than those of ionisation relaxation. Accuracy in the interpretation of diagnostic spectral data can only be maintained through detailed quantitative modelling of the relevant atomic physics. In particular, dynamical plasma models of the solar plasma require matching dynamic atomic models to underpin conclusions drawn from the spectral reduction. The inclusion of important effects such as finite plasma electron density and the influence of metastable levels is essential to reduce the uncertainties associated with equilibrium assumptions.

1. Introduction

Preliminary observations by the CDS and SUMER spectrometers on SOHO have confirmed that even in apparently inactive solar regions there is considerable dynamic activity and spectral variability (see e.g. [1], [2]). Such observations may invalidate the assumption of ionisation equilibrium when the ionisation relaxation timescales are long. Any non-equilibrium has se-

Astrophysics and Space Science is the original source of publication of his article. It is recommended that this article is cited as: *Astrophysics and Space Science* **261**: 91–94, 1999.

rious consequences for many solar studies which have this assumption underlying their analysis techniques ([3], [4]).

If ionisation equilibrium is assumed, a proper treatment should include metastable levels and the density dependence introduced by dielectronic recombination (DR) [5]. Thus the accuracy can be improved and potential mis-interpretation of the spectral data can be avoided (see e.g. [6]).

2. Spectral variability observations by SOHO-CDS

High cadence CDS observations of transition region brightenings ('blinkers') were presented in [2], focusing on three emission lines: He I 537.03 Å , O IV 554.513 Å and Mg IX 368.057 Å with chromospheric ($\sim 2 \times 10^4$ K), transition region ($\sim 1.6 \times 10^5$ K) and coronal ($\sim 1 \times 10^6$ K) formation temperatures respectively. Relative intensity variations from the mean around the region of the 'blinker' events were also examined.

For what matters here, the main results of the analysis were: spectral variability of the He I and Mg IX lines is ~20-30% on timescales of 100-200 s, transition region variations larger than a factor two occur on timescales of 1-30 minutes and while the He emission is spatially related to the 'blinker' the Mg emission is not. Thus it is indicative of 'quiet' coronal variability.

A statistical analysis [7] shows upper limits to the instrumental component of the variability to be ~10%, 4% and 12% for the He, O and Mg lines respectively. With an anticipated 1σ statistical scatter of 4% this implies that a significant part of the variability reflects real variations in the emitting conditions of the the solar plasma.

3. Relaxation timescales

Following [8] and [7] we computed $\tau_{transient}$ which estimates the minimum time for some species to relax into equilibrium in a plasma of fixed electron temperature (T_e) and density (N_e). Its functional form, weighted by N_e, is shown in fig. 1 for some prominent radiators in the solar upper atmosphere.

Table. 1 shows the approximate relaxation estimates for oxygen ions. Emission variability faster than these limits is suggestive of non-equilibrium.

The slope of the curve indicates whether the element is ionising or recombining in the plasma. When fully stripped the ion can only recombine, thus the upward trend at high temperatures. The 'hump' ($\sim 10^5 \rightarrow 10^6$ K) is due to the He-like stage which requires larger than average temperature increases to open the n=1 shell. Note also the broadness of the He-like stage in ionisation balance calculations (e.g. fig. 2). A second 'hump' appears due to the Ne-like stage, i.e. the n=2 shell, for elements of nuclear charge greater than 10. Thus lines emitted by ions of the Li-like and Na-like isoelectronic sequences may be underestimated in non-equilibrium conditions because of

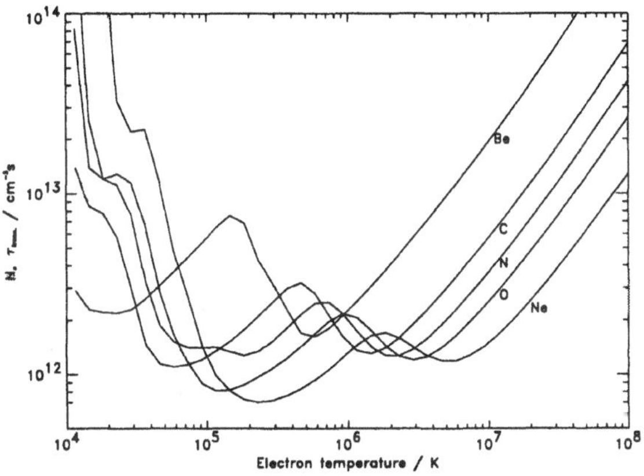

Figure 1. Function $N_e \tau_{transient}$.

TABLE 1. Relaxation Timescales. Solar parameters from [9], boundaries from [10]

	$N_e(cm^{-3})$	$T_e(K)$	$\tau_{transient}$
upper chromosphere	3.7×10^{10}	2×10^4	$\sim 5 \rightarrow 6$ minutes
transition region	$1.3 \times 10^{10} \rightarrow 1.3 \times 10^9$	$5 \times 10^4 \rightarrow 5 \times 10^5$	$\sim 3 \rightarrow 20$ minutes
corona	5.4×10^8	1×10^6	~ 1 hour

the longer recombination times from He-like and Ne-like stages.

Evidence for such discrepancies has been reported in analysis of SOLSTICE data [4]. Transition region lines of Na-like and Li-like isoelectronic sequences were found to be underestimated theoretically by factors between 2.5 and 7. Conversely, despite the longer relaxation timescales, there is yet no clear evidence of non-equilibrium ionisation in the corona. A study of coronal lines observed by SERTS [6] has shown that, even in active regions, variability may be too low to produce detectable non-equilibrium effects.

4. Ionisation equilibrium

In general, for a fixed temperature, increasing the density will shorten the time to reach equilibrium. However, this introduces additional problems e.g. metastable levels can attain populations comparable to the ground level and finite density effects become important.

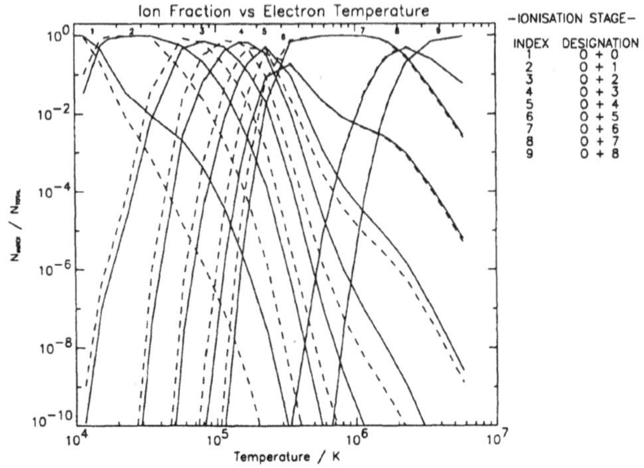

Figure 2. *Unresolved* ionisation balance. solid line - $N_e = 0.0\ cm^{-3}$, dash line - $N_e = 5.0 \times 10^9\ cm^{-3}$.

A treatment of metastable levels improves the ionisation balance calculation by allowing inclusion of the most sophisticated *metastable resolved* ionisation and recombination data. Thus the difference in DR rates to the ground and metastables and their density dependence can be included. Note that the ionisation balance is density dependent at low densities due to DR and at high densities through stepwise ionisation.

To illustrate this, in fig.2 we reconstruct an *unresolved* ionisation balance from its metastable resolved counterpart and contrast the zero density case with that at $\sim 5 \times 10^9$ cm^{-3} representing the upper transition region. This moderate density increase leads to formation temperature shifts of ~30%.

The modelling codes and atomic data of this paper are archived within the Atomic Data and Analysis Structure (ADAS, [11]).

References

1. Carlsson, M., Judge, P.G., Wilhelm, K. (1997), *ApJ Lett.*, **486**, L63
2. Harrison, R.A. (1997), *Sol. Phys.*, **175**, Part 2
3. Lang, J., Mason, H.E., McWhirter, R.W.P. (1990), *Sol. Phys.*, **129**, 31
4. Judge, P.G., Woods, T.N., Brekke, P., Rottman, G.J. (1995), *ApJ Lett.*, **455**, 85
5. Burgess, A. (1964), *ApJ*, **139**, 776
6. Lanzafame, A.C., Brooks, D.H., Lang, J., Summers, H.P., Thomas, R.J., Thompson, A.M. (1998), *A&A*, submitted
7. Brooks, D.H. (1997), *PhD Thesis*, Univ. of Strathclyde, Glasgow.
8. McWhirter, R.W.P. (1960), *Proc. Phys. Soc.*, **75**, 520
9. Gabriel, A. (1976), *Phil. Trans. R. Soc. London A*, **281**, 339
10. Athay, G. (1981), *ApJ*, **249**, 340
11. Summers, H.P. (1994), *JET Internal Report*, JET IR(94)-06

FLARE ACTIVITY OF VISUAL BINARY COU 14

V.S. TAMAZIAN, J.A. DOCOBO
Astronomical Observatory "Ramon Maria Aller", University of Santiago de Compostela, P.O. Box 197, Santiago de Compostela, Spain

AND

N.D. MELIKIAN
Byurakan Astrophysical Observatory, Byurakan, 378433, Armenia

Abstract.
Two flares with an amplitude of more than $0^m.3$ each in B and V colours and another four with less amplitude in U band were detected during photometric and polarimetric monitoring of the nearby visual binary COU14. Irregular long term brightness and colour variations were registered while no significant polarization signal was detected. The variability is attributed to the secondary component whose photometric properties are similar to those of T Tau type stars. For the first time a T Tau type star has been found to be a member of a nearby binary system with well determined orbital parameters.

1. Introduction

Within a framework of the ongoing study of visual binary stars with variable components which are presently being conducted at the Astronomical Observatory "Ramon M. Aller" (see Tamazian et al., 1997; Docobo et al., 1997,) in this article the results of a photometric and polarimetric study of COU 14 (= HD 207652) are presented.

This is a bright (V=5.35, U-B=+0.03, B-V=+0.37; Mermillod, 1987) star classified as F2 III by Nassau and van Albáda (1947) and F2 III-IV by Harlan (1969), both based on the objective prism spectra.

Its duplicity was discovered by Couteau in 1959 (Couteau, 1960) who noted the possible variability of the faint component. Indeed, the magnitude difference between components given by Couteau (1961) is 2^m ($5^m.5$ - $7^m.5$) but careful overlook of all measurements made since then show that at a certain epoch it decreased to $0^m.8$ (Worley, 1978). According to Couteau's most recent observations (Couteau, 1997) the magnitude difference decreased as much as $0^m.4$.

Fernie (1976) measured its brightness V=6.16 allowing COU 14 to be included in NSV Catalogue (Kukarkin et al., 1982) as NSV 13891.

It is noteworthy that no short-period variability has been found by Breger (1969) who observed COU 14 for 3 hours with a high photoelectric accuracy of about $0^m.03$ in order to register short-term δ Scuti type variations.

Astrophysics and Space Science is the original source of publication of his article. It is recommended that this article is cited as: *Astrophysics and Space Science* **261**: 95–100, 1999.

The trigonometric paralax 0."0279 (σ=0."0047) is given in the last edition of the General Catalogue of Stellar Trigonometric Parallaxes (van Altena et al., 1995) being close to the value 0."02991 (σ = 0."00077) measured by HIPPARCOS (1997). The latter gives V=5.38 and magnitude difference between components equal to $1^m.2$. Adopting the paralax value as that measured by HIPPARCOS we obtain 33.4pc as COU 14 distance estimation.

As regards its orbital data, COU 14 is one of the most frequently observed visual binaries and, consequently has the best orbit calculated by Heintz (1973; 1982; 1994), Costa and Morales (1975; 1978), Docobo and Costa (1985), Baize (1986) and Hartkopf et al. (1989). The latter authors, based on precise speckle measurements, have shown that this system has a shorter period (26.1 years) and a considerably smaller semimajor axis (a = 0."3664) than any published before. The orbit inclination they derived is $70^o.3$.

The main purpose of the present study is to conduct the long-term photometric monitoring aimed at directly detectiing the possible variability of COU 14 and investigating its nature. Simultaneously, the polarimetric observations have been performed.

2. Observations and data reduction

The photometric and polarimetric observations in the standard UBVR bands were made in June-September, 1997 with 50cm telescope at the Byurakan Observatory for more than 50 hours of effective exposure time.

The photomultiplier FEU-79 with the maximum spectral sensitivity in the region of 4000-4400 A was used as a detector. In the course of all observations a comparison star SAO 107495 and two background stars were observed.

The accuracy of photoelectric observations in all photometric bands was about $0^m.01$–$0^m.02$ and $0^m.03$ for colour indexes. The standard deviation of polarimetric measurements was estimated at 0.1%. More detailed description of the telescope equipment, observation methodics and data reduction are given by Eritsian and Nersissian (1984).

The polarimetric observations were conducted during the first 8 nights since the measurements only showed a light polarization signal less than 0.3% in all bands.

The regular monitoring programe of the star in U and B bands began on August 8th (and lasted 13 nights) when a flarelike variation of the brightness, an amazing phenomenon for this star, was discovered.

3. The results of UBVR photometry

In Table 1 the results of UBVR photometry are given. In its first column the observation date is given and the next seven columns represent U, B, V, R, U-B, B-V and V-R values, respectively.

As is seen from data given in Table 1, COU 14 showed brightness variations in each colour band for all period of observations.

The maximal rates of the light variation detected for U,B,V and R bands were ΔU=0.23, ΔB=0.30, ΔV=0.36 and ΔR=0.24, respectively.

The brightness maximum in UBV bands was detected on August 10th and September 1st.

It was noticed above that COU 14 monitoring had been conducted for 13 nights. Although its principal aim was to detect short term variations (UV Ceti type flares), each 15-20min. the photometric measurements were also made and accurate data obtained

TABLE 1. Photometric data for COU 14

Date	U	B	V	R	U-B	B-V	V-R
13.06.97	5.65	5.69	5.36	5.48	-0.04	+0.33	-0.12
25.07.97	5.70	5.72	5.32	5.66	-0.02	+0.40	-0.34
27.07.97	5.71	5.70	5.37	5.72	+0.01	+0.33	-0.35
28.07.97	5.75	5.75	5.40	5.71	0.00	+0.35	-0.31
31.07.97	5.78	5.74	5.37	5.61	+0.04	+0.37	-0.24
02.08.97	5.78	5.74	5.53	5.71	+0.04	+0.21	-0.14
04.08.97	5.67	5.76	5.36	5.66	-0.09	+0.40	-0.30
08.08.97	5.67	5.68	5.32	5.51	-0.01	+0.36	-0.19
08.08.97	5.67	5.68	5.32	5.51	-0.01	+0.36	-0.19
09.08.97	5.68	5.70	5.34	5.55	-0.02	+0.36	-0.21
09.08.97	5.68	5.70	5.34	5.55	-0.02	+0.36	-0.21
10.08.97	5.60	5.62	5.17	5.56	-0.02	+0.45	-0.39
10.08.97	5.55	5.63	5.35	5.57	-0.08	+0.28	-0.22
10.08.97	5.64	5.64	5.38	5.58	0.00	+0.26	-0.20
12.08.97	5.74	5.74	5.45	5.68	0.00	+0.29	-0.23
12.08.97	5.71	5.76	5.45	5.68	-0.05	+0.31	-0.23
13.08.97	5.66	5.74	5.34	5.54	-0.08	+0.40	-0.20
13.08.97	5.68	5.72	5.34	5.54	-0.04	+0.38	-0.20
24.08.97	5.73	5.72	5.36	5.63	+0.01	+0.36	-0.27
25.08.97	5.70	5.71	5.35	5.61	-0.01	+0.36	-0.26
25.08.97	5.67	5.72	5.44	5.59	-0.05	+0.28	-0.15
26.08.97	5.70	5.71	5.38	5.58	-0.01	+0.33	-0.20
27.08.97	5.67	5.67	5.37	5.52	0.00	+0.30	-0.15
29.08.97	5.67	5.82	5.50	5.52	-0.15	+0.32	-0.02
29.08.97	5.68	5.77	5.45	5.52	-0.07	+0.32	-0.07
29.08.97	5.68	5.80	5.45	5.54	-0.12	+0.35	-0.09
30.08.97	5.68	5.79	5.41	5.54	-0.11	+0.38	-0.13
30.08.97	5.69	5.67	5.34	5.54	+0.02	+0.33	-0.20
30.08.97	5.68	5.70	5.37	5.48	-0.02	+0.33	-0.11
31.08.97	5.60	5.70	5.37	5.50	-0.10	+0.33	-0.13
01.09.97	5.62	5.62	5.30	5.54	0.00	+0.32	-0.24
01.09.97	5.60	5.52	5.26	5.53	+0.08	+0.26	-0.27

allowing us to follow COU 14's photometric behaviour on the longer time scale without interruption.

The data presented in Table 1 indicate the brightness variation for the time scale of several nights and also one single night. For instance, the mean U magnitudes for August 10th and August 12th are $5^m.60$ and $5^m.72$, respectively. Similarly, for the single night August 10th the brightness in U colour changed from $5^m.55$ to $5^m.64$. The same behaviour can be seen in the other bands (see Table 1).

Taking into account the value of standard deviation ($0^m.01$–$0^m.02$) we can conclude that such variations can not be explained by observational errors and thus, we are dealing with long term brightness variation in the time scale of several days and hours.

As regards the observed colour indexes, the mean colour indexes are U-B= -0.03; B-V=

TABLE 2. Data on detected flares

No.	Date	Tmax(UTC)	Duration	Amplitude
1	28.07.97	21h25m00s	9.5min	0.33 (V)
2	02.08.97	19 22 30	7.0min	0.31 (B)
3	12.08.97	22 18 40	1.5min	0.11 (U)
4	25.08.97	23 21 26	4.5min	0.08 (U)
5	26.08.97	19 55 20	3.3min	0.19 (U)
6	26.08.97	20 18 25	2.1min	0.13 (U)

+0.31 and V-R= -0.21 coinciding rather well with the data given by Mermillod (1987).

It must be noted, however that the observed colours are scattered around these mean values and sometimes clearly exceed $3\sigma = 0^m.09$ as can be seen from the data presented in Table 1.

The character of both slow brightness variation and colour changes seems to be irregular and additional observations are required to verify this trend.

The most interesting result is that besides its slow brightness variation COU 14 showed flare-like rapid variations in time scales of several minutes. The data on these flares are given in Table 2 where in subsequent columns the current number of flare, date and time of maximum, duration and amplitude in corresponding colours are given.

The preliminary results on the first detected flares have already been published (Melikian et al., 1997).

During 50 hours of effective observational time 6 flares were detected in U, B and V bands indicating rather high flare activity (0.12 flare/hour).

4. Discussion

There are certain discrepancies in the luminosity type designation to COU 14. Although all references describe its spectral type as F2, the luminosity class assigned by Nassau and van Albada (1947) F2 III differs from that given later on by Harlan (1969) as F2 III-IV, both based on objective prism spectra. Our recent study of the slit spectra has shown that COU 14 can certainly be classified as F2 IV (Tamazian et al., 1997).

The direct estimation of the absolute magnitude using the paralax value 0."02991 measured by HIPPARCOS (taking into account apparent magnitudes of the components $5^m.5$ and $7^m.5$) lead to M(V) = +2.7 and +4.7 for the main and secondary components, respectively.

It is noteworthy that adopting MK type F2 IV and absolute magnitude M(V) = +2.4 (Schmidt-Kaler, 1982) for the primary we obtain M(V) = +4.4 for the secondary. In both cases the absolute magnitude of the secondary corresponds to the early G0 - G2 type dwarfs.

It must be especially noted that UV Ceti type flare events detected for COU 14 are an unexpected phenomena, since the stars both in the solar vicinity and stellar aggregates possessing flare activity belong to Ke-Me dwarfs with well marked hydrogen emission lines (Krautter,1996). Usually their brightness at the minimum is constant despite typically shown accidental rapid variations in time scales from several seconds to 2-3 hours. They also show strong emission lines of Balmer serie with no significant variability at minimum but during the flare their spectra are very similar to those of T Tau stars.

Based on such similarity Ambartsumian (1954) pointed out that both UV Ceti and T Tau stars represent the red dwarfs at different stage of their evolution. At the last stage of their evolution T Tau stars show UV Ceti type flares (Ambartsumian, 1954). It was ruled out later (Ambartsumian, 1970) that about 25% of T Tau type stars in Orion region show flare activity.

As for now, many T Tau type stars with well detected UV Ceti type flare-like rapid brightness variations are known.

Taking into account all the above mentioned it seems reasonable to suggest that both types of variability are caused by the secondary. In all probability, 1) the estimated spectral type (GO - G2), 2) slow brightness and colour variations of the system, 3) the detected flare-like rapid variations show that the faint component might be T Tau type star, possibly with weak emission lines unseen on the strong background of the primary.

We mentioned above that according to visual estimations made by different observers magnitude difference between components vary from $0^m.4$ to $2^m.0$. In order to compare the homogeneous set of observations Prof. Couteau was asked and kindly overlooked his own data confirming mentioned limits as well as non-variability of the primary (Couteau, 1997). Notice that in such a case the integral apparent magnitude must change roughly from $4^m.9$ to $5^m.3$.

Adopting V=5.5 for primary we can easily estimate the range of secondary's brightness variation needed to satisfy the integral magnitude variation in V colour from $5^m.17$ to $5^m.53$ (see Table 1). So, a roughly estimated variation from $6^m.5$ to $10^m.5$ with an amplitude 4^m is obtained which is typical for T Tau stars.

5. Conclusions

Taking into account the non-variable character of the primary, its belonging to G0-G2 spectral type as well as registered irregular brightness and colour variations and UV Ceti type flares, we may suggest that the secondary component of COU 14 is in all probability a T Tau type star.

It is worth noting that due to obtained results, COU 14 becomes the unique known binary in the solar vicinity with well determined orbital parameters, one of whose components is a T Tau star.

Undoubtedly, further studies of binary and multiple systems with variable components, especially those with known orbital parameters are strongly needed. The increasing number of such systems and careful comparison of their dynamical and astrophysical properties could be very important when discussing many topics concerning the origin and evolution of the stars and stellar systems.

6. Acknowlegements

The authors would like to thank Prof. P.Couteau for providing the recent observational data for COU 14.

One of the authors (N.Melikian) thanks the Administration of the Astronomical Observatory "Ramon Maria Aller" for its hospitality.

This work was supported by the Xunta de Galicia (Spain) under the research project XUGA 23401 B96.

References

Ambartsumian, V.A., Soobsh. Byurakan Obs., 1954, **13**, 3
Ambartsumian, V.A., 1970, Astrofizika, **6**, 31
Baize, P., 1986, A&AS, **65**, 551
Breger, M., 1969, Ap.J.Suppl.Ser., **19**, 79
Costa, J.M., Morales, C., 1975, Urania, **283**, 34
Costa, J.M., Morales, C., 1978, IAU Com. 26 Inf.Circ. No.75
Couteau, P., J. Observateur, 1960, **43**, 1
Couteau, P., J. Observateur, 1961, **44**, 41
Couteau, P., 1997 (private communication)
Docobo, J.A., Costa, J.A., 1985, Com.26 IAU Circ. No.95
Docobo, J.A., Tamazian, V.S., Melikian, N.D., Eritsian, M.A., Karapetian, A.A., 1997, IBVS, 4510
Eritsian, M.A., Nersisian, S.A., 1984, Astrofizika, **20**, 355
Fernie, J.D., 1976, J. R. Astron. Soc. Can., **70**, 77
Harlan, E.A., 1969, A.J., **74**, 916
Hartkopf, W.I., McAlister, H.A., Franz, O.G., 1989, A.J., **98**, 1014
Heintz, W.D., 1973, A.J., **78**, 307
Heintz, W.D., 1982, A&AS, **47**, 569
Heintz, W.D., 1994, A.J., **108**, 2338
"The HIPPARCOS and Tycho Catalogues", 1997, vol. 9, p. 2164, Perryman, M.A.C. (scientific coordinator), ESA Publishing Division, Nordwijk, The Netherlands
Krautter, J., 1996, in **"Light curves of variable stars"**, C.Sterken and C.Jaschek (eds.), Cambridge Univ. Press, Cambridge
Kukarkin, B.N., Kholopov, P.N., Artiukhina, N.M., Federovich, V.P., Frolov, M.S., Goranskij, M.P., Gorynya, M.P., Karitskaya, E.A., Kireeva, N.N., Kukarkina, N.P. et al., 1982,**New Catalogue of Suspected Variable Stars**, Nauka Publ. House, Moscow
Melikian, N.D., Eritsian, M.A., Karapetian, A.A., Tamazian, V.S., Docobo, J.A., 1997, IBVS, 4514
Mermilliod, J.C., 1987, A&AS, **71**, 119
Nassau, J.J., van Albada, G.B., 1947, Ap.J., **106**, 20
Schmidt-Kaler, T, 1982, in Landolt-Bornstein, vol. 2B, part 1
Tamazian, V.S., Docobo, J.A., Chavushian, V.O., Vlasyuk, V.V., 1997, Astrophys. and Space Science Library Ser., **223**, 27, Kluwer Academic Publishers, Dordrecht, Netherland
van Altena, W.F., Lee, J.T., Hoffleit, E.D., 1995, **"The General Catalogue of Trigonometric Stellar Parallaxes"**, Fourth Edition, vol. I, L.Davis Press, Schenectady, New York
Worley, C.E., 1978, Publ. U.S. Naval Obs., **24**, pt.6

X-RAY SPECTROSCOPY OF ACTIVE STARS WITH ASCA AND BEPPOSAX

R. PALLAVICINI, A. MAGGIO AND A. ORTOLANI
Osservatorio Astronomico di Palermo
Palazzo dei Normanni, I-90134 Palermo, Italy

AND

G. TAGLIAFERRI AND S. COVINO
Osservatorio Astronomico di Brera
Via E. Bianchi 46, I-22055 Merate (Co), Italy

Abstract. We discuss representative cases of X-ray spectra of active stars obtained with ASCA and *Beppo*SAX. We focus in particular on our own observations of the RS CVn binaries λ And and UX Ari, of the BY Dra eclipsing binary HD9770, and of the young stars AB Dor and HD35850.

1. X-ray spectroscopy with ASCA

Stellar coronae have optically thin thermal spectra due to line + continuum emission from ionization equilibrium plasmas (Mewe et al. 1995). One-temperature (1-T) and two-temperature (2-T) models are usually assumed to fit these spectra. Somewhat suprisingly, most (but not all) coronal sources observed with ASCA have shown reduced metal abundances (by typical factors of 3 to 5) with respect to *solar* abundances. In many cases, the best fit to the ASCA CCD spectra over the range 0.5 to 10 keV was obtained with a 2-T thermal model with variable non-solar abundances, i.e. by leaving all abundant elements free to vary independently (White et al. 1994, Singh et al. 1996). In other cases, a sufficiently good fit was obtained by leaving only the overall metallicity Z free to vary, i.e. by varying the individual abundances in the same proportion with respect to solar abundances, as we did for λ And (Ortolani et al. 1997a) and for HD35850 (Tagliaferri et al. 1997a).

Astrophysics and Space Science is the original source of publication of his article. It is recommended that this article is cited as: *Astrophysics and Space Science* **261**: 101–104, 1999.

The fact that the coronal abundances measured by ASCA in a given star are typically lower than solar photospheric abundances does not imply necessarily that they are also lower than the *photospheric* abundance of the star. Most stellar coronal sources for which medium resolution CCD spectroscopy has been possible so far with ASCA are bright RS CVn and Algol-type binaries, which are known to have often subsolar photospheric abundances (e.g. Randich et al. 1994). For instance, the RS CVn binary λ And observed by us with ASCA (Ortolani et al. 1997a) was found to have a coronal abundance a factor 5 lower than solar but perfectly consistent with its measured photospheric abundance.

However, there are at least a few cases in which there is an obvious discrepancy between the measured coronal and photospheric abundances. One case is the young rapidly rotating star AB Dor observed simultaneously by us with ASCA and EUVE (Mewe et al. 1996, Ortolani et al. 1997b). The coronal metallicity of this star is $Z \sim 0.3$ in spite of the fact that AB Dor is a ZAMS star with a measured photospheric abundance that is solar. Another case is the young active star HD35850 (Tagliaferri et al. 1997a): again the coronal metallicity measured by ASCA is ~ 0.3 while the measured photospheric metallicity is solar. On the contrary, the active giant β Ceti observed by both ASCA and *Beppo*SAX shows abundances consistent with photospheric values (Maggio et al. 1997), and similarly an analysis of a recent *Beppo*SAX observation of Capella gives a coronal metallicity ~ 0.8 solar, again consistent both with the value derived from EUVE and with its photospheric abundance. Strong metal underabundances for Capella were preliminary reported from ASCA observations by Drake et al. 1996 and White et al. 1996, but the fit was poor and formally unacceptable.

The flaring behaviour of AB Dor as observed by us with ASCA is discussed by Ortolani et al. (1997b).

2. X-ray spectroscopy with BeppoSAX

Although *Beppo*SAX has lower effective area and lower resolution at high energies than ASCA, it has a broader spectral band extending down to 0.1 keV, which is of considerable help for disantangling the emission measure distribution from the elemental abundances (see Favata 1997a, b). Studies of coronal sources with *Beppo*SAX have so far been carried out for Capella (Favata et al. 1997a), VY Ari (Favata et al. 1997b), β Ceti (Maggio et al. 1997) and HD9770 (Tagliaferri et al. 1997b). The temperature structure is generally consistent with that derived from ASCA data, i.e. a 2-T plasma model is sufficient to fit the data satisfactorily, although more complex models (such as a loop-like model as suggested by spatially resolved X-ray observations of the Sun) cannot be excluded. The results on elemental abundances also confirm the ASCA results, showing a large variety of coronal abundances (from nearly solar-like metallicity as in β Ceti to strongly subsolar abundances as in HD9770). The observations are still insufficient to establish whether there are systematic differences between the coronal and photospheric metallicity of stars or between the elemental abundances derived from ASCA and *Beppo*SAX.

Observations of the newly discovered short-period eclipsing binary HD 9770 observed by *Beppo*SAX over four consecutive orbital cycles are discussed in Tagliaferri et al. (1997b).

A very interesting result of *Beppo*SAX has been the detection of hard (> 20 keV) X-ray emission from flares on UX Ari (Pallavicini & Tagliaferri 1997), Algol (Schmitt et al. 1997 as reported in Favata 1997b) and, most recently, AB Dor (Pallavicini & Tagliaferri 1998). This is the first time that hard X-ray emission has been detected from stars and this is a unique result made possible by the high sensitivity and broadband capabilities of *Beppo*SAX.

To illustrate these results in some more detail, we refer to our recent observation of UX Arietis. It was observed by *Beppo*SAX on August 28-30, 1997 for about one day. The observation caught the star during a strong long-lived flare with a peak flux nearly a factor 10 larger than the presumably quiescent emission at the end of the flare. The decay time in the 2 -10 keV band was about 13 hours.

The spectrum around the peak is well fit by a 2-T model with T_{low} = 2.9 keV, T_{high} = 9.6 keV, emission measure ratio $EM_{high}/EM_{low} \sim 1$, and overall metallicity Z = 0.35 solar. During the flare decay both the temperatures and the emission measure ratio change. In the middle of the decay phase, the two temperatures have decreased to 1.0 and 4.0 keV, respectively, the emission measure ratio has increased to $\sim$ 5 and the overall metallicity has slightly decreased to 0.22 solar. At the end of the flare, the two temperatures are 0.9 and 2.6 keV, the emission measure ratio is $\sim$ 3 and the metallicity is still strongly subsolar ($\sim$ 0.2 solar). The quiescent luminosity of the star at the end of the flare is $\sim 1 \times 10^{31}$ erg s^{-1} for an assumed distance of 50 pc, and the total energy released during the flare is $> 5 \times 10^{36}$ ergs.

Hard X-ray emission up to at least 40 - 50 keV was detected from UX Ari by the PDS detector. This emission could be either a non-thermal power-law component (as observed, but at much lower level, during the impulsive phase of solar flares) or a thermal component produced by plasma at $\sim 10^8$ K. An extrapolation to higher energies of a 2-T fit to the 0.1 to 10 keV spectrum indicates that most of the hard X-ray emission is probably thermal, although the presence of a non thermal component cannot yet be excluded. Most interestingly, the hard X-ray emission remains detectable at a level of ~ 0.4 cts/s for the entire duration of the flare till the end of the observation, suggesting the persistence of hard-X emission after major flares.

The hard X-ray emission is about a factor 10 lower than the soft X-ray emission at the end of the flare and nearly a factor 100 less than the soft X-ray flux at the flare peak. Hard X-ray emission is usually observed in solar flares and is attributed to thick-target bremsstrahlung of non-thermal electrons accelerated during the impulsive phase of flares. The ratio of hard to soft X-ray emission in solar flares ($\sim 10^{-5}$) is much lower than for this flare on UX Ari. Moreover, most of the hard X-ray emission in UX Ari is likely to be of thermal origin, contrary to the solar case. Another important difference, if confirmed, is the persistent nature of this emission for many hours after the flare. It is tempting to speculate that this long lasting emission could be related to the radio halos that apparently form after major flares and which are due to non-thermal gyrosynchrotron emission.

References

Drake, S.A., Singh, K.P., White, N.E., Mewe, R., Kaastra, J.S., 1996, in R. Pallavicini and A.K. Dupree (eds.), *9th Cambridge Workshop on Cool Stars, Stellar Systems, and the Sun*, ASP Conference Series 109, 263

Favata, F., 1997a, in R.A. Donahue and J.A. Bookbinder (eds.), *10th Cambridge Workshop on Cool Stars, Stellar Systems and the Sun*, ASP Conference Series: San Francisco, in press

Favata, F., 1997b, in L. Scarsi et al. (eds.), *The Active X-Ray Sky: Results from BeppoSAX and Rossi-XTE*, Accad. Naz. Lincei: Roma, in press

Favata, F., Mewe, R., Brickhouse, N.S., Pallavicini, R., Micela, G., Dupree, A.K., 1997a, A&A 324, L37

Favata, F., Mineo, T., Parmar, A.N., Cusumano, G., 1997b, A&A, L41.

Maggio, A., Favata, F., Peres, G., Sciortino, S., Rosner, R., 1997, A&A , in press

Mewe, R., Kaastra, J.S., Liedhal, D.A., 1995, Legacy, 6, 16

Mewe, R., Kaastra, J.S., White, S.M., Pallavicini, R., 1996, A&A, 315, 170

Ortolani, A., Maggio, A., Pallavicini, R., Sciortino, S., Drake, J.J., Drake, S.A., 1997a, A&A 325, 664

Ortolani, A., Pallavicini, R., Maggio, A., Reale, F., White, S.M., 1997b, in R.A. Donahue and J.A. Bookbinder (eds.), *10th Cambridge Workshop on Cool Stars, Stellar Systems and the Sun*, ASP Conference Series: San Francisco, in press

Pallavicini, R,. Tagliaferri, G., 1997, in L. Scarsi et al. (eds.), *The Active X-Ray Sky: Results from BeppoSAX and Rossi-XTE*, Accad. Naz. Lincei: Roma, in press

Pallavicini, R., Tagliaferri, G., 1998, in preparation

Randich, S., Giampapa, M.S., Pallavicini, R., 1994, A&A 283, 893

Singh, K.P., White, N.E., Drake, S.A., 1996, ApJ 456, 766

Tagliaferri, G., Covino, S., Fleming, T.A., Gagné, M., Pallavicini, R., Haardt, F., Uchida, Y., 1997a, A&A 321, 850

Tagliaferri, G., Covino, S., Cutispoto G., Pallavicini, R., 1997b, in R.A. Donahue and J.A. Bookbinder (eds.), *10th Cambridge Workshop on Cool Stars, Stellar Systems and the Sun*, ASP Conference Series: San Francisco, in press

White, N.E., 1996, in R. Pallavicini and A.K. Dupree (eds.), *9th Cambridge Workshop on Cool Stars, Stellar Systems, and the Sun*, ASP Conference Series 109, 193

White, N.E., Arnaud, K., Day, C. et al., 1994, PASJ 96, L47

X-RAY VARIABILITY AND ROTATION IN THE PLEIADES CLUSTER

G. MICELA AND S. SCIORTINO
Osservatorio Astronomico di Palermo
Piazza del Parlamento 1 - 90134 Palermo - Italy

F.R. HARNDEN JR.
Harvard - Center for Astrophysics
60 Garden Street - Cambridge, MA 02138

AND

R. ROSNER
Department of Astronomy & Astrophysics,
The University of Chicago
640 South Ellis Avenue, Chicago, IL 60637

1. Introduction

The issue of coronal emission variability is one of the current problems of stellar physics. The Sun exhibits large variability in X-rays with amplitude of more than one order of magnitude due to a combination of flares, rotational modulation and 11-year cycle. It is unclear which is the temporal behavior of coronal emission of other stars and how the way in which available observations sample the intrinsic variability affects our view.

Here, we present a preliminary report of a comprehensive variability analysis of X-ray observations of the Pleiades cluster. Taking advantage of IPC observations (obtained in 1979/1981), of pointed PSPC observations (obtained in 1991) and of the more recent HRI observations (1994/1996) we have explored time scales up to the solar cycle length in a substantial fraction of the members. For the IPC observations we have adopted the X-ray luminosities reported by Micela et al. (1990), while, with the aim to minimize systematic effects due to different analysis techniques, we have reanalyzed all the PSPC (originally analyzed by Stauffer et al. 1994, and Micela et al. 1996) and HRI observations in a homogeneous way with a Wavelet based algorithm (Damiani et al. 1997a, 1997b). Furthermore we

Astrophysics and Space Science is the original source of publication of his article. It is recommended that this article is cited as: *Astrophysics and Space Science* **261**: 105–106, 1999.

have analyzed separately the single temporal segments of the PSPC and HRI observations, taken at well separated times.

We find that the maximum likelihood cumulative distribution functions of amplitude variations for all cluster members taken together, and in particular for G stars, are very similar to that obtained for solar flares (Drake 1971). Furthermore, the distribution obtained combining all the *Einstein* and ROSAT data (that probes time scales up to 15 years) does not significantly differ from the distribution obtained from the PSPC data only, that instead probes time scales from days to 18 months (see also Gagné et al. 1995).

Furthermore, taking advantage of recent $v\ sin(i)$ and rotational period determination (mainly for the slowest rotators) we revisit the relation between L_x and rotational velocity for G and K Pleiades compared with the relation for Hyades stars of similar spectral types. We find that in the G star sample is present a relationship between L_x and $v\ sin(i)$ (although with a slope less steep than the canonical Pallavicini et al., 1981, v^2 law), but with a dispersion around this relation of a factor 3 - 10, due exclusively to variability. We note that the Hyades tend to stay in the lower part of the distribution of Pleiades data points. In the Pleiades K star sample the relation between X-ray luminosity and rotation is absent down to $v\ sin(i) \sim 3$ km/sec, with an amplitude variation independent of rotation, well evident also in the "saturated" stars, casting doubt on the interpretation of saturation in terms of complete coverage of stellar surface (filling factor equal to the unity). Next step in the process of understanding the nature of the emission of active young stars will be the removal of the extreme X-ray luminosity values due to the occurrence of flares, with the aim of disentangling the variations due to flares from those due to rotational modulation and cycles.

References

Damiani, F., Maggio, A., Micela, G., Sciortino, S. 1997, *Astroph. J.* 483, 350

Damiani, F., Maggio, A., Micela, G., Sciortino, S. 1997, *Astroph. J.* 483, 370

Drake J. 1971, *Sol. Phis.*, 16, 152

Gagné M., Caillault J-P., Stauffer J.R. 1995, *Astroph. J.*, 450, 217

Micela, G., Sciortino, S., Vaiana, G.S., Harnden, F.R. Jr., Rosner, R., Schmitt, J.H.M.M., 1990, *Astroph. J.*, 348, 557

Micela, G., Sciortino, S., Kashyap, V., Harnden, F.R. Jr., Rosner, R., 1996, *Astroph. J. Suppl.*, 102, 75

Pallavicini, R., Golub, L., Rosner, R., Vaiana, G.S., Ayres, T., Linsky, J.L., 1981, *Astroph. J.*, 248, 279

Stauffer, J., R., Caillault, J.-P., Gagne, M., Prosser, C.F., Hartmann, L.W. 1994, *Astroph. J. Suppl.*, 91, 625

THE SIGNATURES OF STRUCTURED WINDS IN T TAURI STARS

MARIA TERESA V.T. LAGO AND JORGE F. GAMEIRO
Centro de Astrofísica da Universidade do Porto
Rua do Campo Alegre, 823, 4150 Porto - Portugal
Departamento de Matemática Aplicada, FCUP
VITOR M. COSTA
Centro de Astrofísica da Universidade do Porto
Departamento de Matemática, ISEP
AND
LEANDRO MACHADO AND ANTÓNIO PEDROSA
Centro de Astrofísica da Universidade do Porto

1. Introduction

The identification of He I lines λ 4471.5 Å, λ 5875.6 Å, λ 6678.1 Åand He II lines λ 4685.7 Åin emission in the optical spectra of T Tauri stars (e.g. Penston & Keavey 1977, Lago & Penston 1982) was the first clear indication for a temperature structure in the outer layers of these stars. Yet no immediate explanation was proposed for the presence of those high temperature regions in the atmosphere of late type stars.

The ultraviolet spectra later obtained with IUE showing emission lines from metals at various stages of ionization, for example, C I, C II, C III and CIV or Si II, Si III and Si IV, provided further evidence for the temperature structure in the atmosphere. An example is shown in Fig. 1 were the emission at various temperatures received from several T Tauri stars is displayed after normalization for the solar values. Moreover it allowed a quantification in terms of flux or emission measure from the regions contributing to the various lines.

Fig. 1 also shows that the emission from regions of temperature $\geq 10^4$ K far exceeds that observed in the Sun. Furthermore, while some T Tauri stars reach a peak emission just below T $\approx 10^5$ K, for other stars $f_*/f_\odot$

Astrophysics and Space Science is the original source of publication of his article. It is recommended that this article is cited as: *Astrophysics and Space Science* **261**: 107–116, 1999.

clearly keeps on rising until 10^7 K at least. This is an interesting result and a fundamental clue for the structure of the outer layers in T Tauri stars. Simultaneously it provides a tight constraint for the selection of mechanisms powering the activity, which is so very enhanced in these young stars. In this context we are analysing a sample of 25 T Tauri stars using both the IUE Newly Extracted Spectra (INES) and ROSAT Public Data Archives. A first paper focused on three TTS, namely, TW Hya, V410 Tau and CS Cha has been submitted (Costa et al. 1999).

We hope that the study of the UV and X-ray observations of such relatively large sample will help us to understand both the observed differences between the various T Tauri stars and the correlation of such differences with alternative mechanisms operating there.

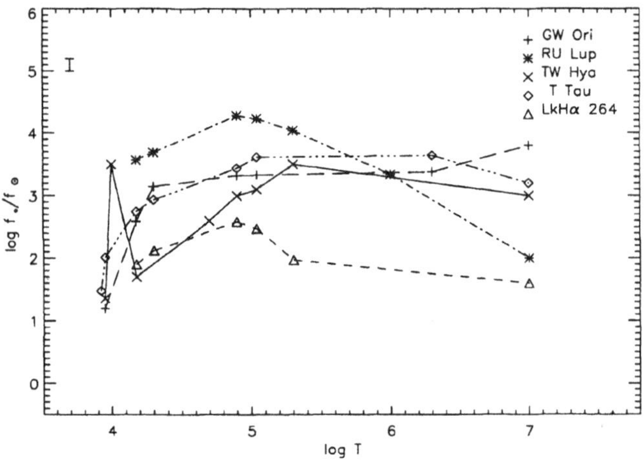

Figure 1. Ratio of stellar to solar surface fluxes for TTS as a function of temperature. For V410 Tau the X-ray flux is an average of two ROSAT observations. A typical error bar is shown in the upper left corner.

An aditional result from the UV data is the possibility to determine the stellar T_{eff} by fitting the observed continuum between ≈ 1200 and 3200 Å. Such fitting is very sensitive to the temperature adopted, mainly in the long wavelength band. It also allows the identification of other contributors such as free-free and free-bound emission from hydrogen and the respective temperature. For most T Tauri stars analysed so far the temperature of the plasma producing such emission is well above 10^4 K, in same cases reaching 5 10^4 K (Lago et al. 1984).

In a few cases, to achieve a good fitting it becomes necessary to introduce a third component. For example, in the case of TW Hya, as shown by Fig. 2,

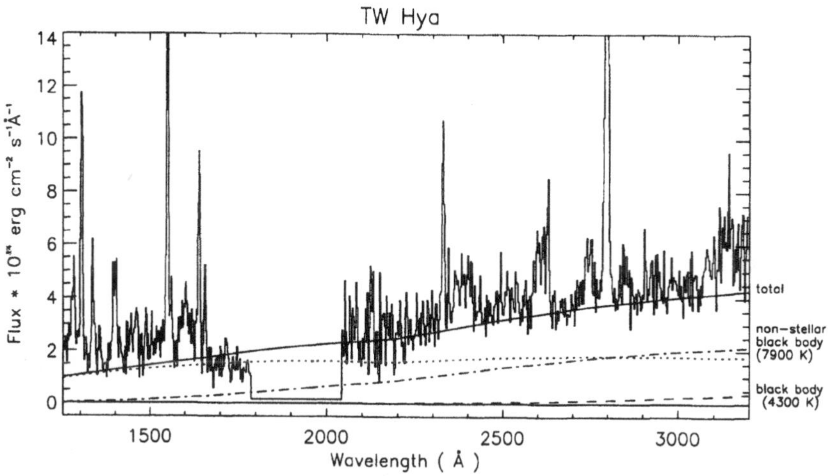

Figure 2. For the star TW Hya the continuum is the sum of hydrogenic free-free plus free-bound emission (dotted line) at 3×10^4 K plus a 4000 K black body emission and a third component - a black body emission at 7900 K covering approximately 5% of the stellar surface. The region around 2000 Å has been supressed due to the very high level of noise.

the third component is well adjusted by blackbody emission at 7 900 K but covering only ≈ 5 % of the stellar surface (Costa & Lago 1998).

2. Rotation

Fast rotation was at some point invoked in connection with the youth of T Tauri stars. However the observations have later shown that these stars are slow rotators.

Yet there are still unanswered questions in connection with the rotation in these early phases of evolution. It is therefore important to have a clear view of what the observations show us, namely the perception of their limitations and uncertainties. For that we have performed a global analysis of the whole set of data on rotation for T Tauri stars (185 stars) available in the literature. We have taken special attention in evaluating the uncertainties involved in the alternative observational techniques: photometry or spectroscopy. Due to the large number of papers involved we have chosen to give no references in this section.

Further to that we have also looked at the statistics in terms of period of rotation and $v \sin i$ versus spectral type. For this analysis confirmed binaries were excluded. The sample contains stars from G2 to M4 spectral types. As

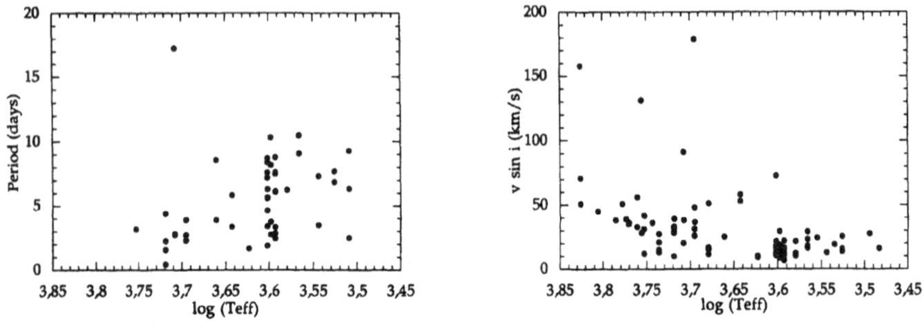

Figure 3. Scatter plots of P_{rot} vs. spectral type (left), and $v \sin i$ vs. spectral type (right).

shown on Fig. 3 there is a clear change of behavior around log $T_{eff} \approx 3.6$ (corresponding to spectral types K7 - M0). For stars of earlier spectral type the period is $\leq$ 4 days, with very few exceptions. For later type stars there is a wider range of values for the period between $\approx$ 2 and 10 days. A similar behaviour is seen in the $v \sin i$ data: stars of $\log T_{eff} \leq 3.6$ display values centered around 20 km s^{-1}, for higher temperature stars the values of $v \sin i$ are systematically higher and seem to vary proportionally to the temperature. Is this an early sign of the well known difference in the rotation rate of main-sequence stars, where stars with a convective envelope rotate much slower than the ones with a radiative envelope?

Another interesting result from this large sample analysis is a clear bimodal distribution in terms of the photometric period, seen in the corresponding histogram. This histogram displays two peaks, one centered around 2 days and another around 8 days. When looking at the histograms relative to the Weak-line T Tauri stars (WTTS) and the Classical T Tauri stars (CTTS), we see that the WTTS show a tendency to rotate faster than the CTTS (see article 'Rotation of T Tauri Stars Revisited' in this volume). However, this is not a clear cut: being a CTTS (WTTS) does not mean that the star will be a slow (fast) rotator.

3. Variability

In T Tauri stars variability is a well established fact. These stars vary in terms of their brightness and of their spectra. Spectral variability affects both the continuum and the lines, the later variations occurring in terms of both the lines intensity and profile.

3.1. THE ULTRAVIOLET

Although limited by the relative scarcity and lack of systematic observations, the variations in the flux of the lines observed with IUE are an important result. This is so because the spectra include lines covering a very wide range of temperatures. In some cases, variations up to 60 % are reported for the strong Mg II lines for observations obtained several months apart. Although no variations were detected on much shorter timescales $\simeq$ 2,5 hours (Penston & Lago 1983). Variations in the flux of higher excitation lines have also been detected, in some cases involving factors as high as 4.

The variations in the continuum, not detectable on scales of one day, are quite remarkable on scales of several months (Lago et al. 1984; Gameiro et al. 1993).

Unfortunately the number of stars for which coherent information is available is very small. However one result is clear: even if variations in the continuum are observed, the variations measured in the lines are real and not just the result of the changes in the continuum.

Of course the limited resolution of IUE does not allow to check whether the variations in the lines were also accompained by any changes in the shape of the profile.

3.2. THE OPTICAL

We selected a sample of 10 T Tauri stars for repeated high resolution observations of Hα, He I and Na I D lines over one week, on time scales ranging from 1 day down to $\approx$ 16 minutes. These lines were selected since they constitute good diagnostic tools for regions of very different temperature. Furthermore, their wavelength proximity allows an almost simultaneous observation. We have analysed the variability of these lines in terms of equivalent width and shape of the profiles.

We will present here the results for one of the stars in the sample, LkHα 264, the closest known T Tauri star (Gameiro et al. 1993, Lago & Gameiro 1998). This star has magnitude $m_V \approx 12.46$ and is located in the Lynds 1457 - 8 dark cloud, at a distance of $\approx$ 65 pc.

Results for other stars in the sample are also presented in another paper (Spectral Variability of T Tauri Stars in the Optical).

3.3. THE Hα LINE PROFILE

LkHα 264 displays a very strong (with equivalent width, $80 \leq W \leq 120\AA$) and broad Hα line (FWHM $\geq$ 300 Km s^{-1}). The line shows a typical III B profile, as do most T Tauri stars (over 33% in the large sample of Reipurth, Pedrosa & Lago, 1996).

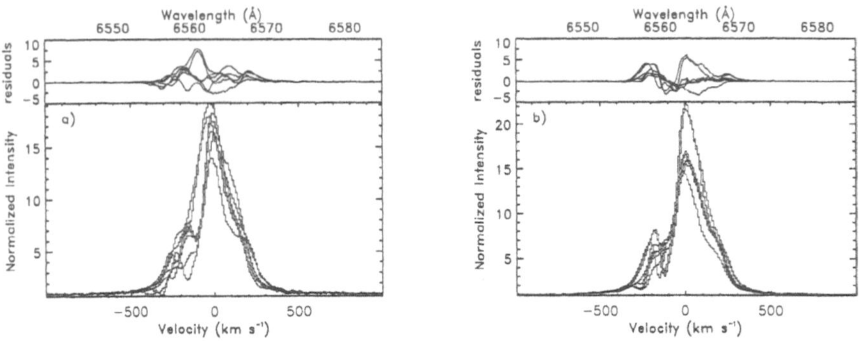

Figure 4. Overlap of the Hα line profiles. The top panel shows the residuals relative to the first profile in the series. a) run of 1990 October, b) run of 1993 November

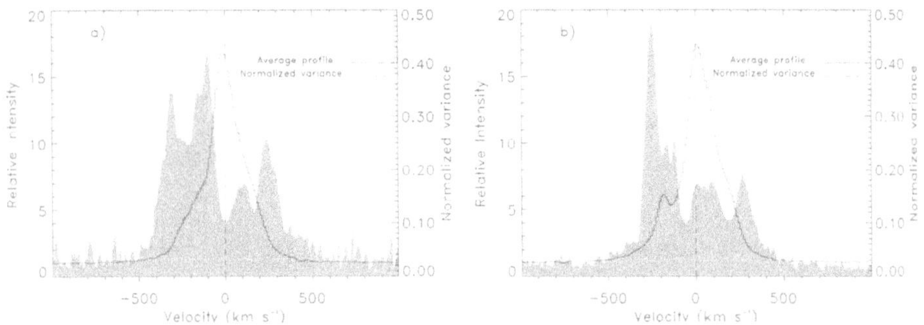

Figure 5. Average profile of Hα and normalized variance. a) Run of 1990 October, b) run of 1993 November.

Fig. 4 shows the variations of the Hα line profile over one week in two different epochs (1990, October 1 to 7 and 1993, November 26 to 30). The lower panels display an overlap of the individual profiles and the upper panels show the residuals relative to the first profile in each set.

The analysis is probably made more clear if we use the normalized variance, as shown in Fig. 5 a) and b). We have also ploted there the average profile for comparison. However, while the data for the first epoch covers only observations $\approx$ 24 h apart, in the second epoch (1993) observations on much shorter time scales are also included.

Fig. 6 shows the normalized variance profiles where we have separated observations on shorter timescales from those 1 day apart. The results show that on longer time scales (1 day) the variations of the Hα line are stronger on the blue side, altough some variability also occurs on the red side. On

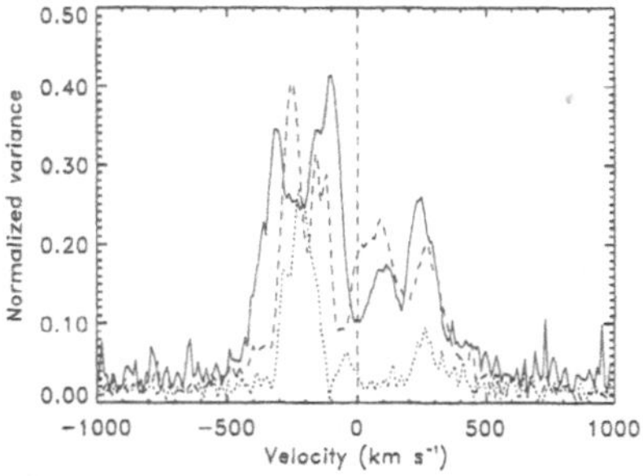

Figure 6. Normalized variance profiles for the Hα observations of 1990 October (solid line), 1993 November (time-scale of 1 d) (dashed line) and the night 1993 November 26 - 27 (dotted line).

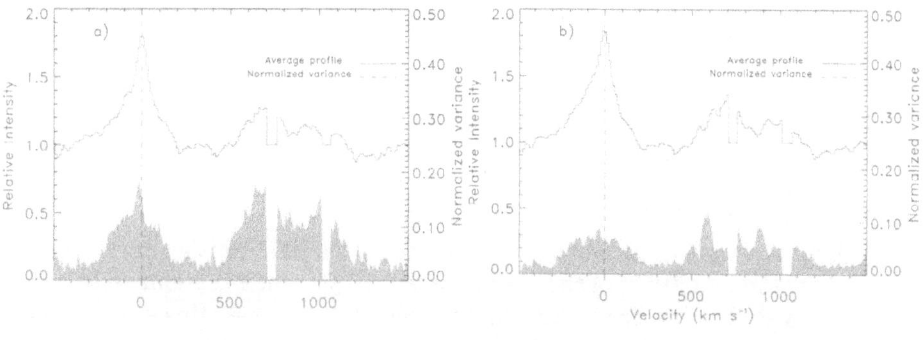

Figure 7. The average and normalized variance profiles of the He I and Na I D lines for (a) 1990 October, (b) 1993 November. The artifact seen on top of the Na I D lines results from the removal of the interstellar components.

short time scales, however, the variations of the Hα line profile are only seen on the blue wing and over a very narrow range of velocities, centered at ≈ -210 Km s^{-1}.

3.4. THE HE I LINE

The He λ 5876 Å line is quite strong ($1.5 \leq W \leq 4.5$ Å) and wide (FWHM$\geq$250 Km s^{-1}). Although the line varies considerably in strength the variations occur over the profile as a whole, the width at the base of the line remains very constant. The variations on short time scales are very small and, once more, take place over the whole profile. The behaviour of the He I line is therefore different from that of the Hα. Fig. 7 displays the average profile for the He I and Na D lines and the normalized variance profile for both runs after removal of the interstellar bands.

3.5. THE NA I λ 5890, 5896 LINE PROFILES

The Na I D lines are more complex and difficult to study, owing to the narrow interstellar bands which have been removed to avoid difficulties while normalizing to the continuum. During the second run the blue side of the normalized variance shows discrete peaks at velocities where absorption features appear at certain times. These peaks, centered at $\approx +600$ Km s^{-1} and $+900$ Km s^{-1} (≈ -150 Km s^{-1} relative to the interstellar components) coincide with the absorption features seen in Hα at similar velocities.

4. Discussion and conclusions

It is interesting to analyse the behaviour of the various optical lines in terms of their equivalent widths as shown in Fig. 8. Fig. 8 a) deals with time scales of ≈1 day. The results show that there is a strong correlation between W_{HeI} and W_{NaI} on both epochs. This implies a common region of origin and a collision dominated regime since the lines differ so much in terms of excitation energy. In previous work we were able to identify this region as being very close to the surface of the star, between 1.1 and 1.2 r_* (Gameiro et al. 1993).

On the first epoch, when the global level of activity in the star as measured by the amplitude of the variations was larger, the Hα seems to behave in a very different way of both the He I and Na I D lines. However, on the second epoch while the global variations are much smaller, the variations in the $W_{H\alpha}$ also correlate well with those of the other two lines.

On shorter time scales the results as shown on Fig. 8 b) are different. There the observations correspond to time scales ranging from less than 20 minutes to hours. The correlation found for $W_{H\alpha}$, W_{HeI} and W_{NaI} seems to hold for time scales of ≈ 1 hour. However, the correlation between W_{HeI} and W_{NaI} seems to hold down to time scales of 1 to 2 hours but not for shorter time scales. On these shorter time scales W_{HeI} changes by ≈ 10 % yet W_{NaI} remains constant. Even on time scales of 1 to 2 hours the variations of the HeI line appear already systematically larger than those of the NaI lines. Therefore, a consistent picture emerging from the global analysis and our conclusion is that the He I and Na I D emission lines are produced in one (same) narrow region, close to the star, very dense and highly variable. On the contrary, the Hα line has a double origin: part of the emission is produced in a relatively small region, close to the star, not too different from the one producing He I and Na I D lines. This can only be seen

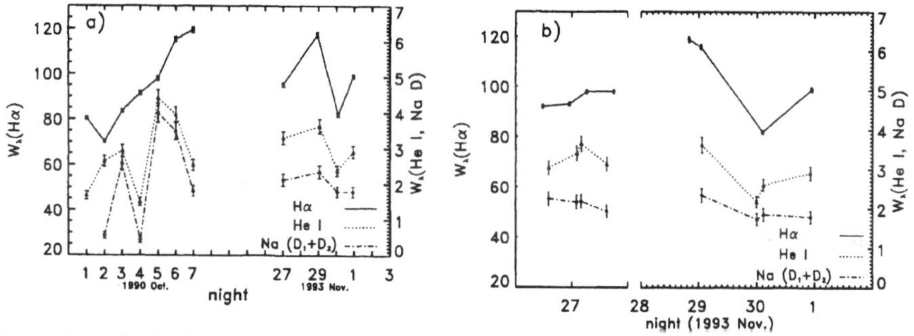

Figure 8. Equivalent width variations of the Hα, He I and Na I (D_1+D_2) lines; **a)** during 1990 October and 1993 November, on time scales of the order of 1 d; **b)** during 1993 November on time-scales shorter than 1 d, ranging from less than 20 min to several hours.

when the global level of activity is low. The atmosphere is then transparent allowing to sample the local behaviour with time. The remaining of the emission in Hα is produced in a more extended region, therefore insensitive to whatever causes the variations in the level of emission of the He I and Na I D lines. This component dominates when the general level of activity of the star is higher and the lower transparency of the upper regions "covers up" the contribution from the inner region.

This is in fact not so different from the results obtained by speckle interferometry for T Tau.

We are pursuing this type of analysis for the remaining stars in the sample. The coherent analysis of such large sample will certainly allow us to clarify several questions still not answered for T Tauri stars:

- the location and characterization of the structure and dynamics of the various temperature regions in the outer layers of TTS,
- through the use of time series analysis to constrain the presence of flows, in terms of direction, structure, dimension, energy and momentum transfer to the circumstellar environment.

The aim is to apply those answers and take advantage of physically sound models of line/continuum formation and transfer of radiation in order to build a comprehensive picture of this early phase of stellar evolution.

References

Costa V.M., Lago M.T.V.T., 1998, Proceedings of the Conference Ultraviolet Astrophysics Beyond the IUE Final Archive, Sevilla, Spain ESA SP-413, 113

Costa V.M., Lago M.T.V.T., Norci L., Meurs E.J.A., 1999, To be submitted

Gameiro J.F., Lago M.T.V.T., Lima N.M., Cameron A.C., 1993, MNRAS, 261, 11

Lago M.T.V.T., Gameiro J.F. , 1998, MNRAS, 294, 272
Lago, M.T.V.T., Penston, M.V., Johnstone, R. (1984), in *Proceedings of Fourth European IUE Conference, Rome, Italy*, p. 233
Penston, M.V., Lago, M.T.V.T., 1983, MNRAS, 202, 77
Reipurth, B., Pedrosa, A., Lago, M.T.V.T. 1996, A&AS, 120, 229

A T TAURI STAR DATABASE

L.P. CARRASQUEIRO
Centro de Astrofísica da Universidade do Porto - Portugal
Rua do Campo Alegre 823, 4150 Porto - Portugal

AND

J.F. GAMEIRO AND M.T.V.T. LAGO
Centro de Astrofísica da Universidade do Porto - Portugal
Departamento de Matemática Aplicada, FCUP - Portugal

Abstract. A T Tauri Star Database was set up with the purpose of designing a user-friendly, graphical database which would provide a quick look into the properties of T Tauri stars.

A T Tauri Star Database was set up as a project leading to the dissertation in "A European Masters Degree in Astronomy".

T Tauri stars are young, low mass stars, whose study is relevant to understand star formation processes.

The FileMaker Pro 3.0 package was chosen to develop the database. This application is available for both Apple Macintosh and IBM PC compatibles and allows the publishing of the database on the internet, through the World-Wide-Web.

The main concerns while structuring and setting up the database were to:

- have a strong graphical component,
- be user-friendly,
- have a hierarchichal organization.

The final goal is for the user to have a global insight of the star in a few minutes.

The database is divided into 3 main sections:

- Basic Characteristics - giving general information about the star: spectral type, age, temperature, magnitude, coordinates, etc. It also includes an image of the sky surrounding the star.

Astrophysics and Space Science is the original source of publication of his article. It is recommended that this article is cited as: *Astrophysics and Space Science* **261**: 117–118, 1999.

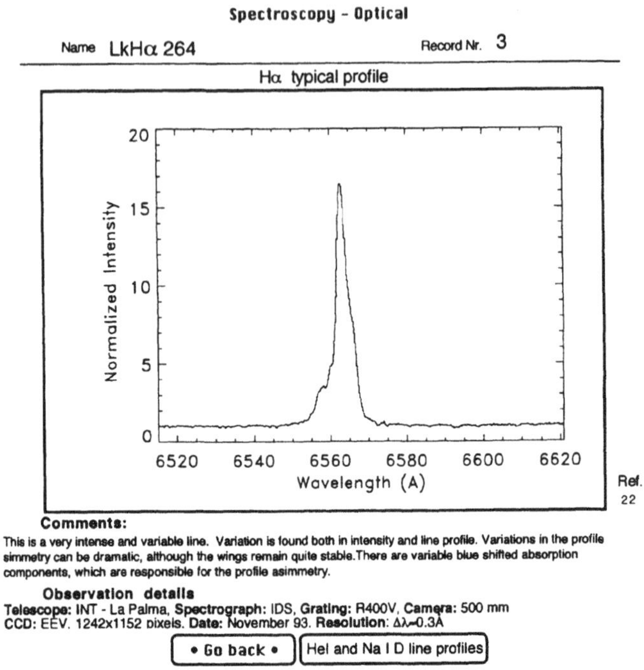

Figure 1. The typical Hα profile field, in this case for star LkHα 264 (Gameiro, 1996).

- Photometry - containing details on colours and indexes, namely information on the photometric variability and period.
- Spectroscopy - subdivided into the major wavebands, from radio to X-rays. It contains line widths, spectra and typical line profiles.

Every field in the database is beside a numerical code which gives the corresponding reference.

Acknowledgements

L.P. Carrasqueiro acknowledges the BM/10615/97 grant which he received from PRAXIS XXI during the dissertation writing period of the M.Sc. course.

References

Gameiro, J.F. (1996), Estudo da Actividade e Rotação em Estrelas de Formação Recente, *PhD Thesis*

T TAURI STARS: THE X-RAY CONNECTION

V.M. COSTA
Centro de Astrofísica da Universidade do Porto - Portugal
Rua do Campo Alegre, 823, 4150 Porto - Portugal
Departamento de Matemática, ISEP

AND

M.T.V.T. LAGO
Centro de Astrofísica da Universidade do Porto - Portugal
Departamento de Matemática Aplicada, FCUP

Abstract. The launch of ROSAT opened a new era in the study of T Tauri stars. Prompted by the IUE results, the new X-ray data confirms the existence of extremely high levels of activity in these young stars, very likely of magnetic nature. We address a detailed analysis of the star TW Hya in the X-ray spectral range. A summary of its properties in the UV will also be included in order to grasp an overall picture of the activity in this star.

T Tauri stars (TTS) are young ($\leq 10^7$ years), low mass stars (< 3 $M_\odot$) still contracting towards the main sequence. Several activity related phenomena, such as X-ray emission, flares and strong chromospheric emission are present in T Tauri stars. Moreover, TTS are also strong UV emitters, 10^3 to 10^4 times in excess of the Sun (e.g. Lago *et al.* 1984).

The observations with the satellites *Einstein* and ROSAT revealed a high number of previously unrecognized stars in the class, the Weak T Tauri stars (WTTS). In general TTS also have X-ray luminosity 10^3 times in excess that of the Sun (e.g. Neuhäuser *et al.* 1995).

For the Classical T Tauri star TW Hya we have performed a detailed spectral analysis from the UV to the X-ray band using IUE re-calibrated spectra and ROSAT data. In the UV band, emission lines characteristic of temperatures between 1.5 and 2×10^5 K have been identified. The continuum can be explained by the sum of the stellar component (a black body at $T_{eff} \simeq 4300$ K) plus hydrogenic free-free and bound-free emission at $T \simeq 3 \times 10^4$ K and a non-stellar black body at $T \simeq 7900$ K (Fig. 1). This

Astrophysics and Space Science is the original source of publication of his article. It is recommended that this article is cited as: *Astrophysics and Space Science* **261**: 119–120, 1999.

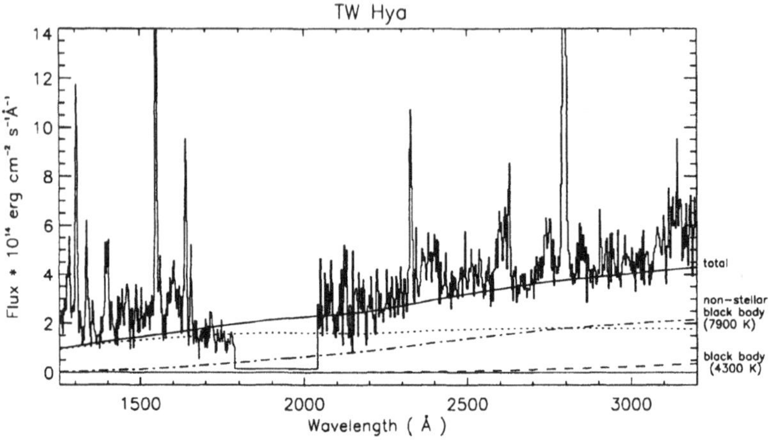

Figure 1. For the star TW Hya the ultraviolet continuum is the sum of hydrogenic free-free plus free-bound emission (dotted line) at 3×10^4 K, black body and also requires a third component, a black body at 7900 K covering approximately 5% of the stellar surface.

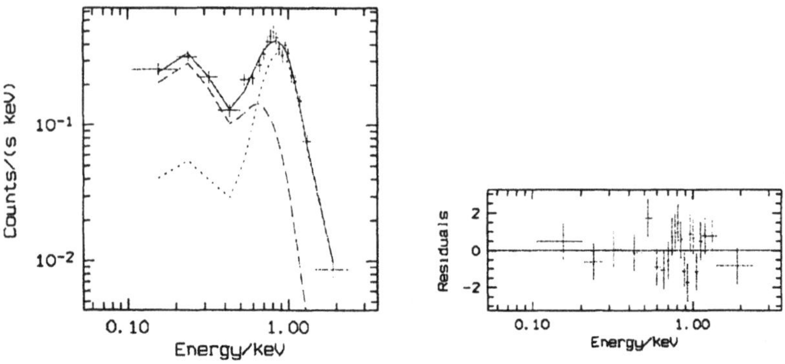

Figure 2. ROSAT spectrum of TW Hya where the best fit requires a two-component model. The residuals are also shown.

non-stellar emission is likely to result from accretion or flaring activity. Further out regions at coronal temperatures $\simeq 2.2 \times 10^6$ K and 1.1×10^7 K seem to be the responsible for the thermal emission (Raymond & Smith, 1977) in the 0.1 - 2.4 keV ROSAT spectral band with a total luminosity of 2.97×10^{30} erg s^{-1}.

References

Lago, M.T.V.T., Penston, M.V., Johnstone, R. (1984), in *Proceedings of Fourth European IUE Conference, Rome, Italy*, p. 233

Neuhäuser, R., Sterzik, M.F., Schmitt, J.H M.M., Wichmann, R., Krautter, J. (1995), *A&A*, **297**, 391

Raymond, J.C., Smith, B.W. (1977), *ApJS*, **35**, 419

A DEEP X-RAY SURVEY OF THE PMS POPULATION OF THE UPPER SCO-CEN ASSOCIATION

S. SCIORTINO, F. DAMIANI AND G. MICELA
Osservatorio Astronomico di Palermo, "G. S. Vaiana"
Piazza del Parlamento 1, 90134 Palermo, Italy

AND

F. FAVATA
Astrophysics Division - Space Science Dep. of ESA, ESTEC
Postbus 299, NL-2200 AG Noordwijk, The Netherlands

We have searched for low-mass pre-main-sequence stars in the Upper Sco-Cen star-forming region using ROSAT pointed observations analyzed with an innovative source detection method based on wavelet transforms (cf. Damiani et al. 1997a,b). In a ROSAT PSPC image we find 32 sources down to a limiting flux of 3.9 10^{-14} erg cm^{-2} s^{-1}. In two HRI observations that cover with greater sensitivity approximately one-third of the same region we find 17 sources, down to a limiting flux of 1.4 10^{-14} erg cm^{-2} s^{-1}. Half of the HRI sources where not detected either in the PSPC pointed observation nor are reported in the ROSAT Bright Source Catalog, nor in the final analysis of Einstein IPC data of the same region, yielding 18 sources down to a limiting flux of 9.1 10^{-14} erg cm^{-2} s^{-1}. In our survey, we have a total of 50 distinct X-ray sources, of which only 13 were known as Einstein IPC sources in a sky region of nearly the same area as that we have surveyed. Our new analysis allows us to largely increase the census of likely PMS stars in the Upper Sco-Cen region and shows that the IPC survey of PMS stars in this region by Walter et al. (1994, hereafter WVMBM) is more incomplete than previously reported. A recent X-ray surveys, based on the ROSAT All-Sky Survey (RASS) data, studied a much larger areas of the sky, but with the same limiting sensitivity as the older *Einstein* observations and have identified several ($\simeq$ 100) new low-mass WTTS in the region (Preibisch et al. 1997). To put on a more quantitative base the bias due to the shallow limiting sensitivity we compared the Log N - Log S (cf. Fig. 1) derived from the *Einstein* IPC results and that built from our ROSAT survey. For fluxes greater than 10^{-13} erg cm^{-2} s^{-1} the

Astrophysics and Space Science is the original source of publication of his article. It is recommended that this article is cited as: *Astrophysics and Space Science* **261**: 121–122, 1999.

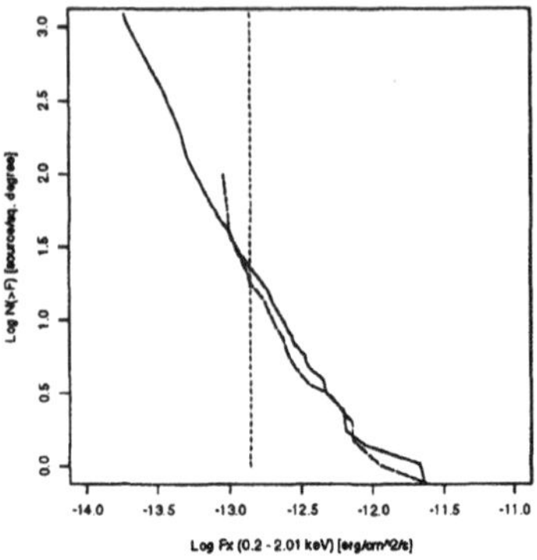

Figure 1. The Log N - Log S deduced from the sample of sources detected with the pointed ROSAT observations (solid line) and with the Einstein IPC (long-dashed line). The vertical line marks the flux of the weakest IPC source reported by WVMBM. Note that the larger size of the ROSAT source sample allows us to better define the shape of the Log N – Log S, in particular below the limit of the WVMBM survey. Since the Log N – Log S does not seem to flatten at weaker flux, the total number of X-ray emitters, most of which are likely associated with stars of the Upper Sco association, is significantly greater than that obtained with the IPC and reported by WVMBM.

"*Einstein*" and the "ROSAT" Log N – Log S are in very good agreement. The ROSAT Log N - Log S shows that the number of sources continues to increase (with a slope of approximately 1.5) toward weak fluxes and indicates the incompleteness of the sample of the sources detected with the *Einstein* IPC. The likely PMS nature of the stellar counterparts of the faint X-ray sources is supported by follow-up optical spectroscopy of some of these sources, as well as by the presence of variability of X-ray emission in a substantial fraction of them, and by the characteristics of available X-ray spectral information. A complete account of this work can be found in the paper by Sciortino et al. (1998). We are currently in the process of identifying the optical counterparts of a larger fraction of the X-ray sources detected in our survey. Photometric observations have been obtained and spectroscopic observations have been scheduled at ESO in June 1998.

References

Damiani F., Maggio A., Micela G., Sciortino S., 1997a, Astroph. J. 483, 350
Damiani F., Maggio A., Micela G., Sciortino S., 1997b, Astroph. J. 483, 370
Preibisch T., Zinnecker H., Guenther E. et al. 1997, Cool Stars, Stellar Systems and the Sun 10, R. Donahue and J. Bookbinder eds., in press
Walter F. M., Vrba F. J., Mathieu R. D., Brown A., Myers P. C., 1994, Astron. J. 107, 692 (WVMBM)
Sciortino S., Damiani F., Favata F., Micela G. 1998, A&A, in press.

THE ULTRAVIOLET VARIABILITY OF THE T TAURI STARS

A. I. GOMEZ DE CASTRO, M. FRANQUEIRA AND N. HUELAMO
Instituto de Astronomía y Geodesia CSIC-UCM
Facultad de CC. Matemáticas, Universidad Complutense de Madrid
28040-Madrid, Spain

AND

E. VERDUGO
IUE Observatory-VILSPA,P.O.Box 50727,28080-Madrid,Spain

1. Introduction

T Tauri stars (TTSs) show enhanced ultraviolet (UV) continuum and line emission with respect to main sequence stars of similar spectral types (Imhoff & Appenzeller 1989; Gómez de Castro, 1998). There is mounting evidence of this excess being produced by the release of gravitational energy from infalling material (Bertout et al 1988, Simon et al 1990, Gómez de Castro & Fernández, 1996). In particular, Bertout et al (1988) have shown that the UV excess of the TTSs can be produced at the boundary layer between the accretion disk and the star assuming accretion rates of ~ 3 $(10^{-7}$ - $10^{-9})$ $M_{\odot}yr^{-1}$. However, it is now evident that some properties of the TTSs cannot be explained with this simple boundary-layer model since some TTSs exhibit periodic photometric variability which is most likely produced by the presence of *hot spots* on the stellar surface.

The detection of hot spots has made rise the suspicion that the infalling material could be chanelled by strong dipolar fields onto the stellar surface. Some theoretical models have been developed mainly addressing the implications of magnetically channelled accretion in the spin-down of PMS stars and the generation of mass outflows (Tout & Pringle 1992, Cameron & Campbell, 1993, Shu et al 1994, Pearson & King 1995) however, key observational tests have not been carried out until recently. If the material falls onto the stellar surface channelled by the field lines, the UV flux (the accretion flux) variation should be rotationally modulated. The Final

Astrophysics and Space Science is the original source of publication of his article. It is recommended that this article is cited as: *Astrophysics and Space Science* **261**: 123–126, 1999.

Archive of the International Ultraviolet Explorer (IUE) provides a unique data base to study the UV variability of the TTSs. In this contribution we present a brief summary of our main findings.

2. UV monitorings of TTS with the IUE

Around a ~ 20% of the TTSs included in the Herbig & Bell Catalogue (1988) have been observed with the IUE; a summary list can be found in Gómez de Castro & Franqueira 1997a. However only 14 of them have been monitored during time scales comparable with their rotation periods; these are: V410 Tau, BP Tau, RY Tau, HD 283572, T Tau, DF Tau, UX Tau, DN Tau, DR Tau, SU Aur, GW Ori, AB Dor , TW Hya and DI Cep. The presence of hot spots has been reported without ambiguity for: DN Tau and BP Tau (Vrba et al 1986), DF Tau (Bouvier and Bertout 1989) and DI Cep. Detailed studies of the UV variability during the rotation period have been carried out only for two of these sources: BP Tau (Simon et al 1990, Gómez de Castro & Franqueira, 1997b) and DI Cep (Gómez de Castro & Fernández).

DI Cep is a CTTS classified as G8 IV with a hot spot (T ~ 8500 K) covering 1-3 % of the visible hemisphere. The monitoring campaign was carried out with the Short Wavelength spectrograph (1200-2000 Å) and the optical FES Camera of the International Ultraviolet Explorer (IUE) from July 12 to July 26, 1992 to study the temperature structure of the spot. As a result, it was shown that the far UV spectrum of DI Cep is dominated by strong emission lines of O I, C IV, Si IV, Si II and Si III with typical surface fluxes of ~ 10^6 erg cm^{-2} s^{-1}. The UV fluxes (lines and continuum) vary in phase and reach the maximum when the optical flux (FES) does. The light curves are similar in all the lines indicating that there is a broad range of temperatures in the spot (from 10^4 to 10^5 K) that is similar to that observed in the plages of magnetically active cool stars (e.g. II Peg). However, in DI Cep the *light curves of the UV lines and continuum are correlated with the optical light curve* (see Gómez de Castro & Fernández, 1996). These observations support the theories in which the accreting material is magnetically channelled onto the star heating its surface at the impact regions. Variations in the temperature of the spot between observations taken 1 year appart suggest that the infalling material is more likely channelled by a transient loop structure attached to the star rather than by a strong stellar dipolar field.

BP Tau is a CTTS classified as K7 having a hot spot with temperature 8211 K covering 0.36 % of the visible hemisphere (Vrba et al 1986). Simon et al (1990) showed that the UV continuum variations are correlated with the optical variations. The line variations are best studied from the IUE monitoring of BP Tau carried out from the 5th to the 19th of January of 1992, when the star was monitored during 2 rotational periods. The analysis of the data shows that lines which can be excited by radiative recombination processes, such as those from O I and He II have periodic-like light curves, whereas lines that are only collisionally excited do not follow the same trend (Gómez de Castro & Franqueira 1997b). The detection of an absorption component redshifted by 81 km/s in the high resolution Mg II profiles obtained with the HST provides *direct evidence of the presence of warm material (T$\sim$ 10^4 K) falling onto BP Tau* and suggests that the various contributions to the line by the accretion flow, the wind and the magnetosphere may blur the Mg II (and Si II) light curves.

Finally, we present in Fig.1 an analysis of the variations of the Mg II profiles of **SU Aur**, which has been monitored with the IUE for $\sim$ 2 rotation periods in the high dispersion mode. The signal-to-noise ratio of the spectra is low and the variations in the profile are best sorted out using cross-correlation methods.Note that the line centroid shifts back and forth during the rotation period.

3. Summary

UV monitorings of TTSs are instrumental to study the mass inflow and separate the various contributions from the atmosphere, the wind and the accretion flow. The analysis of the data available in the IUE Archive indicates that UV continuum and lines emission is concentrated in some hot regions which most likely trace the location of the magnetically channelled inflow. There are still, major questions to be answered as why the spectroscopic tracers display so different behaviours in BP Tau and DI Cep. These differences could be related with the geometry of the field (dipolar, cuadrupolar, tangled magnetic field), with the relative relevance of fossil to dynamo-generated fields and with accretion rates.

Acknowledgments: We thank the VILSPA staff for their support during this work. This research was partly supported by the Ministerio de Educación y Cultura (MEC) of Spain through the research grant PB93-491.

References

Bertout, C., Basri, C., Bouvier, J., 1988,ApJ,330,350
Bouvier, J., Bertout, C., 1989,A&A,211,99
Cameron, A.C., Campbell, C.G., 1993,A&A,274,309
Gómez de Castro, A.I., 1998; *UV Astrophysics beyond the IUE Final Archive*, Ed. ESA Publications Division, ESTEC, Noordwijk, The Neetherlands, in press.
Gómez de Castro, A.I., Fernández, M., 1996,MNRAS,283,55
Gómez de Castro, A.I., Franqueira, M., 1997a, *ULDA Access Guide to T Tauri Stars observed with IUE*,ESA SP-1205, Ed. ESA Publications Division, ESTEC, Noordwijk, The Neetherlands.

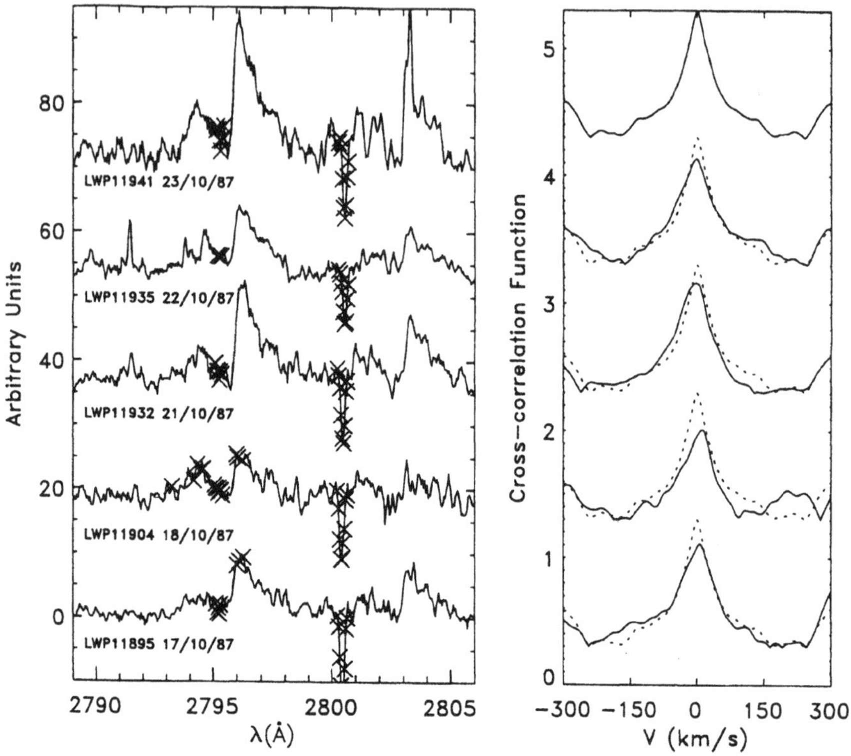

Figure 1. **Left:** *Mg II(UV1) profiles in SU Aur obtained with the IUE (high dispersion mode) in 1987 (rotation period 3.4d).* **Right:** *Comparison of the profiles using a cross-correlation method via FFT. The autocorrelation of LWP11941 produces a symmetric peak centered at 0 velocity (top). The cross-correlation of the LWP11941 profile with the LWP11935, LWP11932, LWP11904 and LWP11895 are plotted below (continuous line) on top of the autocorrelation of the LWP11941 profile (dashed line).*

Gómez de Castro, A.I., Franqueira, M., 1997b,ApJ, 482, 465

Imhoff C.L., Appenzeller, I., 1989 in *Exploring the universe with the IUE satellite* Ed. Y.Kondo Reidel Publ. Comp.,295

Pearson, K.J., King, A.R., 1995,MNRAS,276,1303

Simon, T., Vrba, F.J., Herbst, W., 1990,AJ,100,1957

Tout, C.A., Pringle, J.E., 1992,MNRAS,256,269

Vrba,F.J., Rydgren,A.E., Chugainov,P.F., Shakovskaya, N.I., Zak,D.S., 1986,ApJ,306,199

X RAYS AND UV EMISSION LINES IN T TAURI STARS

Based on observations by IUE

N. HUÉLAMO, A.I. GÓMEZ DE CASTRO AND M. FRANQUEIRA
Instituto de Astronomía y Geodesia
Facultad de CC. Matemáticas (UCM), Avda. Complutense s/n, 28040 Madrid, Spain

1. Introduction

The T Tauri stars (TTS) are low-mass, pre-main-sequence stars. It is well known that the UV emission lines fluxes of the TTS are significantly higher than the observed in late-type main sequence stars. In this work we make use of the IUE Final Archive data to study this problem.

2. The Data

The sample of TTS observed in the low dispersion mode with the IUE satellite has been studied in detail by Gómez de Castro & Franqueira (1997; hereafter GF). We have studied the UV spectrum of the TTS which have good quality spectra in the whole IUE range (1200-3000 Å). We have analized the lines of CIV(UV1), HeII(1640Å) and MgII(UV1). The line fluxes have been measured by gaussian fitting to the line profiles in the mean spectrum[1]. The fluxes have been corrected using the visual extinctions of each star *Av*.

The X-rays fluxes for the T Tauri stars have been calculated from the EINSTEIN satellite data. We have assumed that the underlying spectrum is a coronal-like thermal spectrum (Raymond-Smith spectrum) with a typical temperature of 0.86 KeV (10^7K).

3. Flux-Flux diagrams

We have studied the location of the TTS in flux-flux diagrams in order to compare them with late-type main-sequence and giants stars. We have used

[1]Note that these fluxes are average fluxes of all the observations carried out by the IUE for each star. It is well known that the line fluxes may vary by a factor of ~ 2

Astrophysics and Space Science is the original source of publication of his article. It is recommended that this article is cited as: *Astrophysics and Space Science* **261**: 127–128, 1999.

as standars those included in the RIASS Coronathon Sample (RCS)(Ayres et al.,95). We show the most relevant flux-flux diagrams in Figures 1a and 1b. We have divided the emission line fluxes by the bolometric fluxes of each star. These quantities are represented by R_{line}. In Fig.1a , we represent the X-ray flux versus the CIV flux. The average CIV flux of the TTS is 2 orders of magnitude larger than the observed in late-type stars, however the X-rays flus is just 1 magnitude larger. In Figure 1b we observe that the TTS show an X-rays defficiency (or a MgII excess) when compared with late-type stars. There is a weak correlation between MgII and X-rays fluxes with *a negative slope* , just the opposite to the observed in late-type stars; In Fig. 1c we show the flux-ratio diagram CIV vs. X-rays to HeII, in which the TTS display a different behaviour than MS stars and giants.

4. CONCLUSIONS

The strong UV emission lines fluxes cannot be explained within the context of standard atmospheric models for late-type stars since the X-rays fluxes are far too low. We have found a weak correlation between the "chromospheric" MgII and the "transition region" CIV lines, suggesting a connection between the mechanism of line formation; however the slope of the correlation is different than the observed in late-type stars. This may be caused, in part, by the contribution of the wind to the Mg II line. The X-rays excess of the TTS with respect to late type stars could be responsible of the photoionization of some of these species, as Fig.1c suggests. However, the source of the X-rays emission is still uncertain.

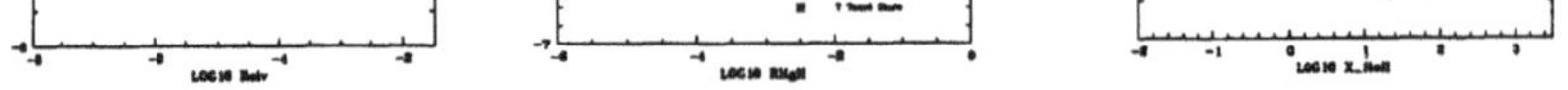

Figure 1. Flux-flux diagrams: a)X-rays vs. CIV, b)X-rays vs. MgII; Flux-ratio diagram: c) CIV vs. X-rays to HeII

References

Ayres, T.R., Fleming T. A., Simon, T. et al. (1995) , *ApJS*, **96. no. 223**

Gómez de Castro, A.I., Franqueira, M. (1988) Access Guide n.8, SP-ESA, 1205 . ESA Publication Division, ESTEC, Noordwijk, The Neetherlands.

SPECTRAL VARIABILITY OF T TAURI STARS IN THE OPTICAL

J.F. GAMEIRO AND M.T.V.T. LAGO
Centro de Astrofísica and Departamento de Matemática Aplicada da Faculdade de Ciências da Universidade do Porto Rua do Campo Alegre, 823, 4150 Porto, Portugal

Abstract.

We present the results of time variability analysis of the Hα, He I (λ5876 Å) and Na D (λ5890, 5896 Å) lines in two active T Tauri stars, DL Tau and DR Tau. The aim is the study of time behaviour of these lines, during one week, as well as possible correlations among them. The observed profiles of some lines are very complex, often showing characteristics of outflow and accretion.

1. Introduction

DR Tau and DL Tau are active classical T Tauri stars, of spectral type K7–M0 (Herbig & Bell, 1988) with strong emission and few absorption lines. The spectra of these two stars are characterized by a strong veiling, ranging from 6 to 20 times the level of the continuum in DR Tau (Hartigan *et al.*, 1995) and values larger than 2 in the optical band for DL Tau (Basri & Batalha, 1990). In DR Tau, Kenyon *et al.* 1994 observed variations in the V and R bands larger than 1 magnitude and in DF Tau, Chavarria-K 1979, reported on strong photometric varibility. The spectroscopic variability is also strong in both stars, and occurs on different time scales in DR Tau, ranging from days to hours. These fluctuations, seem to be related with the varibility of the veiling level (Guenther & Hessman, 1993).

2. Profiles variations

The spectroscopy was carried out at the La Palma Observatory, with the Isaac Newton Telescope, using the Intermediate Dispersion Spectrograph ($\lambda/\Delta\lambda > 25\,000$). The data includes repeated observations on time scales of one day, during 1990 October 1 to 7. The data of October 3 are not included owing to bad weather and DL Tau was not observed on October 5. The spectral reduction has been done in the usual way.

Astrophysics and Space Science is the original source of publication of his article. It is recommended that this article is cited as: *Astrophysics and Space Science* **261**: 129–132, 1999.

The choice to look at those profiles, was made with a view of probing the conditions prevailing both in the cooler and hotter regions of the atmosphere. The strength and shape of these emission lines and their behaviour over the week, provide important constraints for modeling the star and understanding other related phenomena such as mass loss. A similar study has been done for the T Tauri star LkHα 264 (Gameiro *et al.*, 1993).

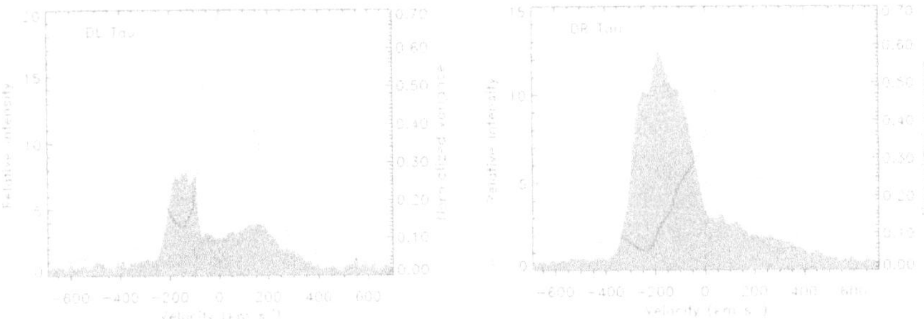

Figure 1. Average profile of Hα and normalized variance, for DL Tau and DR Tau.

2.1. Hα

Fig. 1 displays the average profile and normalized variance of Hα obtained during the week. The profile is highly variable in both stars.

In DL Tau, the profile maintains its shape during the week. It is very intense and centered at $\approx$ 0 km s^{-1}, with a blueshifted absorption, similar to the average profile plotted on Fig. 1. The variations occur on the very localized absorption component, over velocities in the range -200 to -100 km s^{-1}.

In DR Tau, the profile is centered at 100 km s^{-1} with the variations occuring over a large band of velocities in the blue part of the line. The peak of variance occurs at ≈ -200 km s^{-1}, but the variations are important over velocities in the range -300 to 0 km s^{-1}. The variations seem to be caused by a transient absorption. The evolving characteristic time scale of this blue absorption seems to be less than one day, which makes impossible its fit with the present data.

Independently of the differences in the line shape, the far wings on both stars, are seen to extend to the same high velocities and are quite constant during the week.

2.2. HE I AND NA I D

For both stars, the emission lines in this band show weak variations during the run. The peak of variance, is less than 0.1, still very small when compared to the level obtained for the Hα line. In DR Tau, the peak of variance for He I occur in the red wing of the profile and could be associated with the intensity variation of a possible inverse P Cygni profile. However the data is incomplete to search for any kind of periodicity.

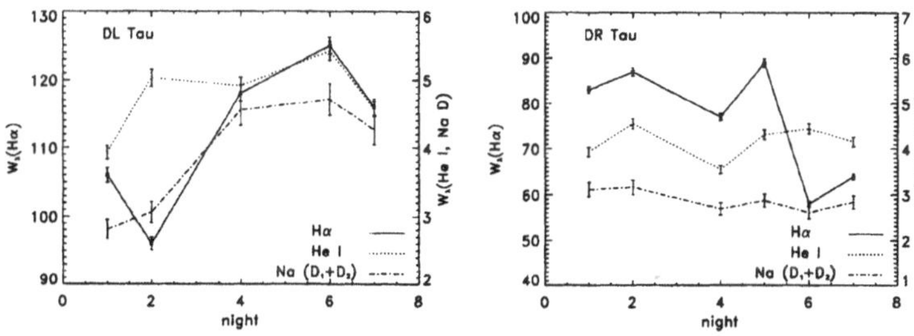

Figure 2. Equivalent widths variations.

3. Correlations

The measured equivalent widths of Hα, He I and Na I (D_1+D_2) are displayed as well as the corresponding error bars (Fig. 2). There is no detection of periodicity in the equivalent width behaviour of these emission lines. Our temporal coverage is too short, in DL Tau, to check for any correlation between the profile variation and the photometric period of 9.4 days. For the behaviour of the equivalent width of Hα and Na D, we found a coefficient of correlation of 0.88. In DR Tau, no correlation has been measured, however if the two last nights are excluded (a decrease of equivalent width in Hα occurs in those nights) the correlation coefficient between Hα and He I would increase to 0.92. Despite the short temporal coverage, the equivalent width of He I presents a quasi-periodic variation of $\approx 4-5$ days, lower than the photometric period (larger than 5 days, Kenyon *et al.* 1994

Fig. 3 shows the correlation matrix for the Hα line in both stars. The range of velocities corresponds to the portion of the line with large variations. The contour plot range from 0.9 to 1, with the lowest contour representing a real correlation at the 99 % confidence level. This is useful to analyze possible correlations in the variations occurring at different parts of the profile.

In DL Tau some parts of the line are relatively well correlated, namely the red wing up to 250 km s^{-1}. The blueshifted absorption component appears uncorrelated with the rest of the line profile. These results are very different from what we found for other T Tauri stars, namely DR Tau, where the red wing of Hα is well correlated for high velocities. In our

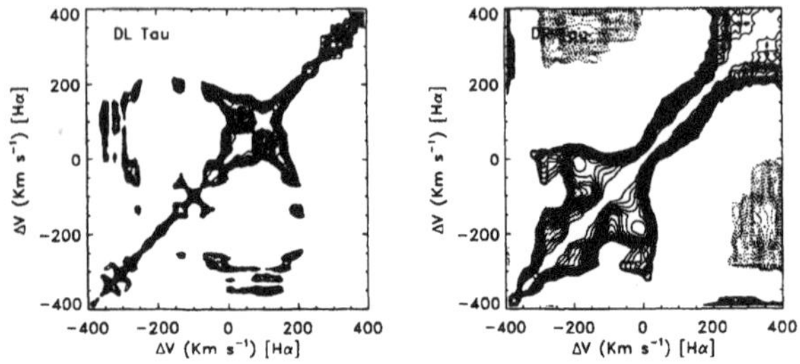

Figure 3. Correlation matrices, Hα versus Hα, for DL Tau and DR Tau.

sample of T Tauri stars under study (10 stars), DR Tau is the only one which presents such strong inverse correlation. The anti-correlation occurs between the velocity band $[-300, 0]$ km s^{-1} and $[+200, +400]$ km s^{-1}.

The search for correlations He I versus He I and between two different lines (He I versus Hα) was also tried, but the results were found to be unreliable for a conclusion due to the small variations of the He I line.

4. Conclusions

The equivalent width variability of emission lines of Hα, He I and Na D, in DL Tau, is mainly caused by variations on the relative intensity. The lines of Na D and Hα seem to be well correlated, but He I presents a different behaviour. A blueshifted absorption at velocity $[-200, -100]$ km s^{-1}, characterize the Hα line. The depth of this absorption changes during the week, but is not correlated with emission, suggesting different regions of formation.

The results obtained for DR Tau suggest that, although the star presents a large veiling with strong fluctuations, the Hα line profile variability cannot be explained only by variations in the veiling. The registered variations are associated with a transient blueshifted absorption, but the present data is incomplete to fit the evolution of this feature.

References

Basri, G., Batalha, C. (1990), *ApJ*, **363**, 654
Chavarria-K C. (1979), *A&A*, **79**, L18
Gameiro, J.F., Lago, M.T.V.T., Lima, N.M., Cameron, A.C. (1993), *MNRAS*, **261**, 11
Guenther, E., Hessman, F.V., (1993), *A&A*, **268**, 192
Hartigan, P., Edwards, S., Ghandour, L. (1995), *ApJ*, **452**, 736
Herbig, G., Bell, K. (1988) Lick Obs. Bull., **1111**
Kenyon, S.J., Hartmann, L., Hewett, R., Carrasco, L., Cruz-Gonzales, I., Recillas, E., Salas, L., Serrano, A., Strom, K.M., Strom, S.E., Newton, G. (1994), *AJ*, **107**, 2153

RAPID SPECTRAL VARIABILITY OF DR TAU.

K.W. SMITH
Physics Department, QMW,
Email: K.W.Smith@qmw.ac.uk

G.F. LEWIS
Dept. of Physics and Astronomy, University of Victoria

I.A. BONNELL
Institute of Astronomy, Cambridge

P.S. BUNCLARK
Royal Greenwich Observatory, Cambridge

AND

J.P. EMERSON
Physics Department, QMW

1. Introduction

The Classical T Tauri stars (CTTS) are pre-main sequence stars surrounded by an accretion disk. Material from the disk is being accreted by the central object, probably by a strong magnetic field anchored to the star (Königl 1991). The accreted material is channelled down the magnetic field lines and free falls onto the star's surface, forming strong shocks. This gives rise to excess optical continuum emission and spectral emission lines (Bertout 1989).

The accretion-driven emission features are known to be variable (Guenther & Hessman 1993). In many cases this variability is thought to be periodic, with periods corresponding to known photometric periods of a few days. The implication of this is that variable accretion signatures arise due to rotation of the system. Such variability is to be expected if the accretion streams and shock regions are not axisymmetric.

Astrophysics and Space Science is the original source of publication of his article. It is recommended that this article is cited as: *Astrophysics and Space Science* **261**: 133–135, 1999.

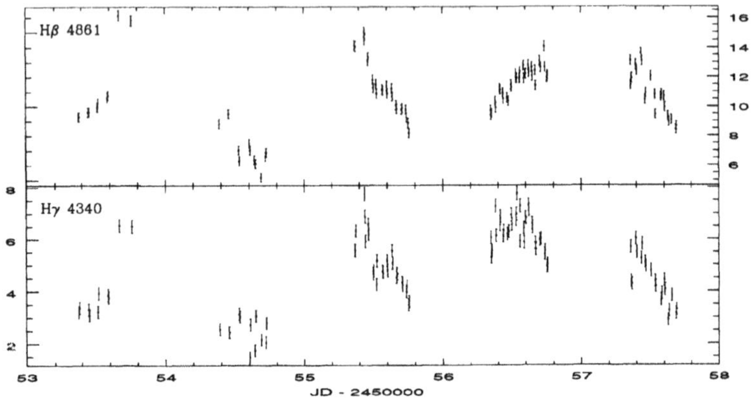

Figure 1. Equivalent width time series' for H_β and H_γ in the spectrum of DR Tau.

2. Observations and analysis

We obtained time series spectroscopy of six CTTS with a time resolution of approximately one hour. Here, we present the time series' of the lines in the spectrum of DR Tau, the source with the most activity on short timescales.

Figure 1 shows the equivalent width time series of two of the Balmer lines, H_β and H_γ. These were amongst the strongest lines in the DR Tau spectra.

The variability of the different lines is complex. No convincing evidence of a periodicity less than the 5 night time span of the data was found. The activity of the different lines is at times closely correlated. At other times the activity is entirely different.

We used a cross correlation technique to test the resemblance of the activity of the different lines to one another. Examples of cross correlation functions (CCF's) are shown in Figure 2. The CCF's for night 3 show strong correlation between the different lines. The CCF's on night 4 show evidence of a correlation between certain lines which is offset from zero time difference. This indicates that the two time series' resemble one another when offset by some time lag.

3. Conclusions

Interpretation of this behaviour is problematic in the absence of detailed radiative transfer models of the inner regions. The complex line behaviour suggests that the environment of the star may be morphologically complex, possibly comprising magnetically rigid accretion streams, orbiting conden-

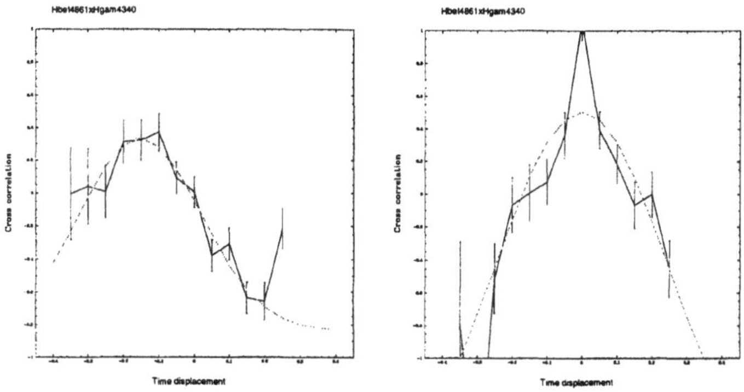

Figure 2. Cross correlation functions between pairs of lines on night 3 and night 4. The functions have been fitted with Gaussians, to find the most likely time lag. Error bars $\sqrt{\frac{1}{N}}$, where N is the number of pairs of points in each bin. This is the expected standard error if there is no correlation between the two time series'. On the left, H_β and H_γ are shown to be strongly correlated at zero time lag with the night 3 data. On the right, the same two lines show an apparent time lag of approximately 0.1 daysd on night 4, H_γ varying in advance of H_β.

sations of material, accretion shocks on the stellar surface and so on. The variability itself could be the result either of a variable process, such as variable accretion, or could be caused by the rotation of the magnetically rigid inner system. One possibility is that the apparently time-lagged activity of the different lines on night 4 might be caused by an accretion shock passing over the stellar limb.

A fuller account of this work can be found in Smith et al (1997) and Smith et al (1998).

References

Bertout C., 1989, Ann. Rev. Astron. & Astroph., 27, 351.
Guenther & Hessman, 1993, A&A, 268, 192.
Königl A., 1991, ApJ, 370 L39.
Smith K.W., Bonnell I.A., Lewis G.F., Bunclark P.S., 1997, *MNRAS* 289, 151.
Smith K.W., Lewis G.F., Bonnell I.A., Bunclark P.S. & Emerson, J.P. 1998, *MNRAS*, Submitted.

THE STRUCTURE OF THE SHOCK WAVE IN THE CASE OF AN ACCRETION ONTO LOW MASS YOUNG STARS

S.A. LAMZIN
Seccion Departamental de Astronomia y Geodesia
Facultad de Ciencias Matematicas, Universidad Complutense de Madrid, Spain

AND

Sternberg Astronomical Institut
Moscow V-234, 119899 Russia

The problem of calculation of the accretion shock wave structure in the case of young stars is formulated physically and mathematically. Some results of its numerical solution are presented and dependence of the shock wave structure on the parameters of the problem is demonstrated. It was found that relative intensities and profiles of lines in T Tauri stars spectra depend generally speaking on velocity and density of infall gas as well as on geometry of the accretion. It is concluded that results of the calculations can be used to solve the problem of young stars activity nature. See for details Lamzin (1998).

We try to explain the nature of Classical T Tauri Stars (CTTSs) activity as a result of disk accretion of a circumstellar matter onto the surface of a magnetized young star. We connect the region of CTTSs emission spectrum formation with the shock wave, which appear when infalling matter collides with the surface of the star. Before to calculate the structure of the accretion shock wave some physical assumptions were made to simplify general system of radiation hydrodynamic equations.

Let us assume that from the qualitative viewpoint the structure of the shock wave is the following (Lamzin 1995). Inflowing gas, compressed at the shock front smoothly decelerates and cools from $T \sim 10^6$K down to some thousands degrees, transiting finally into hydrostatically equilibrium stellar atmosphere. Gas cooling occur due to volume emissivity , such that the main portion of energy is concentrated in the quanta with energy > 13.6eV.

Astrophysics and Space Science is the original source of publication of his article. It is recommended that this article is cited as: *Astrophysics and Space Science* **261**: 137–140, 1999.

The half of quanta from cooling zone moves upward and is absorbed by matter before the shock front, heating and ionizing gas in this region, which is however transparent for the continuum at $\lambda > 912$ Å. The second half of quanta moves downward and is absorbed in the transition region between cooling zone and stellar atmosphere. Finally their energy is reemitted by high density gas predominantly in continuum, such that Lyman quanta carry away only negligible portion of the total energy, so the radiation of the transition zone practically has no influence on the structure of the upper layers. We shall assume also that the shock front is at rest relative to the stellar surface.

In the frame of these assumptions infall gas velocity V_0 and density N_0 far before the shock front should be in the range 150–450 km/s and $10^{10} - -10^{13}(\mathrm{cm}^{-3})$ respectively in the case of CTTSs (Lamzin 1995). Then the characteristic lengths of post-shock cooling zone and H II zone before the shock front are $<< R_*$ as well as the width of the accretion zone in the direction perpendicular to the gas velocity. We have solved therefore 1-D radiative hydrodynamic equations supposing $(\partial/\partial t)_z = 0$ everywhere.

Equations of continuity, momentum and energy conservation as well as ionization balance where considered. We have supposed that kinetic temperatures of all ions and atoms are the same, but in general differs from that of electrons. It was assumed also that chaotic magnetic field is freezen into accreted plasmas, strength of which is much less then that of large scale magnetic field of the star, but large enough to suppress electron conductivity. Radiative and dielectronic recombination, excitation and ionization via electron collisions as well as photoionization (including possibility of Auger effect), charge-transfer reactions with H and He of more than 80 ions of H, He, C, N, O, Ne, Mg, Si, S and Fe were considered. Contribution of ~ 200 spectral lines as well as free-free and free-bound continuum emission into cooling function were taken into account. See Lamzin (1998) for details.

TABLE 1.

V_0, km/s	200	300		400
$\lg N_0$, cm^{-3}	12	11	12	11
z_{HII}, cm	$1.3\,10^8$	$2.1\,10^{10}$	$7.0\,10^8$	$2.4\,10^{10}$
T_1, K	$1.7\,10^4$	$1.8\,10^4$	$2.0\,10^4$	$1.8\,10^4$
x_H	0.0027	0.0008	0.0007	0.0005
x_{C4}	0.096	0.17	0.16	0.093
T_e^{max}, K	$5.6\,10^5$	$1.3\,10^6$	$1.3\,10^6$	$2.3\,10^6$
z_{pst}, cm	$1.4\,10^6$	$9.3\,10^7$	$9.4\,10^6$	$2.0\,10^8$

One can find in the Table some parameters of the accretion shock wave for different values of V_0 and N_0 : pre-shock HII zone length, gas temperature as well as H I and C IV ion abundances immediately before the shock front, maximal post-shock electron temperature and the size of this zone.

Some ions have significant abundance in 3 regions: in the pre-shock H II zone, just after the shock front and in the recombination post-shock zone. Thus generally speaking line profiles of these ions should have 3 separate peaks because gas velocity in these regions are equal to V_0, $V_0/4$ and $\simeq 0$ respectively. Relative intensities of these peaks depend significantly on V_0 and N_0, such that in different way for different ions. For example when $lgN_0 = 12$ the ratio of CIV 1550 Å doublet fluxes from regions before and after the shock front is equal to 1:17 at $V_0 = 200$ km/s, but to 2:1 at $V_0 = 300$ km/s. C III 1908 Å intercombination line flux from post-shock zone is negligible in all cases, while in the case of O VI 1031 Å doublet the flux from pre-shock zone is always small.

To derive line profiles from the accretion zone as a whole one should summarize local line profiles over all its points, which are oriented at different angles relative to the observer. It means that the real profiles depend on the geometry of the accretion zone and can differ significantly from 1-D onces. Only lines of ions, which form at $T \geq 2\,10^4$ K are suitable for the comparizon of the theory with observations because "low temperature" lines can form in stellar wind as well. Apparently "high temperature" lines of CTTSs are redshifted and indeed have multicomponent structure (Gomez de Castro et al. 1994).

Some lines e.g. C IV 1550 Å have optical depth more than 1, so "limb darkening" should be different for different lines. It means that line ratios generally speaking should depend on the geometry of an accretion zone due to the same reasons as line profiles.

Gas of the post-shock zone is a source of X-ray emission. In the pure 1-D approch external X-ray luminocity of the shock $L_x^{obs} \equiv 0$ because (by definition!) one have infinity absorbing matter before the shock front. But in reality some part of X-ray quanta can escape through the walls of an accretion column. Then the observed X-ray radiation is much less intensive and has significantly more hard spectrum than the spectrum of post-shock zone.

If stellar magnetic field is a dipol one, such as the angle α between magnetic and rotational axes is 0, then the accretion zone has a form of uniform accretion belt $\theta_1 < \theta < \theta_2$ (in each hemisphere). One can see from the Table below how much L_X^{obs} differs from that of post-shock zone for different positions of the accretion zone and accretion mass rate when the inclination angle $i = 0$. If $\alpha > 0$ then variations of X-ray flux and its spectrum should occur during stellar rotation.

TABLE 2.

θ_1	θ_2	$\dot{M}_{ac}$	L_{ac}	L_x^{obs}	L_x^{obs}/L_x^{pst}
20	40	$2.1\,10^{-8}$	$6.2\,10^{32}$	$1.0\,10^{30}$	$6.9\,10^{-3}$
30	50	$2.7\,10^{-8}$	$8.0\,10^{32}$	$2.2\,10^{30}$	$1.5\,10^{-2}$
50	70	$3.7\,10^{-8}$	$1.1\,10^{32}$	$7.6\,10^{30}$	$5.1\,10^{-2}$

The main conlusions can be summarized as follows:

1) Numerical calculations have confirmed *a'posteriori* the validity of the main physical assumptions adopted to simplify the general system of radiative hydrodynamic equations, i.e. the problem was solved in selfconsistent way.

2) Relative intensities and profiles of UV lines depend strongly on V_0 and N_0, so it looks possible in principle to find these parameters from comparison of the theory with observations.

3) The model predicts that
a) *relative intensities and profiles of UV lines*
b) *UV continuum intensity and spectral energy distribution*
c) *X-ray luminocity and spectrum*
depend on the geometry of the accretion zone and vary during stellar rotation if $\alpha > 0$.

References

Gomez de Castro A.I., Lamzin S.A., Shatskij N.I. (1994), *Astron. Reports*, **71**, p. 609
Lamzin S.A. (1995), *Astron. and Astrophys*, **295**, p. L20
Lamzin S.A. (1998), *Astron. Reports*, in press

EXTENDED ($15 \times 6\,AU$) Hα EMISSION AROUND T TAU N

P. J. V. GARCIA
CRAL/Observatoire de Lyon, CNRS UMR 5574, France
Centro de Astrofísica da Universidade do Porto, Portugal
AND
E. THIÉBAUT
CRAL/Observatoire de Lyon, CNRS UMR 5574, France

1. Abstract

We present an image of T Tauri N reconstructed from speckle interferometry observations made at the 3.6 m Canada-France-Hawaii telescope (CFHT). Two spectral windows, at Hα [6540-6580 Å] and red continuum [6580-6840 Å] were used.

The star is unresolved in the continuum ($\theta \leq 0''038$), setting an upper limit for its size of $\leq$ 5.3 AU (assuming a 140 pc distance). At Hα (continuum subtracted) two components are observed, an unresolved one with 70 % of the Hα flux and a resolved $0''106 \times 0''043$ (15×6 AU) with 30 %. The extended component is elongated in the North-South direction.

The extended Hα component scale, it's non-detection in the continuum, and TTau Hα profile shape suggest that the extension is a disk corona.

2. Discussion

Fig. 1 shows a preliminary image restored from the Hα measurements with an imposed resolution of $0.8 \times D/\lambda$ Because our deconvolved image uses also the *phase* information (which isn't present in the visibility) we can further constrain the visibility models of Devaney et al., 1995: *Model 1 with a* $15 \times 6\,AU$ *extended component (1/3 Hα flux) plus an unresolved one with (2/3 Hα flux) is clearly favored.*

Several ideas have been proposed to explain the Hα emission from CTTSs: Alfvén wave driven winds , stochastic winds, inner disk winds and magnetospheric boundary layer accretion. All these models predict emis-

Astrophysics and Space Science is the original source of publication of his article. It is recommended that this article is cited as: *Astrophysics and Space Science* **261**: 141–142, 1999.

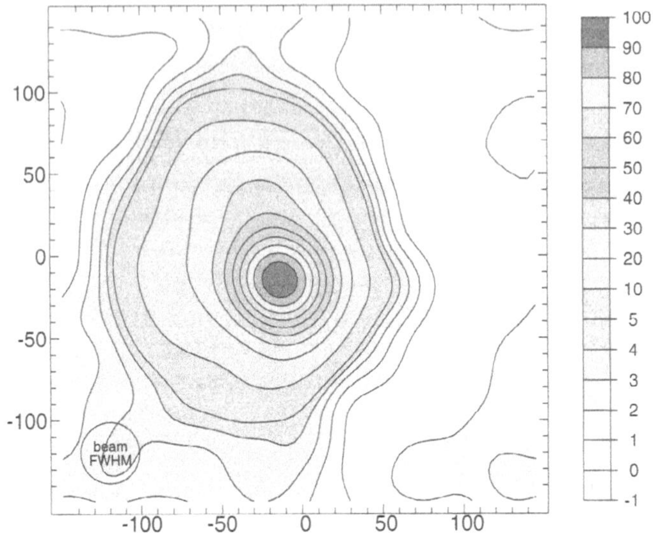

Figure 1. Deconvolved image of T Tau N Hα , scale in marcsec., North is up and East is to the left. The imposed resolution is $0.8D/\lambda$.

sion regions $r < 0.2\,AU$ and hence can't explain TTau N's extended Hα component. Furthermore these models assume that only one physical region/mechanism is responsible for the Hα while our observations suggest that the Hα profile has contributions from 2 distinct components.

The published Hα profile for TTau (Reipurth et al., 1996) combined with the assumption that the disk is optically thick, the fact that no redshifted forbidden lines are observed in T Tau spectra and that strong blueshifted absortions are very rare and only observed for FU Orioni objects, which isn't the case of TTau N leads us to conclude that the extended component has a low velocity.

This low velocity and extension agree with the disk corona scenario proposed by Kwan and Tademaru 1988 and could be the Hα counterpart of the forbidden emission coronae detected by Hirth et al, 1997 for several CTTSs.

Acknowledgements: P. Garcia acknowledges financial support from FCT via the PRAXIS XXI/BD/5780/95 grant.

References

Devaney, N., Thiébaut, E., Foy, R., Blazit, A., Bonneau, D., Bouvier, J., de Batz, B., & Thom, Ch. (1995), *A&A*, **Vol. no. 300**, pp. 181

Hirth, G. A., Mundt, R., Solf, J. (1997), *A&A S.*, **Vol. no. 126**, pp. 437

Kwan, J., &Tademaru, E. (1995), *Ap.J*,**Vol. no. 454**, pp. 382

Reipurth, B., Pedrosa, A., & Lago, M. T. V. T. (1996), *A&A S.*, **Vol. no. 120**,pp. 229

BALMER LINE PROFILES VARIATIONS IN SU AURIGAE FROM THE MUSICOS 96 MULTI-SITE CAMPAIGN

J. M. OLIVEIRA
ESA Space Science Department, ESTEC
Centro de Astrofísica da Universidade do Porto
Y. C. UNRUH
Institute of Astronomy, University of Vienna
B. H. FOING
ESA Space Science Department, ESTEC
AND
THE MUSICOS 96 COLLABORATION

1. The Hα variability

SU Aur is a G2 classical TTauri star. The Balmer lines in this star show very large intensity fluctuations, and a persistent pattern of simultaneous gas inflow and outflow around the star has been reported (Petrov *et al.*, 1996).

Our data set was obtained during the MUSICOS 96 campaign and includes 126 echelle spectra from 5 observatories: Xinglong, OHP, INT, McDonald and CFHT (Unruh *et al.*, 1997).

The variability of the Hα profiles in SU Aur is clearly visible in this data set. The more striking variability appears in the blue wing. Two redwards moving features can be easily tracked for more than 1.5 days, respectively starting at HJD 2450408 and HJD 2450412.

Fig.1 (a) shows the average and normalized variance profiles. It can be seen in the variance profile that the main variability comes from velocity ∼-80kms^{-1}. This velocity corresponds to where Johns & Basri (1995) detected a marginal 7.7 day period. In the periodogram of our data set, we find a strong peak at 2.8 day between -250 to -200 kms^{-1}, consistent with the average 3 day periodicity, usually accepted (e.g. Giampapa *et al.* 1993 and Johns & Basri 1995). There is a strong indication of a 5 day period, both in the blue and in the red wings of the profile, indicating that these

Astrophysics and Space Science is the original source of publication of his article. It is recommended that this article is cited as: *Astrophysics and Space Science* **261**: 143–144, 1999.

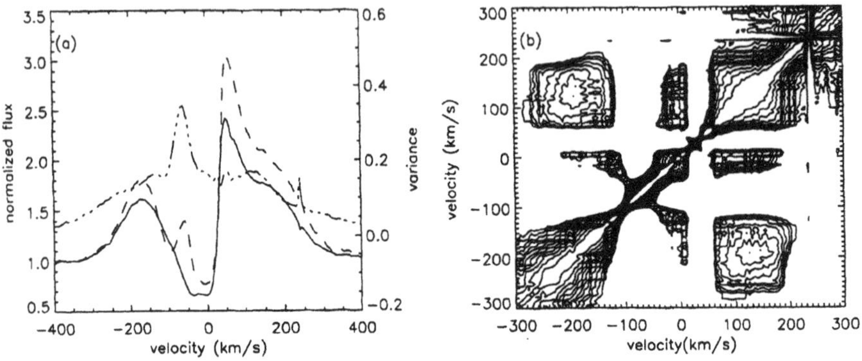

Figure 1. (a) The Hα average (full line) and normalized variance (dotted dashed line) profiles as well as the profile that most strongly deviates from the average (dashed line). (b) The autocorrelation matrix of the Hα line profiles. No global correlation is found, indicating that continuum variations are not responsible for the profile variations.

variations might be related. To further investigate this relation, we plotted the correlation matrix of the Hα profiles (Fig.1 (b)). The immediate conclusion from these correlations is that the profiles do not vary uniformly, as no strong global correlation can be seen, excluding continuum variations as the cause of the profile variability. The variations in the red wing seem very well correlated, better that what is observed in the blue wing. The most interesting point is that there is a relatively strong correlation between the two wings, indicated by the contours centered at $v_x \sim$-200 kms^{-1} and $v_x \sim$+150 kms^{-1}. This occurs at the velocity range where the 5 day period was detected. The main absorption does not seem to be at all correlated with any other part of the profile.

Acknowledgements

JMO acknowledges the support of the *Fundação para a Ciência e Tecnologia* (Portugal) under the grant BD9577/96. The authors wish to thank the MUSICOS 96 collaboration (see names in Unruh *et al.*, 1997).

References

Giampapa M.S. et al. (1993), *ApJ*, **89**, 321
Johns C.M. & Basri G. (1995), *ApJ*, **449**, 341
Petrov P.P. et al. (1996), *A&A*, **314**, 821
Unruh Y.C. et al. (1997), *The 10th Cambridge Workshop on Cool Stars, Stellar Systems and the Sun*, ed. R.A. Donahue & J.A. Bookbinder, in press

IS IT POSSIBLE TO OBSERVE RADIATION OF FUORS BOUNDARY LAYER?

S.A. LAMZIN
Sternberg Astronomical Institute
Moscow V-234, 119899 Russia

M. TEODORANI, L. ERRICO, A.A. VITTONE
Osservatorio Astronomico di Capodimonte
I-80131 Napoli, Italy

F. GIOVANNELLI
Istituto di Astrofisica Spaziale
I-00044 Frascati, Italy

AND

C. ROSSI
Istituto Astronomico
Universita La Sapienza di Roma,I-00161 Roma, Italy

We have estimated the optical depth of stellar wind of 3 FUORs (V1057 Cyg, FU Ori, Z CMa) in Balmer and Paschen continuum and conclude that it seems plausible that it exceeds unity. If so it is not possible in principle to observe optical and UV radiation from the innermost and the hottest regions of these objects, i.e. boundary layer of the accretion disk.

It is widely assumed now that high luminocity of FUORs is due to the disc accretion of circumstellar envelope onto the central young star with an accretion rate $\dot{M}_{ac} \sim 10^{-4}\ M_{\odot}/yr$. Boundary layer should give some contribution into the radiation of FUORs, but Kenyon et al.(1989) could not detect it during IUE observations of FU Ori, V1057 Cyg and Z CMa. They have concluded that angular velocity of the central stars outer regions is close to critical one or that the surface area of boundary layers is much larger than in Lynden-Bell and Pringle (1974) theory. Petrov and Herbig (1992) consider negative results of boundary layer emission search as an argument against the disk accretion theory.

Astrophysics and Space Science is the original source of publication of his article. It is recommended that this article is cited as: *Astrophysics and Space Science* **261**: 145–146, 1999.

FUORs have a strong outflowing envelopes and if they are opaque in Balmer and Paschen continuum it is not possible in principle to observe radiation of boundary layer. To estimate continuum optical depth of FUORs stellar wind we have compared it with that in lines of Balmer and Pashen series. Asuuming that if Balmer (Paschen) line has blueshifted absorbtion feature then its optical depth in the wind exceeds unity, one can estimate low limit for column density of hydrogen atoms along the line of sight with first (second) excited level and to obtain low limit of optical depth in respective continuum – see Errico et al. (1997) for details.

We have found that it seems plausible that continuum optical depth of stellar wind of FU Ori, V1057 Cyg and Z CMa exceeds 1 in 1800-8400 Å spectral band, i.e. outflowing gas can screans radiation of these FUORs boundary layers.

Kenyon et al.(1988) and Welty et al.(1992) have succeded to reproduce continuum SED of FU Ori, V1057 Cyg and Z CMa in the frame of thin accretion disk model assuming that the spectrum of elementary disk ring of radius r and gas temperature $T(r)$ is identical to the spectrum of a supergiant with effective temperature $T_{ef} = T(r)$. But in fact it looks strange that this model gives resonable agreement with observations because at so high accretion rate one can not expect that disk temperature distribution is equal to $T(r)$ of Lynden-Bell and Pringle model.

If outflowing envelopes of FUORs are indeed opaque then continuum emission of stellar wind should also be taken into account. One can assume that mentioned above agreement of observed and theoretical SEDs is by chance: two disadvantages of the model (the difference of real temperature distribution from the adopted one and the ignorance of stellar wind emission) have compensated each other. Respective numerical calculations are necessary to check this hypothesis.

References

Errico L., Lamzin S.A., Vittone A.A. et al. (1997), *Astron. Letters*, **23**, p. 604
Kenyon S.J., Hartmann L., Hewett R. (1988), *ApJ.* **325**, p. 231.
Kenyon S.J., Hartmann L.W., Imhoff C.L., Cassatella A. (1989) *ApJ.* **344**, p. 925
Lynden-Bell D., Pringle J.E. (1974), *MNRAS* **169**, p. 603
Petrov P.P., Herbig G.H. (1992), *ApJ* **392**, p. 209
Welty A.D., Strom S.E., Edwards S. et al. (1992) *ApJ.* **397**, p. 26

NEAR INFRARED LINES AS DIAGNOSTICS OF ACCRETION IN T TAURI STARS

D.F.M.FOLHA
Physics Department, Queen Mary & Westfield College
London E1 4NS, United Kingdom
Centro de Astrofísica da Universidade do Porto
Rua do Campo Alegre, 823, 4150 Porto, Portugal

AND

J.P.EMERSON
Physics Department, Queen Mary & Westfield College
London E1 4NS, United Kingdom

Abstract We present results from a spectroscopic survey of Paβ & Brγ emission in a sample of T Tauri stars, mostly in the Taurus-Auriga complex. Velocity resolved spectra ($\Delta v \sim 15$ km s^{-1}) show that, of the 41 stars observed in Paβ emission, 34% display IPC redshifted absorptions, while, of the 29 stars observed in Brγ emission, 21% display IPC redshifted absorptions. The lines are broad (FWHM ~ 200 km s^{-1}) and lack blueshifted absorptions. These observations are consistent with the near infrared hydrogen lines forming predominantly in infalling material, as suggested by magnetospheric accretion column models. Radiative transfer calculations of the Paβ and Brγ line profiles in a very simplified, spherically symmetric situation are discussed.

1. Introduction

The currently favoured accretion model in T Tauri stars, the magnetospheric accretion model, invokes the presence of a dipolar stellar magnetic field which disrupts the circumstellar disk out to a few stellar radii. From there material falls freely onto the star along magnetic columns (e.g. Camenzind 1990). This model provides a mechanism that dissipates angular momentum, therefore keeping the star rotating well below break-up velocity, as it is observed. This model also provides an explanation for the presence of the winds observed in these stars.

Strong Hydrogen line emission in T Tauri stars has traditionally been interpreted in terms of mass loss but there were severe problems in explaining all the features in the profiles (e.g. Hartmann *et al.* 1990). More recent modelling of the Balmer line profiles in a magnetospheric accretion scenario

Astrophysics and Space Science is the original source of publication of his article. It is recommended that this article is cited as: *Astrophysics and Space Science* **261**: 147–150, 1999.

(Hartmann *et al.*, 1994) suggests that most of the underlying emission is produced by *infalling* material, and this interpretation has been supported by the results from a velocity resolved optical spectroscopic survey of 15 classical T Tauri stars (Edwards *et al.*, 1994).

Near infrared Hydrogen lines are expected to form in the denser parts of the circumstellar envelope of a T Tauri star and their line profiles reflect the dynamics of the region where the lines are formed. These profiles constitute one of the few observational tests that can be made of the magnetospheric accretion model. Najita *et al.* (1996) showed that the Brγ line profiles in 5 Young Stellar Objects were broadly consistent with an origin in infalling material. The Paβ lines have different optical thickness from the Balmer and Brackett lines and provide further tests of the magnetospheric infall (or any other) model. We present results based on observations of Paβ and of Brγ profiles for a sample of T Tauri stars from the Taurus-Auriga complex.

The observations were carried out at the United Kingdom Infrared Telescope (UKIRT) between 15-16 December 1995 where velocity resolved (R$\sim$20,000) spectra of Paβ and of Brγ were obtained for a large sample of T Tauri stars (49 stars at Paβ and 34 stars at Brγ).

2. Results

Paβ and Brγ emission is not present in all the observed T Tauri stars. Paβ is in emission in 41 out of 49 stars and Brγ in 29 out of 34 stars. The emission lines were classified according to the classification scheme developed by Reipurth et al. (1996). 51% of the Paβ profiles and 72% of the Brγ profiles are generally symmetric, suffering no or little influence from absorption features (Type I). The bulk of the remaining profiles, 34% at Paβ and 21% at Brγ, display inverse P Cygni profiles (IPC), while the others are classified either as Type II R or as Type III R. The typical velocities at which the redshifted absorption occurs in the IPC line profiles are $100-300$ km s^{-1}. Only one star displays a blueshifted absorption feature (only at Paβ) superimposed on the emission line.

The observed emission line profiles are very broad, with a typical FWHM of $\sim$ 200 km s^{-1} and with line wings extending to around $\sim$ 300 km s^{-1} in the blue and $\sim$ 200 km s^{-1} in the red. The line peaks tend to be slightly blueshifted but very rarely redshifted and the asymmetry factor (ratio of the equivalent width blueward to equivalent width redward of the rest velocity) is typically between 1 and 2 for the Paβ line with somewhat larger values for the Brγ line.

3. Discussion

Wind models are clearly not capable of producing IPC profiles in the Hydrogen lines. The observed velocities of the absorption features in the IPC Paβ and Brγ profiles are typical of those expected for material falling freely onto a T Tauri star from a few stellar radii out, indicating that these lines are formed in an accretion scenario.

While an infall origin of the lines displaying IPC profiles does not seem to be in doubt, that of the more frequent symmetrical lines (Type I profiles) is more debatable. The fact that some T Tauris certainly show infall can be used to argue that probably all do, however one must also ask whether wind models are able to produce Paβ and Brγ lines with Type I profiles. In reality both infalling material and a wind no doubt exist in most T Tauris, but the relative importance of the two in producing H lines may be different in different T Tauri stars

The Paβ and Brγ lines in T Tauri stars have seldom been modelled. Hartmann et al. (1990) present results for Brγ in the context of wind models. The model profile is unlike any of our observed Brγ profiles. Muzerolle *et al.* 1998 present Brγ profiles computed in the context of the magnetospheric accretion scenario. While qualitatively there is a good agreement between the model and our observations, quantitatively there are some discrepancies. The peak intensities of model line profiles tend to be considerably larger than those observed and the model profiles are too narrow compared to the observations. The ability of this model to produce so many Type I profiles is not clear.

4. Radiative Transfer Modelling

In view of the paucity of Paβ and Brγ line profile modeling found in the literature, line profiles were computed using simple spherically symmetric wind and accretion models, constructed following the procedures described by Johns & Basri 1995. A slightly modified CLOUDY radiative transfer code (Ferland, 1993) was used. Of course this is too simplistic to be expected to describe magnetospheric accretion, but is nevertheless instructive. After calculating the level populations at each radii with CLOUDY, a two-level Hydrogen atom was used to compute optical depths and source functions, from which the line profiles are calculated.

For typical parameters used to model Balmer line profiles (see Hartmann *et al.* 1990), the wind models computed tend (as expected) to produce normal P Cygni profiles at Paβ and Brγ, unlike those observed. Accretion models from which matter falls radially and freely from several stellar radii produce, as expected, IPC profiles similar to some of those observed.

The strength of the line peaks can be matched by tuning the mass accretion rate and the velocities at which the line peaks occur are usually consistent with those observed. Also, the observed line widths can be reasonably well explained by choosing appropriately the size of the modeled region, i.e. where free fall begins ($3R_* < R_{out} < 10R_*$). The redshifted absorption resulting from these calculations is however too deep when compared to the observations. Furthermore, the physical reason why free fall might start only at R_{out} between 3 and 10 stellar radii is unclear in a spherically symmetric scenario.

While spherically symmetric wind and accretion models can produce Type I profiles for very high densities (high wind mass loss/accretion rates), they fail in producing this type of profiles with peak intensities similar to those observed. High mass loss/accretion rates ($\dot{M} \sim 10^{-6}\ M_\odot\ yr^{-1}$) imply Pa$\beta$ and Brγ line peak intensities which are more than one order of magnitude higher than those observed, as well as being themselves too high when compared to the results from other observations.

It will be intresting to examine further the ability of profiles computed in the context of the magnetospheric accretion scenario (eg Muzerolle *et al.*1998) to explain the Type I profiles, assuming a random range of viewing angles to the T Tauri stars.

5. Conclusion

The observations presented here suggest that quite a large percentage of Paβ and Brγ line profiles, those displaying inverse P Cygni profiles, are formed in an accretion scenario. The observed infall velocities are similar to those expected from the magnetospheric accretion model. Simple radiative transfer modeling, predicts IPC profiles of depths similar to those observed are present in these lines even when a radial spherically symmetric infall is considered. However the observed range of emission peak strength to IPC depth is not reproduced, an indication that there is not truly spherical symmetry. Neither the spherically symmetric accretion nor wind models explored were not able to produce Type I line profiles with peak intensities similar to those observed and also have difficulties in explaining the observed line widths of all profiles. Thorough radiative transfer calculations in a more realistic accretion model, such as that of Muzerolle *et al.* 1998 are needed to address these problems.

Acknowledgements

We thank Christopher Johns-Krull for giving us access to his modified version of CLOUDY and for providing the code to compute the line profiles. The United Kingdom Infrared Telescope is operated by the Observatories on behalf of the UK Particle Physics & Astronomy Research Council. D.F.M. Folha acknowledges financial support via the CIENCIA/BD/2799/93 grant within the PRAXIS XXI program.

References

Camenzind M. 1990, *Rev. Mod. Astron.* **3**, 234
Edwards S., Hartigan P., Ghandour L., Andrulis C. 1994, *AJ* **108**,1056
Ferland G. 1993, Univ. Kentucky Dept. Physics and Astronomy, *Internal Rept.*
Hartmann L., Calvet N., Avrett E.H., Loeser R. 1990, *ApJ* **349**, 168
Hartmann L., Hewett R. & Calvet N. 1994, *ApJ* **426**, 669
Johns C. M. & Basri G. 1995, *ApJ* **449**, 341
Muzerolle J., Calvet N., & Hartmann L., 1998, *ApJ* **492**, 743
Najita J., Carr J.S. & Tokunaga A.T. 1996, *ApJ* **456**, 292
Reipurth B., Pedrosa A. & Lago M.T.V.T. 1996, *A&AS* **120**, 229

M.H.D. MODELS FOR STELLAR WINDS

C. SAUTY
Observatoire de Paris - Université Paris 7, DAEC, F-92190 Meudon, France

K. TSINGANOS
University of Crete, Dept. of Physics, GR-71003 Heraklion, Greece

AND

E. TRUSSONI
Osservatorio Astronomico di Torino, Strada dell'Osservatorio 20, I-10025 Pino Torinese, Italy

1. Introduction

In situ measurements of the Solar Wind in the interplanetary medium (e.g., reviews by C. Parnell and J-L. Bertaux in this volume) and observations covering almost all stellar types (e.g., reviews by T. Lago, H. Zinnecker and M. Smith in this volume) provide evidence that the external atmospheric layers of stars are accelerated to tens, hundreds, or, thousands of kilometers per second (with even higher speeds in compact stellar objects and AGN). Thus, plasma outflows seem to provide a link between stellar interiors and the interstellar medium. As such, outflows seem to participate actively to stellar evolution by removing mass, angular momentum and magnetic flux from the central object.

Although all these facts are well known, a complete understanding of the acceleration mechanism of outflows still remains an open problem. A second intriguing question is the variety of shapes that these outflows can take. They appear almost conical or bipolar around evolved stars while they focus into extremely collimated jets around young stars. A sketch of these two extreme configurations is shown in Figure 1. The presence of jets is probably linked to the existence of accretion discs. Nevertheless, accretion alone does not seem to explain the physical processes that take place in the outflowing gas to eventually form a very well collimated jet.

Astrophysics and Space Science is the original source of publication of his article. It is recommended that this article is cited as: *Astrophysics and Space Science* **261**: 151–160, 1999.

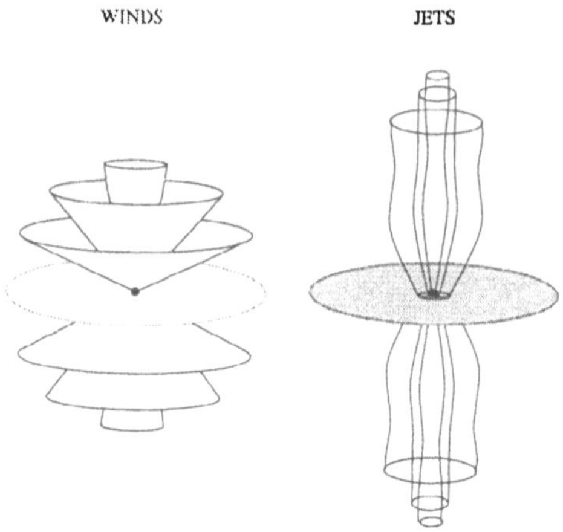

Figure 1. Two sketches showing a theorist's view of the morphologies of winds and jets as inferred from observations. Winds (left) are thought to roughly expand conically while jets (right) cylindrically.

The propagation of jets, their destruction by instabilities and their interaction with the interstellar medium deserve a separate study by themselves and are also treated in this volume (e.g., Schmidt's review for the theoretical part and Zinnecker's for the observations). The same is true for the stellar (solar) wind and its termination shock in the interstellar medium. The point that we want to address here is more specifically what happens between the source and the terminal shocks and how we can explain the various observed morphologies. From the technical point of view, one may start by solving the familiar MHD equations for a heated plasma at rest and rotating with the stellar magnetosphere and show how it eventually escapes out of the gravitational potential with rather high terminal velocities and some degree of collimation.

The very knotty structure of jets, as well as their measured proper motions are invoked in many studies as an evidence that the outflows have a strongly time-dependent nature. This point of view can be pushed to the limits by arguing that jets are formed as ballistic bullets. All this may probably be partially true in the sense that some violent events like Fu-Ors in YSOs may be strong and sudden ejections. However this argument is also invoked from a theoretical point of view because numerical simulations deal more easily with rapid variations and time dependent MHD equations than with the stationary problem (See Kudoh & Shibata 1997, Ouyed & Pudritz 1997). Yet, the timescale of the outflow variation inferred from the proper motions and the spacing of the knots can be as long as months or years. Some analogy can be drawn with the solar wind where a global steady outflow is observed in quiet periods with some violent events like Coronal Mass Ejections and some long term variations due to solar activity. This is strikingly similar to the pulsating jet model proposed for protostars in the simulations of M. Smith (this issue).

In that respect, the study of the steady MHD equations is equally important. In this domain analytical treatments have already been successfull in providing some physical understanding (Heyvaerts & Norman 1989, Léry 1997). A particular set of analytical models uses the so-called self-similar approach which enables the conversion of the partial MHD differential equations into nonlinear ordinary ones. This is achieved by separating the variables and scaling all functions with one of the spatial coordinates. The so-called radially self-similar class of models are thought to better apply to disc winds ejected from the external part of an accretion disk (e.g., Blandford & Payne 1982). A second class is the so-called meridionally self-similar models which seem to better describe the stellar part of the ejection and/or the internal part of the accretion disc.

The *combined* effects of the gas pressure and magnetic fields in accelerating, collimating and confining jets have not been discussed adequately. Various models, especially disc-wind models, assume for simplicity a cold plasma such that the magnetic field is the crucial and only ingredient, for both the acceleration and confinement of the outflow. An MHD approach combining the magnetic and the thermal effects has been proposed recently in a series of papers (Trussoni et al. 1997). In Sauty & Tsinganos (1994) a criterion for the collimation of winds into jets has been proposed but assuming that gas pressure played no role in the jet confinement itself. We will discuss at the end some extensions of these results including both ingredients and how this can suggest some scenario for the evolution of stellar jets into stellar winds.

2. Basic Principles of Steady Flows

Thus in our working frame winds and jets are outflows of a fully ionized plasma flowing at a bulk flow speed **V** and carrying a magnetic field **B** in the gravitational field of a central body of mass $\mathcal{M}$. The perfect gas has a mean density ρ, pressure P and temperature T. The familiar ideal MHD equations are employed for a physical description of these phenomena. In particular, under steady and axisymmetric conditions free integrals exist, i.e., functions which remain constant on the magnetic surfaces generated by the revolution around the magnetic/flow symmetry axis of the system of a poloidal magnetic line $A(r,\theta) \doteq$ constant, (Tsinganos 1982). Specifically, on the surface of such a flux tube $A =$ const., the following physical quantities remain invariant:

- Ψ_A(A), the ratio of the mass and magnetic fluxes,
- $L(A)$, the *total* specific angular momentum carried by the flow *and* the magnetic field,

- $\Omega(A)$, the corotation frequency or angular velocity of each streamline at the base of the flow.

Furthermore, it is well known that the poloidal (p) and azimuthal (φ) components of the magnetic field and the velocity can be expressed in terms of these free integrals and the poloidal Alfvén Mach number $M = \sqrt{4\pi\rho(V_p^2/B_p^2)}$, using spherical (r, θ, φ) or cylindrical (ϖ, φ, z) coordinates (for details see Sauty & Tsinganos, 1994). On the other hand, the two integrals $L(A)$ and $\Omega(A)$ are not independent if the flow is trans-Alfvénic. In such a case, at the cylindrical distance ϖ_a of the Alfvén point ($M = 1$) from the field/flow axis of a flux tube labeled by A, they are related as $\varpi_a^2 = L/\Omega$. Moreover the product of the two $L\Omega = \varpi_a^2\Omega^2$ is known as the energy of the magnetic rotator (Belcher & MacGregor 1976).

3. Critical points and limiting characteristics

The major difficulty with the steady MHD equations is that they are of mixed elliptic/hyperbolic type. A realistic solution describing a plasma which is initially at rest and is accelerated up to a terminal shock must go through the various elliptic and hyperbolic domains. In other words, in order to obtain an accelerated steady flow the boundary conditions on the source (star or disc) must adapt themselves such that the flow becomes super fast-magnetosonic. This is done mathematically by crossing correctly some fixed points called critical points, similar to the sonic point in Parker's classical solar wind theory (1958).

In magnetized fluids, waves can propagate with three characteristic speeds: the Slow Magnetosonic (SM), Alfvenic (A) and Fast Magnetosonic (FM) speeds. In a steady flow, a terminal shock may form in the flow without propagating upwind, only when such waves cannot propagate backwards. In one dimensional flows, such as equatorial winds (Weber & Davis 1967) or models with a fixed poloidal fieldline geometry, this is achieved when the poloidal component of the velocity becomes successively larger than the three wave speeds in the poloidal direction. Thus the critical surfaces correspond to the locations where $V_p = V_{SM,\,p}$, $V_{A,\,p}$, $V_{FM,\,p}$. Conversely, in self-similar models the same happens when $V_\perp = V_{SM,\,\perp}$, $V_{A,\,\perp}$, $V_{FM,\,\perp}$, where $\perp$ indicates the component in the poloidal direction perpendicular to the direction in which self-similarity is assumed.

This apparent contradiction is solved when it is realized that in both cases the critical surfaces correspond to limiting characteristics in the flow (Tsinganos et al. 1996). The conclusion is that limiting characteristics are the boundaries between causaly disconnected domains. This crucial connection of the limiting characteristics with the critical points has been extended to the general axisymmetric case by Bogovalov (1997). In order that a solu-

SELF-SIMILAR JETS

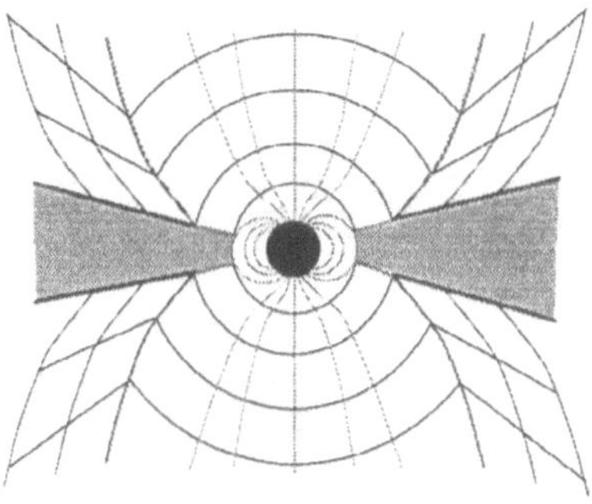

Figure 2. Sketch of the central engine showing the star and its magnetosphere surrounded by the accretion disk. Dotted lines in the poloidal plane correspond to streamlines and solid lines to surfaces of constant M. The outer part of disk outflows may be better described by r-self-similar models while θ-self-similar models are more appropriate for winds from the inner and stellar region.

tion be physically acceptable, it must cross all limiting characteristics such that waves from the terminal shock cannot propagate upwind and destroy the steady configuration.

4. Self-similar models

The basic assumptions of the two main classes of radial (r) and meridional (θ) self-similar models are summarized in Table 1.

TABLE 1. Main assumptions used in self-similar models. Those marked with a (*) can be more complicated to include higher multipole components.

r-self-similar	θ-self-similar
$A = f(\theta)r^x$	$A = f(r)\sin^2\theta$ (*)
$M^2 = M^2(\theta)$	$M^2 = M^2(r)$
cold plasma ($P = 0$)	Fixed streamlines, $P \propto P_o(r) + P_1(r)\sin^2\theta$
or	or
polytropic, $P = K(A)\rho^\gamma$.	Free streamlines, $P \propto P(r)(1 + \kappa A)$

In radial self-similar disc-winds, surfaces of constant poloidal Alfvén Mach number are cones while the magnetic flux scales with a power of the spherical radius. Conversely in meridional self-similar stellar winds surfaces of constant poloidal Alfvén Mach numbers are spheres while the magnetic flux scales with the colatitude ($\sin^2\theta$ is the lowest order while Lima et al. 1997, Vlahakis & Tsinganos 1998 use higher order multipoles). We shall discuss in the next sections results from meridionally self-similar models.

The main reason for using models of winds ejected from a disc is the correlation between observed accretion and mass loss rates and also the efficiency of the extracting mechanism (Blandford & Payne 1982). However these models fail to describe the internal part of the jet so they are in some sense complementary to the second class of θ-self-similar models. Also, the absence of an exact crossing of all critical points in the solutions presented in r-self-similar models prevents from considering their conclusions as definitive. Solutions that connect to a laminar accretion flow do not cross the last critical point and recollimate towards the axis (Ferreira 1997). Within the same set of assumptions, cylindrically collimated jets do not connect to a laminar accretion structure and do not either reach the last limiting characteristic. Yet, solutions that cross all critical points exist (Contopoulos 1995, Tsinganos et al. 1996). However they seem to be too efficient in extracting mass and angular momentum from the disk (Ferreira 1997). In any case they recollimate towards the axis so they necessarily terminate by a shock as suggested by Gomez de Castro & Pudritz (1993). However, the role of the inhomogenity in the gas pressure distribution has not yet been taken into account in these models (Ferreira 1997) and a full parametric study of this extra variable is yet to be performed.

In a series of studies, solutions of the MHD equations that are self-similar in the meridional direction have been also analysed (e.g. Sauty & Tsinganos 1994, Trussoni et al. 1997, Lima et al. 1997). Such a treatment allows to study the physical properties of the outflow close to its rotational axis. As in this region the contribution to acceleration of the magnetocentrifugal forces is small, the effect of a thermal driving force is important. This implies also that the structure of the gas pressure in the flow is essential.

5. Application to the Solar Wind

The first application of meridional self-similar models was to model stellar winds and, of course, the quiet solar wind (Tsinganos & Low 1989). The initial model was refined and successfully applied to coronal holes and the fast solar wind (Tsinganos & Sauty 1992). This was further extended by including magnetic field and higher multipole components of the magnetic field by Lima et al. (1997) but assuming a radial geometry. They showed that such a model could nicely reproduce the latitudinal dependence of the solar wind mass efflux, velocity, density and radial magnetic field, in agreement with the differential rotational law of the Sun. Comparison of the model with the Ulysses' data is shown in Fig. 3.

6. Application to the Collimation of Winds into Jets

Another application of meridional self-similar flows is to model the acceleration and collimation of jets. This can be done by imposing the geometry of

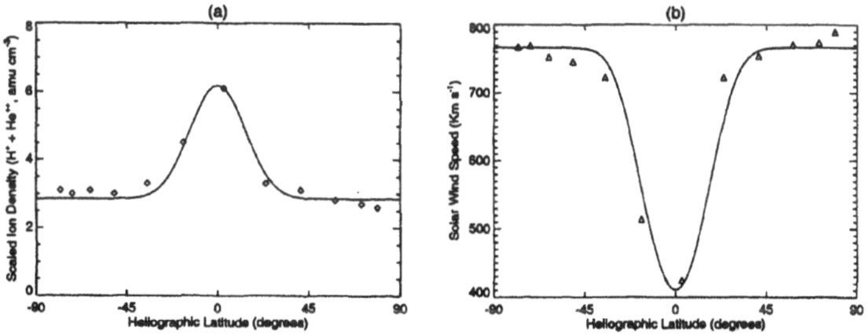

Figure 3. Self-similar model for the Solar wind proposed by Lima et al. (1997). It shows a satisfactory agreement with the Ulysses' data (triangles and squares). The ion density as a function of colatitude is plotted in (a) while the radial velocity in (b).

the flow (Trussoni et al. 1997; see Table 1). In such a case it has been shown that cylindrical jets could form only in the presence of both rotation and toroidal magnetic fields, i.e., only in Magnetic Rotators. Under-pressured jets (i.e. jets with a pressure along the polar axis lower than the pressure at their outer edge) can be confined either by gas pressure or, by magnetic pinching. While over-pressured jets (i.e. jets with a pressure along the axis higher than the pressure at their outer edge) are necessarily magnetically confined.

Within the same self-similar class but with a free geometry and related meridional and polar pressure (see Table 1), a relation between the terminal morphology of the outflows and the energetic distribution across the fieldlines can be drawn. In such models, each variation with colatitude of the physical quantities is easily controlled by one parameter due to the self-similar assumptions. The asymptotic conditions in the outflows are actually governed by four parameters (see Sauty & Tsinganos, 1994, for details):

- $\delta \propto \Delta\rho/\rho$ which controls the increase ($\delta > 0$), of the density with colatitude,
- $\kappa \propto \Delta P/P$ which controls the increase ($\kappa > 0$) or decrease ($\kappa < 0$) of the pressure with colatitude,
- λ which controls the rotation, and
- $Q \propto P(r = \infty)$ which is in dimensionless form the asymptotic value of the pressure.

Δ indicates the difference between a non polar line and the polar axis. λ^2 is a crucial parameter as it is proportional to the so called *magnetic rotator energy* ΩL (Belcher & MacGregor 1976). In a cold plasma ΩL is the "reservoir" that provides the energy to magnetically launch and confine the plasma.

The efficiency of the magnetic rotator to confine the jet independently of the pressure term depends on how much the same energy is used to accelerate the jet. It is measured by the following parameter evaluated at the stellar surface and normalized for conveniency to the magnetic rotator energy,

$$\frac{\epsilon}{2\lambda^2} = \frac{E_{\mathrm{MR}} - E_{\mathrm{R},o} + \Delta E_{\mathrm{G}}^{*}}{E_{\mathrm{MR}}}, \tag{1}$$

where $E_{\mathrm{MR}} = L\Omega$ is the energy of the magnetic rotator, $E_{\mathrm{R},o}$ is the centrifugal energy of the footpoints and

$$\Delta E_{\mathrm{G}}^{*} = -\frac{\mathcal{GM}}{r_o}\left[1 - \frac{T_o(\alpha)}{T_o(\mathrm{pole})}\right] \propto -(\delta - \kappa), \tag{2}$$

is the transverse variation of the gravitational energy (per unit mass) corrected from the thermal compensation. Let us briefly explain this point. Along the polar axis the acceleration of the wind is done by thermal processes alone. On a non polar line, if the relative increase of the density is not compensated by an identical relative increase of the pressure ($\delta - \kappa > 0$) then the thermal driving is less efficient along this line. This must be supplied by magneto-centrifugal driving. Thus part of E_{MR} is used to accelerate the non polar flow. In that sense, $\epsilon/2\lambda^2$ is the relative amount of the energy of the magnetic rotator that is not used to accelerate the flow out of the gravitational well or to make the flow rotating. In other words $(\epsilon/2\lambda^2) < 0$ corresponds to <u>Inefficient Magnetic Rotators</u> and $(\epsilon/2\lambda^2) > 0$ to <u>Efficient Magnetic Rotators</u>.

Once $\epsilon/2\lambda^2$ is fixed on the source boundary, the asymptotic cylindrical radius of a flux tube in units of the cylindrical radius at the Alfvén surface, $G_\infty = \varpi_\infty/\varpi_a$, can be calculated using the force balance across the jet. For an asymptotically cylindrical jet, the force balance across the flux tube becomes $F_C + F_B + F_P = 0$. F_C is the centrifugal force, F_B is the total magnetic transverse force (magnetic tension + magnetic pressure gradient) and F_P is the transverse pressure gradient. The result depends on $\epsilon/2\lambda^2$, κ and Q and is plotted in Figs. 4. For a given set of these parameters, we may have zero, one or more solutions for G_∞, with zero meaning that cylindrical collimation cannot be achieved. We also see that there are three different domains where two of the three forces dominate the transverse momentum balance:

• $F_C + F_P \approx 0$, the thermally confined regime where the magnetic contribution is negligible. The jet is pressure confined and centrifugally supported *(left part of Figs. 4b,c,d)*.

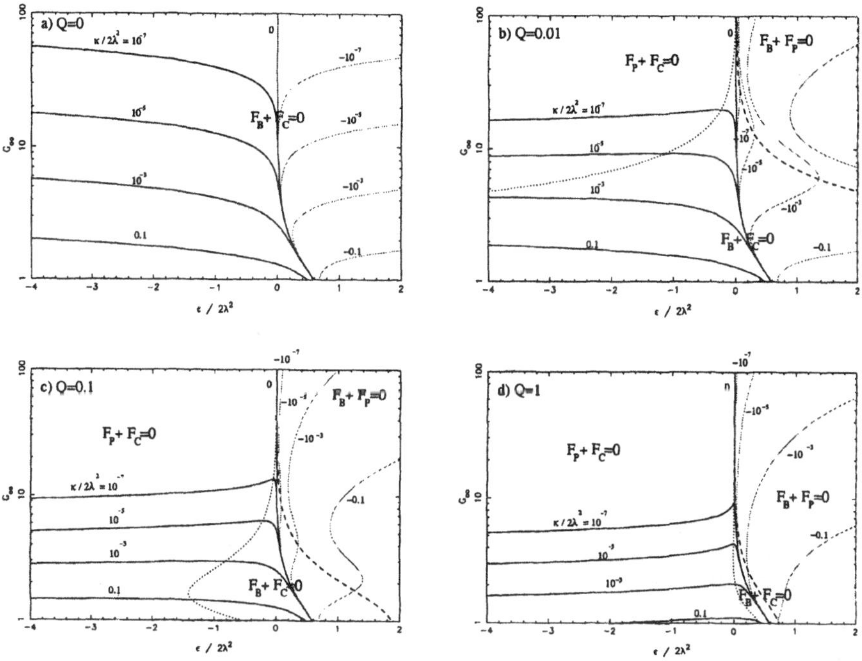

Figure 4. Plots of the asymptotic radius of cylindrically collimated outflows in units of the Alfvénic radius, $G_\infty = \varpi_\infty / \varpi_a$, versus $\epsilon/2\lambda^2$. Each curve corresponds to a constant value of $\kappa/2\lambda^2$ and Q.

• $F_C + F_B \approx 0$, the magnetocentrifugal regime where the pressure gradient is negligible. The jet is magnetically confined and centrifugally supported *(Fig. 4a and central part of Figs. 4b,c,d).*
• $F_B + F_P \approx 0$, the thermally supported regime where the centrifugal force is negligible. The jet is magnetically confined and pressure supported.

Underpressured jets ($\kappa > 0$) are always in one of the two first regimes while overpressured jets ($\kappa < 0$) are in one of the two last regimes. The limit between the various domain is drawn with a dashed line and a dotted line. As the asymptotic pressure Q increases from Figs. 4a to 4d, the magnetocentrifugal domain shrinks around the curve $\kappa = 0$.

7. Evolution of Jets into Winds

Sauty & Tsinganos (1994) (see Table 1) have shown that, in the case of iso-pressured outflows (i.e. $\kappa = 0$), the outflows collimate into a cylindrical jet only if the magnetic rotator is efficient, $\epsilon/2\lambda^2 > 0$, as shown in Figs. 4 (curve $\kappa = 0$). Otherwise the flow is asymptotically radial. The same is also true for over-pressured jets ($\kappa < 0$), except that the threshold value of

$\epsilon/2\lambda^2 > 0$ is now positive. Qualitatively, it means that such outflows should evolve from a cylindrical jet-like shape towards a bipolar radial wind as the star evolves and looses angular momentum ($\epsilon/2\lambda^2$ decreases).

Under-pressured outflows ($\kappa > 0$) behave quite differently. In this case cylindrical collimation is always achieved. So, as the star changes from beeing an efficient magnetic rotator to an inefficient one ($\epsilon/2\lambda^2$ decreases from positive to negative values), its outflow switches from magnetically confined to pressure confined ones. However in a realistic scenario, as the jet widens the asymptotic pressure Q also decreases, and presumably κ too, so we may expect to go from Fig. 4d to Fig 4b. Thus as the star evolves the jet radius becomes larger and larger. Last but not least, an exact equipartition between pressure and magnetic confinement (along the dotted line) correspond to a compromise to have the most effective thermal acceleration and magnetic braking. It practically corresponds to a maximal asymptotic radius. Thus it is very attractive to conjecture that the jet gets wider and wider following this condition of equipartition and that magnetic fields and pressure must be taken equally into account. This raises a last but interesting question: in such conditions one may wonder if the solar wind itself does tend to be collimated far from the Sun, at least around the polar axis.

References

Belcher, J.W., MacGregor, K.B., 1976, ApJ, 210, 498
Blandford, R.D., Payne, D.G., 1982, MNRAS, 199, 883
Bogovalov, S.V., 1997, A&A, 323, 634
Contopoulos, J., 1995, ApJ., 450, 616
Ferreira, J., 1997, A&A, 319, 340
Gomez de Castro, A.I., Pudritz, R.E., 1993, ApJ, 409, 748
Heyvaerts J., Norman C.A., 1989, ApJ, 347, 1055
Kudoh, T., Shibata, K., 1997, ApJ, 474, 362
Léry, T., 1997, Thèse de Doctorat, Université de Strasbourg.
Lima, J.J.G., Priest, E.R., Tsinganos, K., 1997, Proceedings of the Fifth SOHO Workshop, ESA SP-404, 521-526
Ostriker, E., 1997, ApJ, 486, 291
Ouyed, R., Pudritz, R.E., 1997, ApJ, 482, 712
Parker, E.N., 1958, ApJ, 123, 664
Sauty, C., Tsinganos, K., 1994, A&A, 287, 893
Trussoni, E., Tsinganos, K., Sauty, C., A&A, 1997, 325, 1099
Tsinganos, K., 1982, ApJ, 253, 775
Tsinganos, K., Low, B.C., 1989, ApJ, 342, 1028
Tsinganos, K., Sauty, C., 1992, A&A, 255, 405
Tsinganos, K., Sauty, C., Trussoni, E., Contopoulos, J., Surlantzis, G., 1996, MNRAS, 283, 811
Vlahakis, N., Tsinganos, K., 1998, MNRAS, (submitted)
Weber, E.J., Davis, L.Jr., 1967, ApJ, 148, 217

ON THE SOURCE OF OPTICAL VARIABILITY AND OUTFLOW IN Z CMA SYSTEM

S.A. LAMZIN
Sternberg Astronomical Institut
Moscow V-234, 119899 Russia

M. TEODORANI, L. ERRICO, A.A. VITTONE
Osservatorio Astronomico di Capodimonte
I-80131 Napoli, Italy

E.A. KOLOTILOV
Sternberg Astronomical Institute
Crimean Laboratory, 334413 Nauchny, Ukraine

AND

A.S. MIROSHNICHENKO, R.V. YUDIN
Pulkovo Astronomical Observatory
196140, Saint-Petersburg, Russia

We discuss our new (Teodorani et al. 1997) and already published photometric and spectroscopic data regarding the binary system Z CMa and conclude that these data, in contrast to wide spread opinion, do not contradict the hypotesis that a FU Ori type secondary rather than the IR primary is the main source of the observed optical variations and outflow. In the frame of this hypothesis, we explain observed larger polarization in the emission lines than in the continuum as a result of Thomson scattering of line quanta inside the region of their formation.

Z CMa belongs to FU Ori type objects (FUORs) and has an IR companion at projected separation of 0.1". The companion (or Z CMa NW) is dominant in SED at $\lambda > 2\,\mu m$ and has $\sim$ 2 times larger L_{bol} than its optical counterpart Z CMa SE. The system is a sourse of powerful outflow. Whitney et al. (1993) discovered Z CMa emission lines are much more polarized than the continuum and have concluded finally that IR companion is responsible for gas outflow and 1987 yr. outburst. Here we discuss possibility of the opposite view point.

Astrophysics and Space Science is the original source of publication of his article. It is recommended that this article is cited as: *Astrophysics and Space Science* **261**: 161–162, 1999.

We belive that FUOR Z CMa SE was the reason of 1985 and 1987 outbursts because:

1) they have occured just at the end of a $\sim$ 15 yrs. period of nearly constant Z CMa brightness, and then followed by a decay of brightness. It seems resonable to ascribe all these variations to the same component of the system and we argue that it is Z CMa SE.

2) the amplitude of these outbursts subsequently decreased from U to K spectral bands, what looks strange if the reason of the flash was the rise of the IR companion luminosity.

3) we argue that the flash could not be due to the temporary clearing of Z CMa NW dusty envelope along the line of sight.

We have analised profiles of forbidden lines in the spectra of Z CMa and its jet and have concluded, that the line of site passes within the cone of the outflowing gas. It is why we belive that IR companion surrounded by opaque dust shell can not be the source of the outflow. We have presented arguments that forbidden lines and permitted lines with P Cyg profiles form in different regions of the same outflow.

We have estimated an optical depth of the ionized component of Z CMa stellar wind relative to Thomson scattering and found $\tau_{Th} \geq 0.03$. The lower limit of τ_{Th} we derived refers to quanta coming from the central source, which reach the observer along the line of sight, such as quanta of Pashen continuum with free-bound optical depth $\tau_{fb}(R)$ not too much larger than unity if at all. But quanta of strong emission lines pass the distance $L \gg R$ inside the region of line formation due to multiple resonance scattering. Thus their optical depth relative to Thomson scattering $\tau_{Th}^{line} \simeq L\sigma_{Th}N_e$ should be much larger than our lower limit. This is why we believe that Thomson scattering of emission lines quanta inside the densest part of the stellar wind can be responsible for the excess polarization mentioned above. Numerical calculations are necessary in order to estimate the effect quantitatively.

References

Teodorani M., Errico L., Vittone A. et al. (1997), *Astron. Astrophys Supl.* **126**, p. 91
Whitney B.A., Clayton G.C., Shulte-Ladbeck R.E. et al. (1993), *ApJ*, **417**, p. 687

MODELS IN T TAURI STARS TESTED THROUGH Hα

ANTÓNIO PEDROSA, M.T.V.T. LAGO
Centro de Astrofísica da Universidade do Porto
Rua das Estrelas S/N , 4150 Porto,
Portugal

1. Introduction

Since the pioneering works of Joy(1945) and Herbig(1970), the knowledge of T Tauri stars (TTS) has been steadily building up. It is now well established that these stars are young, with ages not exceeding 3×10^6 yrs, low mass, variable, with a late type spectrum with emission lines superimposed.

One *class* of T Tauri stars, the so-called Classical T Tauri stars (CTTS), displaying strong Balmer, emission lines ($W(H_\alpha)$ often exceeding 100Å) and exhibiting in many cases a Balmer jump (Valenti et al. 1993).

Many models have addressed the question of Hα line formation on CTTS, some of which based on outflow. The existence of a wind is suggested by the presence of a blue absorption, although only a small fraction of the stars the absorption goes below the continuum. Many of these models are powered by Alfvén waves (Lago 79, Hartmann et al. 82), with mass loss rates on the order of 10^{-8} $M_\odot$ yr^{-1} (DeCampli 1981), according to mass loss estimates of 10^{-8}-10^{-7} $M_\odot$ yr^{-1}(Kuhi 1964). Some of them (Hartmann 90, Johns & Basri 1995) use turbulence as the major broadening mechanism, to produce the broad lines observed.

On section 2 we present the Hα line observations, with a discussion of their features, followed by a section where a model is presented based on an expanding chromosphere, that reaches several hundred km/s at a distance of a couple stellar radius from the star. No turbulence was introduced, although the stark effect was taken in account.

2. The Hα line observations

We have performed extensive *observations* on a group of 43 Southern T Tauri stars (Reipurth et al. 1996). Using the CAT+CES at ESO, La Silla, we have

Astrophysics and Space Science is the original source of publication of his article. It is recommended that this article is cited as: *Astrophysics and Space Science* **261**: 163–166, 1999.

obtain a set of high resolution (R = 50 000), high signal to noise Hα line spectra, some of which are displayed on Figure 1.

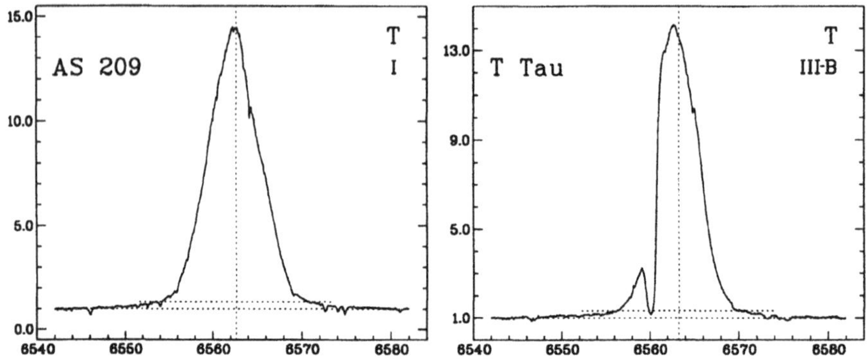

Figure 1. Typical symmetric (left) and P Cygni (right) Hα line profiles.

Some major *features* could be found from the spectral analysis. They showed that Hα is a strong line in emission, with large equivalent width, sometimes greater than 200 Å, with the profiles dominated by symmetrical type, 25% , and P Cygni type, 55% . The lines showed highly symmetrical wings, with the velocity distribution at $I_{max}/40$ similar in red and blue with a peak close to 300 km/s, see Figure 2. At the base the wings can reach 900 km/s.

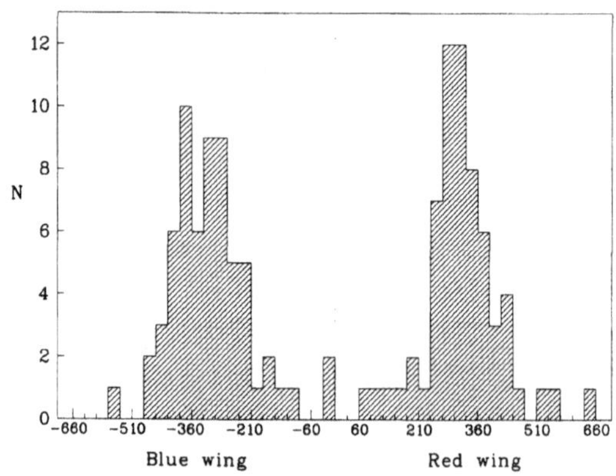

Figure 2. Histograms with the velocity distributions at $I_{max}/40$ for both wings.

3. Modeling

3.1. THE MODEL

We constructed a model to try to reproduce the observed line profiles (Pedrosa & Lago 1998). Similarly we wanted to understand the role played by several parameters, like temperature of the gas, mass loss rate, velocity law, on the physical conditions of the emission region. We also wanted study their influence on the profiles, to seek for a line broadening mechanism other than turbulence, still paying attention to other parameters like the atomic model (statistical distribution) and to the extent of the atmosphere.

The *model* is based on an spherically symmetric, isotermic, expanding wind. Only statistical equilibrium of hydrogen is computed, and a modified Sobolev method to compute the radiation field at the line frequencies was used. The continuum is computed by dividing the atmosphere in layers, and the stellar continuum is consider to be Plankian.

As *inital conditions* we had to supply the gas temperature, that we have taken between 8000 and 12000 K, a velocity law with a form $v(r) = v_0 + v_1(1-(\frac{r}{R_*})^\alpha)$, with v_0 between 1 and 10km/s, v_1 between 300 and 400 km/s, and the exponent α ranging from 0.5 to 2. The density at the base of the wind was consider to be in the interval $10^{11} - 10^{13}$ p/cm^3 and the extent of the atmosphere as 10 R_*.

The *line profiles* were computed by a formal integration of the equation of radiative transfer, with a Doppler+Stark profile as the intrinsic line profile, with complete redistribution.

3.2. THE RESULTS

For the *degree of ionization*, th results showed that under the initial range of conditions considered, the atmosphere is in a threshold between being fully ionized and fully neutral, see Figure 3.

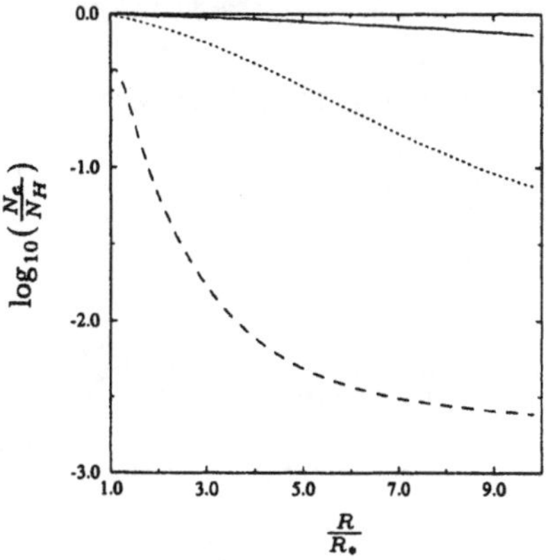

Figure 3. Models with different states of ionization. **Solid line** - ionized wind when $\rho_0 = 10^{12}$ p/cm^3, $T_e = 12000$ K, $T_{ef} = 4000$ K, $v_0 = 5$ km/s, $v_1 = 350$ km/s, $\alpha = 1$; **Dotted line** - partially ionized wind when $\rho_0 = 10^{11}$ p/cm^3, $T_e = 10000$ K, $T_{ef} = 5000$ K, $v_0 = 10$ km/s, $v_1 = 350$ km/s, $\alpha = 0.5$; **Dashed line** - neutral wind when $\rho_0 = 10^{11}$ p/cm^3, $T_e = 8000$ K, $T_{ef} = 4000$ K, $v_0 = 10$ km/s, $v_1 = 350$ km/s, $\alpha = 1$.

The Hα lines are presented in Figure 4, for several initial conditions. Very broad and intense lines are formed mainly when the density at the base is increased. The presence of a blue absorption is also very noticeable.

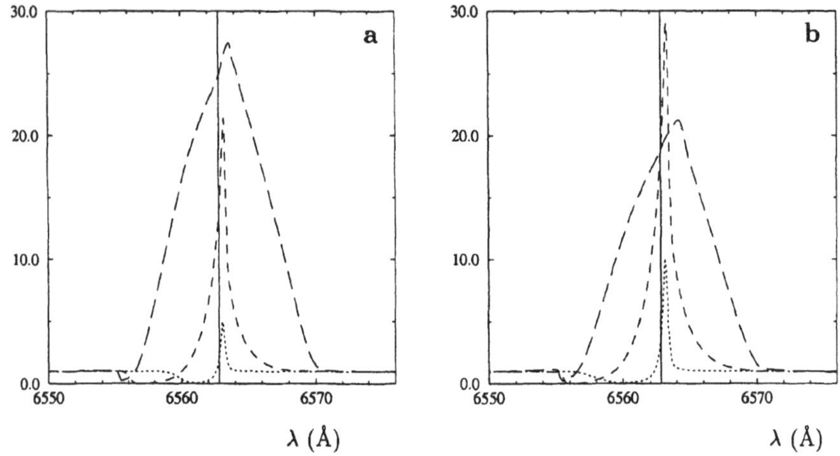

Figure 4. H_α lines generated by models with different electron temperatures: **a)** $T_e = 8000$ K, $T_{ef} = 4500$ K, $v_0 = 10$ km/s, $v_1 = 350$ km/s, $\alpha = 1$; **b)** Same as **a)** except for $T_e = 10000$ K; **Dotted line**, $\rho_0 = 10^{11}$ p/cm^3; **Dashed line**, $\rho_0 = 10^{12}$ p/cm^3; **Long Dashed line**, $\rho_0 = 10^{13}$ p/cm^3.

These results point towards the existence of a region at a temperature close to 10000 K that is responsible for the line emission. Such region must be narrower than the one considered here, in order to lower the line intensity. A temperature structure must also be present in order to avoid the presence of the blue absorption and to produce other lines of different excitation energies such as NaI D, Ca II H & K and HeI among others.

References

DeCampli, W.M., 1981, ApJ 244, 124
Hartmann, L., Edwards, S., Averett, E.H., 1982, ApJ 261, 279
Hartmann, L., 1990, ApJ 349, 168
Herbig, G.H., 1970, Men. Soc. Roy. Sci. Liège, 52 Ser., 19, 13
Jonhs, C.M., Basri, G., 1995, ApJ 449, 341
Joy, A.H., 1945, ApJ 102, 168
Lago, M.T.V.T., 1979, DPhill. thesis Univ. Sussex, Engl.
Kuhi, L.V., 1964, ApJ 140, 1409
Pedrosa, A., Lago, M.T.V.T., 1998, A& A (Submitted)
Reipurth, B., Pedrosa, A., Lago, M.T.V.T., 1996, A& AS, 120, 229
Valenti, J.A., Basri, G., Johns, C.M., 1993, ApJ 106, 2024

ON THE ECLIPSES IN GW ORI DOUBLE SYSTEM

LAMZIN S.
Sternberg Astronomical Institute
Moscow V-234, 119899 Russia

AND

SHEVCHENKO V., GRANKIN K., MELNIKOV S.
Astronomical Institute of Academy of Science of Uzbekistan, Tashkent

GW Ori is a single line spectroscopic binary CTTS with orbital period $P \simeq 242^d$ and nearly circular ($e < 0.05$) orbits of the companions (Mathieu et al.1991). Radius of the primary according to these authors is $R_1 \simeq 5.6\,R_\odot$.

Shevchenko et al. (1993) and Grankin (1995) have found Algol-like decreasing of GW Ori brightness during 1988, 1990 and 1992 observational seasons. Their duration were near 20^d and they were symmetrical relative to spectroscopic phase 0.0. Such events were not observed before 1988 yr and disappeared after the 1992 yr. Shevchenko et al. (1993) have interpreted this phenomenon as a result of primary eclipse by dusty envelope surrounding the secondary.

The duration of the dips was constant during 1988-1992 yrs, but their amplitude subsequently decreased. This means probably, that the size of the eclipsing region did not vary, while the envelope opacity monotonically decreased. The parameters of the system will be estimated below assuming that the size of the envelope is equal to the radius r_2 of the secondary Roshe lobe, which we consider as a sphere.

Let a be the distance between the centers of stars, $q = m_2/m_1 < 1$ – mass ratio, i – the angle between normal to the orbital plane and the line of site. According to Mathieu et al. (1991)

$$\frac{a\,q\,\sin i}{1+q} = 1.5\,10^{12}\ \text{cm}. \qquad (1)$$

Astrophysics and Space Science is the original source of publication of his article. It is recommended that this article is cited as: *Astrophysics and Space Science* **261**: 167–168, 1999.

If the total duration of the eclipse is $2\psi \simeq 0.1$ in phase units, then one can derive:

$$\sin^2(2\pi\psi) + \cos^2 i = \frac{R_1 + r_2}{a}. \quad (2)$$

According to Antokhina (1989) $\xi = r_2/a$ value is connected with q as follows:

$$q = \frac{\xi^3(\xi^2 - 3\xi + 3)}{(1-\xi)^2(1-\xi^3)}. \quad (3)$$

The system of equations (1)-(3) contains 5 unknown values: i, q, a, R_1, r_2. We have solved it as a function of i for $R_1 = 5.6$, 6.6 and 7.6 $R_\odot$. Mass function of Mathieu et al.(1991) was used to derive masses of the components.

If to assume that the age of the system is $< 3\,10^6$ yrs. then the mass of the primary m_1 should be $> 2\,M_\odot$ as far as its $\log T_{ef} = 3.75$ – see Fig.8 in Mathieu (1994). Thus only i-values, for which $m_1 \geq 2\,M_\odot$ are suitable in our case. We have found finally that the distance between components is 1.05 a.u. with the accuracy 5%, and $i = 83^o$. The mass of the secondary is close to $0.25\,M_\odot$.

GW Ori system is surrounded by an opacque gas-dust envelope with characteristic size $\sim$ 300 a.u. (Mathieu et al. 1991). If orbital plane is indeed inclined at 83° to the line of site, then the envelope is a thin circumbinary disk. We suppose that non-stationary accretion of the disk matter onto the double system is the consequence of the interaction of the system with the disc (Artymowicz,1992).

Average brightness of GW Ori has decreased initially and then increased during 1985-1996 yrs period. Let us assume that significant amount of matter felt onto the double star from the circumbinary disk and has appeared finally inside the Roshe lobe of each star. Dust inside the primary Roshe lobe was the reason of general decreasing of GW Ori brightness, while matter around the secondary become the reason of the eclipses. Captured matter has fallen finally onto the stars, so GW Ori brightness has returned back to the initial state and eclipses have disappeared.

References

Antokhina E.A. (1988), *Astron. Reports* **65**, p. 1164
Artymowicz P. (1992), *PASP* **104**, p. 769
Grankin K.N. (1995), *PhD Thesis*, Tashkent, Uzbekistan
Mathieu R.D. (1994), *Ann. Rev. Astron. Astrophys.*, **32**, p. 465
Mathieu R.D., Adams F.C., Latham D.W. (1991), *AJ* **101**, p. 2184
Shevchenko V.S., Vitrichenko I.A., Grankin K.N. et al. (1993), *Astron. Letters* **19**, p. 334

THE EVOLUTION OF YOUNG STARS, PROTOSTELLAR JETS & BIPOLAR OUTFLOWS - A UNIFICATION SCHEME

MICHAEL D. SMITH
Astronomisches Institut der Universität Würzburg,
Am Hubland, 97074 Würzburg, Germany

Abstract. Close links between jet evolution and protostellar evolution are beginning to be understood. Firstly, stellar jets are reviewed here, establishing the accretion-outflow connection. Then, outflows from young stars are reviewed, suggesting a synchronised development in the star and outflow. This yields a unification scheme in which rising molecular jets dominate the early protostellar epoch, followed by a jet-driven outflow stage and, finally, a bow-driven ballistic stage. This scheme is quantified, yielding the systematic changes in the bolometric, mechanical and shock luminosities and the cross-over phase from dense molecular jets to light atomic jets.

1. Introduction

Highly collimated outflows are associated with very young stars. We study the outflows not only to understand interstellar physics, chemistry and dynamics, and not only for the immaculate beauty of youth, but to understand how stars form. So what information should we be looking for to test the formation models? Wish list: (1) jet rotation, (2) jet speed, (3) polar magnetic field, (4) azimuthal field, (5) shear, (6) violent outbursts, (7) precession, (8) oscillations in power, (9) jet entrainment (i.e. core disruption), (10) focussing and (11) raw material (e.g. atoms, dust).

We must attempt to recognise these properties by interpreting the emission lines of molecules and atoms, and radio continuum emission images. The problem addressed here is that we observe jets and young stars in many vastly different evolutionary stages and that we lack a scheme in which to intepret the derived properties.

Jets are not restricted to the pre-main sequence stage - they are associated with many types of star, as shown in Table 1. An important

Astrophysics and Space Science is the original source of publication of his article. It is recommended that this article is cited as: *Astrophysics and Space Science* **261**: 169–178, 1999.

Class	Accretion	System	Examples	Ref.
Pulsar	-	Neut. Star	Crab, Vela	1
HMXB	Y	Neut. Star	SS 433, CygX-3	1
LMXB	Y	Neut. Star	CirX-1	1
Super Soft Sources	Y	W. Dwarf	RXJ05139-6951	1
Symbiotic	Y	W. Dwarf	R Aqr	1
(Pre-)Planetary Neb.	-	AGB/WR	NGC 7354, Hb 4	2
BHXT	Y	Black Hole	GRS 1915+105	1
CTTS	Y	T Tauri	HL Tau, DG Tau	3
YSO-LM (low mass)	Y	Protostar	HH 47, HH 111	4
HAeBe	Y	Protostar	Lk234, 1548C27	5
YSO-HM (high mass)	Y	Protostar	DR 21, W,75N	6

TABLE 1. Stellar Jet Systems. See (for example): [1]Livio (1997),[2]Hajian et al (1997), [3]Stapelfeldt et al (1997), [4]Reipurth & Heathcote (1997), [5]Mundt & Ray (1994), [6] Davis et al (1998).

Type	Example	Comment	Review/Ref.
Microjets	DG Tau	atomic (rotation?)	Solf (1997)
Radio Jets	Cep A	thermal free-free	Rodriguez (1997)
HH Flows	HH 1/2	atomic, shocks	Reipurth et al 9(1997)
Parsec-Jets	HH 24	atomic, shocks	Eislöffel & Mundt (1997)
Giant HH Flows	HH 111	atomic, shocks	Bally & Devine (1997)
Turbulent Jets	HH 110	at./mol shocks	Noriega-Crespo (1997)
Molecular Jets	HH 212	H_2, shocks	Zinnecker et al (1998)
	HH 211	CO, bullets, bows	Guilloteau et al (1997)
Bipolar	RNO 43	CO rot. outflows	Padman et al (1997)

TABLE 2. Protostellar Collimated Outflows

clue to any underlying scheme is the accretion-outflow connection, as emphasized by Ray (1996) and Livio (1997). Table 1, as well as Table 2, show also that we cannot classify all jets from young stars together without careful consideration.

Why do jets from young stars differ so widely? This is neatly seen

Reference	Dim.	Meth.	gas	n/cm^3	v (km/s)
Raga et al '95	2D	2nd	at-mol	50	130
Stone & Norman '94	3D/2	PPM	atomic	60	235
de GDP & Benz '93	3D	SPH	atomic	4 10^4	400
Cliffe et al '96	3D	2nd	isoth	-	-
Cerqueira & de GDP	3D	SPH	at./MHD	600	400
Würzburg Jets	3D	2nd	mol-mol	10^5	100

TABLE 3. Numerical Jet Propagation Models (formation models excluded).

by examining the parameters assumed by various groups undertaking numerical simulations (Table 3). Most simulations have concentrated on atomic outflows of low density. However, the Würzburg group are investigating very dense molecular jets. These arise at a very early stage in star formation (Class 0), involving a much higher mass ejection rate albeit for a shorter time.

The Würzburg hydrodynamic simulations (Suttner et al 1997) assume no outflow evolution, but involve variations in many of the parameters, including shear, precession, opening angle, oscillations and outbursts. Several general predictions have ensued from these simulations (Smith et al 1997b). Davis et al (1998) employ the results as a framework for analysing a specific outflow.

The latest 'Hammer Jet', involving strong short-term pulses superimposed on long-term outbursts, produces noteworthy gaps before the jet infrared emission knots appear (Fig. 1). Numerically, the gaps are a result of the 'hammer structure' i.e. the imposed time variations in the jet velocity, and are only present for a fraction of the time, clearly some time after a strong outburst sequence has ended. Gaps are indeed observed in several sources. Extinction could well be important. The jet outburst model, however, is favoured here due to faint knot structures in the gap (e.g. HH 34, Ray et al 1996) and proper motion of the innermost knot (HH 47, Micono et al 1998).

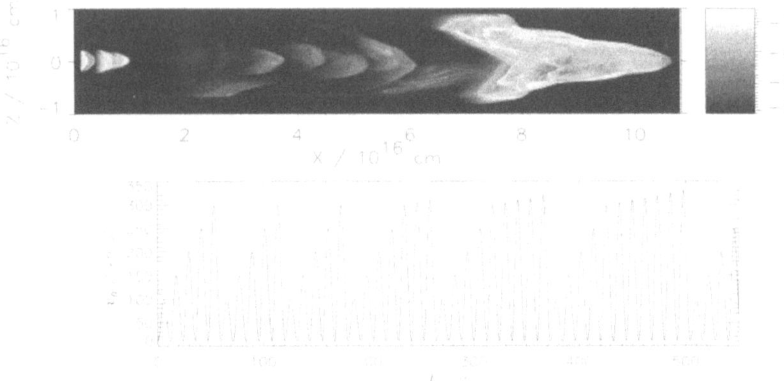

Figure 1. The latest hydrocode simulation of a pulsating and outbursting jet molecular jet (see Völker et al 1998), demonstrating (at 414 years) a gap followed by complex knot structures. the infrared H_2 1-0 S(1) emission is displayed. The imposed velocity variations are displayed.

2. The Unification Model: Framework

This model will assume the following.

• Certain stars, perhaps all stars, begin generating a collimated outflow in their early embedded phase (early Class 0), reaching a peak mass outflow rate, $\dot{M}_o$, in their late Class 0 phase, and then slowly decaying, finally becoming unobservable in the Class 1-2 phase. The idea that both outflows and protostars have comparable lifetimes is derived from the recent discovery of numerous giant bipolar outflows associated with LM-YSOs (see Table 2). We shall take a law of the form

$$\dot{M} = \dot{M}_o(e/\alpha)^\alpha (t/t_o)^{-\alpha} \exp(-t_o/t), \tag{1}$$

with α providing the decay rate, and the peak mass outflow occuring at time $t = t_o/\alpha$. An alternative plausible model, would involve an extended plateau in $\dot{M}$ for a time of several t_o (to simulate the often-modelled constant accretion rate).

• We relate $\dot{M}$ to the stellar accretion rate $\dot{M}_a$ by

$$\dot{M} = \epsilon(t)\, \dot{M}_a, \tag{2}$$

and take here $\epsilon = \epsilon_o$, a constant of order 0.1 i.e. we assume a model in which some fraction of the mass reaches the stellar surface and the rest

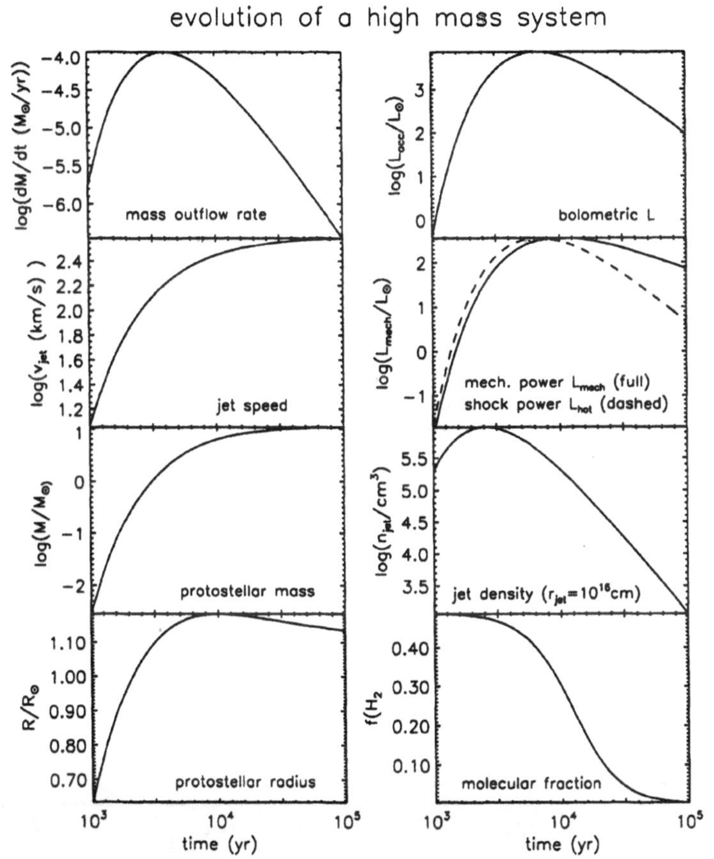

Figure 2. The formation of a 11 $M_{\odot}$ star, with $\alpha = 2.5$, and $t_o = 2.8t_1 = 10^4$ yr.

is ejected along with a greater part of the magnetic field and angular momentum (Shu et al 1988). Then the total accreted mass, i.e. the stellar mass is given by

$$M_* = h(t)\dot{M}_o t_o \tag{3}$$

where $h(t) = (1/\epsilon)(e/\alpha)^{\alpha}\Gamma(\alpha - 1, t_o/t)$ approaches a constant for $t >> t_o$ (Γ is the incomplete Gamma function which approximates to a constant for $t >> t_o$).

• The jet model contains five physical parameters: $\dot{M}$, atomic nuclei density n_{jet}, velocity v_{jet}, molecular fraction f, and dust (mass) fraction d. The jet radius is then a dependent variable. Evidence for a suitable

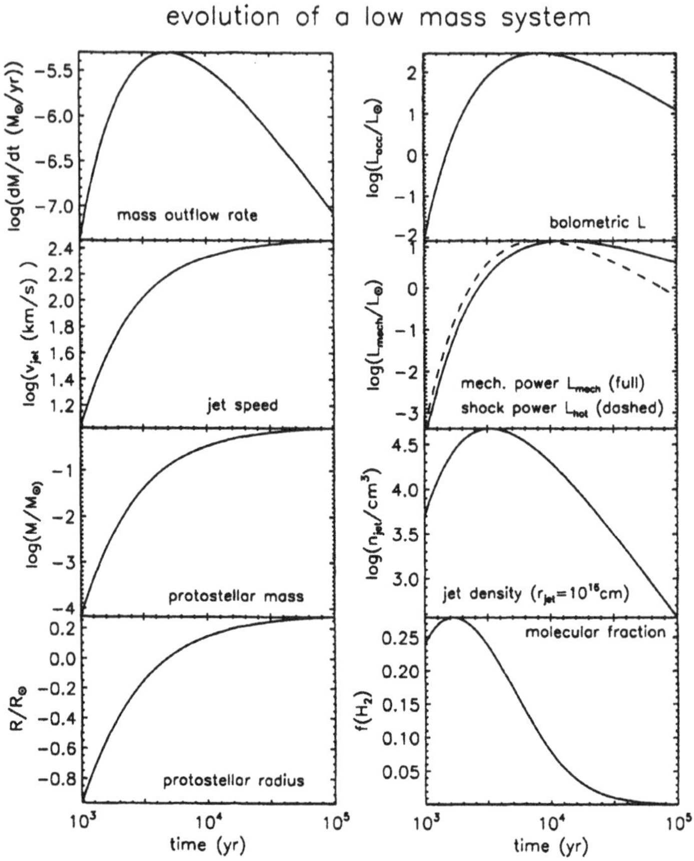

Figure 3. The formation of a $0.9\,M_\odot$ star, with $\alpha = 2$, and $t_o = 3t_1 = 10^4$ yr.

velocity law is found in Table 3. We take here

$$v_{jet} = v_f \exp(-t_1/t) \tag{4}$$

with v_f a constant, mainly for analytical convenience, with a suitable early 'acceleration' phase', and a late approach to a constant-speed proportional to the stellar escape speed.

• If we assume a jet model in which $v_{jet} = \chi\,(GM_*/R_*)^{1/2}$, then the stellar radius is also given through v_{jet}. (A more complex path would involve calculating R_* explicitly from M_a and assumptions about the collapse process as well as the early pre-main sequence evolution, requiring a numerical analysis.) Here, the implicit function for R_* is

$$R_* = \left[\frac{\chi^2 G\dot{M}_o t_o}{v_f^2}\right] h(t)\, \exp(2t_1/t). \tag{5}$$

At this point, we consider suitable parameters for low-mass and high-mass star formation. For the peak mass outflow rate per jet, we take $\dot{M}_o = 10^{-4}\, M_\odot/yr$ (YSO-HM) & $\dot{M}_o = 5\times10^{-6}\, M_\odot/yr$ (YSO-lM). The accretion rate onto the star is $10\times \dot{M}_o$. The fall-off is taken as $\alpha = 2.5$ (YSO-HM) & $\alpha = 2.0$ (YSO-LM), consistent with what we infer about accretion (Hartmann 1997). The important timescale will be taken as $t_o = 10^4$ yr. The resulting stellar and outflow evolution are displayed in Figs. 2 & 3.

3. The Unification Model: Predictions

The stellar bolometric luminosity, (twin-)jet mechanical and (twin-)jet shock luminosities displayed in the figures are defined by:

$$L_{bol} = L_{acc} = \frac{GM_*\dot{M}_a}{R_*} \tag{6}$$

$$2L_{mech}(t) = \left[\int_0^t \dot{M}\, v_{jet}^2 dt\right] /t \tag{7}$$

$$2L_{hot}(t) = \dot{M}\, v_{jet}^2. \tag{8}$$

Note that the total instantaneous jet power and the bolometric luminosity are linearly related by definition: $2L_{hot} = \epsilon L_{bol}$. Observationally, all three luminosities are constrainable (given some rather big assumptions) from continuum radio-infrared spectra, CO low-J rotational flux and H_2 infrared emission, respectively.

In non-evolving outflows, the ratio of CO and H_2 fluxes does not provide a measure of the outflow age (Smith et al 1997a). In evolving outflows, the situation is different. The mechanical luminosity, on integrating (7), is

$$2L_{mech}(t) = \dot{M}_o v_f^2 \frac{t_o + 2t_1}{t}\left(\frac{e}{\alpha}\right)^\alpha \left(\frac{t_o}{t_o + 2t_1}\right)^\alpha \Gamma\left[\alpha - 1, (t_o + 2t_1)/t\right]. \tag{9}$$

Hence $L_{mech}/L_{hot} \propto t^{\alpha-1}$ at late times, i.e. $t >> t_o, t_1$). This implies relatively high CO/H_2 flux ratios for fast evolution (α high) as suggested for W 75N (Davis et al 1998).

A transition from jet-driven to bow-driven (i.e to a snowplough) outflow will also occur for $\alpha > 1$ (applying the above formalism to the momentum). The details, sensitive to the external medium, will be presented elsewhere. Note that Giant HH Flows show evidence for deceleration in the outer bow shocks (Bally & Devine 1997) - a sign that the snowplough phase has been reached.

The outflow density (see 7th captions) is a strongly evolving parameter. The displayed quantity is

$$n_{jet} = \frac{\dot{M}}{\pi m_p}\left(\frac{10^{16}\,cm}{r_{jet}}\right)^2 \tag{10}$$

i.e. the density at a distance of $10^{17}\,cm$ for a jet with a 12° opening angle. This explains the different choices in density discussed in section 2. Note the high peak jet densities predicted in both the YSO-HM and YSO-LM cases.

The velocity of rotation in the jet is small on resolvable scales. Given that all the angular momentum is extracted through the jets, and at the stellar radius the (disk) rotation speed is $\sqrt{(GM_*/R_*)}$, then

$$v_{rot} = \frac{R_* v_{jet}}{2\chi\epsilon r_{jet}}. \tag{11}$$

This formula and the two evolution figures make it clear that rotation is best looked for at the *later* stages (maximising v_{jet} and R_*). Thus 'microjets' (see Table 2) probably offer the best opportunity to study angular momentum extraction.

Molecular hydrogen formation (bottom-right captions) in the jet requires (1) the presence of dust on which to form and (2) the time to cool (so as not to dissociate). To cool below $\sim 4000\,\mathrm{K}$ takes about $t_d = 2\,10^6$ seconds, (the inverse of the average Einstein coefficient for vibrational transitions). This implies that H_2 jets may not be detectable until a distance of $\mathrm{D} = 2\,10^6\,v_{jet}$. Given a reformation timescale of $t_r = 10^{17}/(n\,d)$ then yields the molecular hydrogen number fraction as

$$f = \frac{t_d}{t_r + 2t_d} \tag{12}$$

where we shall take $r_{jet} = 10^{13}$ cm (i.e. the jet is very weakly collimated on this scale) and a full compliment of dust (d = 1). The result, as shown in Figs. 2 & 3, is a gradual transformation from H_2-dominated to H-dominated jets. Note that for the YSO-LM example, a large fraction of atomic H is always present in the jet: the maximum H_2 number fraction reached is 0.29 (fully molecular being 0.5). Clearly, many processes are simply ignored here (shock dissociation, photo-destruction). Nevertheless, an interesting result is the persistance of at least a few per cent H_2 to quite late times.

4. Conclusions

The unification model presented here results from a collection of well-motivated ideas and well-hidden assumptions.

The observed correlation of CO-derived mechanical luminosity with bolometric luminosity (see Padman 1997) is here predicted through L_{bol} & L_{mech}. The model yields $L(CO) \sim 0.1 - 0.03\, L_*$ (except in the bow-driven stage where the full CO flux becomes difficult to measure). Note that L_{shock} (or its observational couterpart $L(H_2)$) should be even better correlated with L_{bol}.

Eruptions on the timescale of 1000-3000 years, superimposed on the systematic mass outflow, have also been investigated (numerically). These produce similar variations in L_{bol}, L_{shock}, n_{jet} and the molecular fraction, f. *Molecular* bullets may then be generated at quite late times in the protostellar evolution. (Other variability effects have been discussed by Cabrit et al (1997) & Smith et al (1997b).)

The model produces a strongly-evolving jet density and molecular chemistry, which helps us to understand how we might improve numerical simulations of outflows. A more detailed analysis should consider the external medium, which has been ignored here. However, jets will be propagating into different environments, and this will certainly produce variation in evolution.

The size of the outflows predicted here are of order 1 parsec at the age of 10,000 years, but rise to 30 parsecs at 10^5 years (this is the maximum size as given by the speed v_{jet}). Very large outflows are being found (e.g. Padman et al 1997), even though the luminosities are predicted to fall quite fast (Figs 2 & 3). Furthermore, the outflows become hard to detect as the velocity of the CO gas is gradually reduced by mixing.

Very young outflows (under 3000 years, on the rising part of the $\dot{M} - t$ curve) will be recognisable not only by their size (0.12 pc at 3000 years for the low-mass example) but by particular infrared and submillimeter emission line signatures (Völker et al, 1998, in prep).

Last, but not least, Henriksen et al (1997) present a unification model which contrasts with the present scheme - based on a very strong Class 0 outburst, and assuming a constant velocity outflow. Alternatively, rather than smooth evolution, a sequence of FUor outbursts could shape events.

Acknowledgements

I thank the Deutsche Forschungsgemeinschaft (Schwerpunkt Programm 'Physics of Star Formation'), and Roland Völker, Chris Davis, Hans Zinnecker & Hal Yorke.

References

Bally, J., Devine, D., 1997, in *IAU Symp. 182, Herbig-Haro Flows*, p. 29
Cabrit, S., Raga, A., Gueth, F., 1997, in *IAU Symp. 182, Herbig-Haro Flows*, p. 3
Cerqueira, A.H., Gouveia Dal Pino, E., 1998, preprint.
Cliffe, J.A., Frank, A., Jones, T.W., 1996, MNRAS, 282, 1114
Davis, C.J., Smith, M.D., Moriarty-Schieven, G., 1998, MNRAS, submitted.
Eislöffel, J., Mundt, R., 1997, AJ, 114, 280
Gouveia Dal Pino, E. Benz, W., 1993, ApJ, 410, 686
Guilloteau, S., Dutrey, A., Gueth, F., 1997, in *IAU Symp. 182, Herbig-Haro Flows*, p. 365
Hajian, A.R., Balick,B., Terzian, Y., Perinotto, M., 1997, ApJ, 487, 304
Hartmann, L., 1997, in *IAU Symposium 182, Herbig-Haro Flows*, p. 391
Henriksen, R., André, R., Bontemps, S., 1997, A&A, 323, 549
Livio, M., 1997, in *IAU Colloquium 163, ASP Conf. Series* **121**, p. 845
Micono, M., et al 1998, MNRAS, in press.
Mundt, R., Ray, T.P., 1994, in *ASP Conf. Series* **62**, p. 237
Noriega-Crespo, A., 1997, in *IAU Symposium 182, Herbig-Haro Flows*, p. 103
Padman, R., Bence, S., Richer, J., 1997, in *IAU Symposium 182, Herbig-Haro Flows*, p. 123
Raga, A.C., Taylor, S.D., Cabrit, S., Biro, S., 1995, A&A, 296, 833
Ray, T.P., 1996, in *Solar & Astrophysical MHD flows*, eds. Tsinganos, K.C. et. al. (Kluwer) p. 539
Ray, T.P., et. al. 1996, ApJL, 468 L103
Reipurth, B., Heathcote, S., 1997, in *IAU Symp. 182, Herbig-Haro Flows*, p. 3
Rodriguez, L.F. 1997, in *IAU Symposium 182, Herbig-Haro Flows*, p. 83
Shu, F.H., Lizano, S., Ruden, S.P., Najita, J., 1988, ApJL, 328, L19
Smith, M.D., Suttner, G.,Yorke, H.W., 1997a, A&A, 323, 223
Smith, M.D., Völker, R., Suttner, G.,Yorke, H.W., 1997b in *IAU Symposium 182, Herbig-Haro Flows*, p. 303
Solf, J., 1997, in *IAU Symposium 182, Herbig-Haro Flows*, p. 63
Stapelfeldt, K.R., et. al., 1997, in *IAU Symposium 182, Herbig-Haro Flows*, p. 355
Stone, J.M., Norman, M.L., 1994, ApJ, 420, 237
Suttner, G., Smith, M.D., Yorke, H.W., Zinnecker, H., 1997, A&A, 318, 595
Völker, R., Smith, M.D., Yorke, H.W., Suttner, G., 1998, preprint
Zinnecker, H. et al., 1998, preprint

EXCITATION OF H_2 IN STELLAR OUTFLOWS: SHOCKS VERSUS FLUORESCENCE

AMADEU J.L. FERNANDES
Centro de Astrofisica da Universidade do Porto
Rua do Campo Alegre 823, Porto, Portugal

AND

PETER W.J.L. BRAND
Institute for Astronomy
Blackford Hill, Edinburgh, EH9 3HJ, Scotland

1. Introduction

The high-velocity molecular gas produced in the early stages of star formation generates energetic interactions with the surrounding molecular gas cloud. The resulting shock waves excites the abundant H_2 molecules in to higher rotational and vibrational energy states. Herbig-Haro objects, commonly associated with supersonic jets arising from young embedded stars, are a common result of this shock interaction.

It has been suggested that photo-dissociation of the molecules may also play an important role in the excitation of molecular hydrogen lines near regions of active star formation (Wolfire and Königl, 1991). For example, work by Burton *et al.* (1992) has shown that the UV radiation field from early-type stars in a central H II region induces the nearby molecular cloud surfaces to show intense emission from atomic, molecular and ionized species such as CO, H_2, [O I], [C II] and [Si II]. However, the broad velocity profiles observed at certain positions within the flow also suggest the presence of shocks.

The work by our group has highlighted the possibility that the UV radiation field necessary to excite the higher energy levels of H_2 may be produced in the dissociated apex of a bow-shaped C-type shock (Fernandes and Brand, 1995). We have suggested that the excess H_2 emission observed in HH7 for the higher excitation lines ($v \geq 3$) arises from H_2 fluorescence which is produced by H I Lyman-α pumping of the lower density ($\sim 10^3$–

Astrophysics and Space Science is the original source of publication of his article. It is recommended that this article is cited as: *Astrophysics and Space Science* **261**: 179–182, 1999.

10^4) pre-shocked gas. Moreover, the bulk H_2 emission is excited in a bow C-shock. In a recent work we have shown that the same phenomena may be occurring both in high- and low-mass star-forming regions; DR21 is one such example of a high-mass region where we also found indirect evidence for H_2 fluorescence in the observed near-infrared H_2 emission spectra (Fernandes *et al.*, 1997).

2. Discussion

Many recent high-resolution images of shocked flows have provided some evidence for clumpy structure associated with the emission from these regions. The bow geometry for each clump which we assume here, is consistent with the structure observed for many HH objects (Allen and Burton, 1993) and molecular outflows (Davis and Smith, 1996). These bow shocks are modeled by finite segmentation of the bow surface into smaller surface elements, each one with a different normal component of the shock velocity which determines the local H_2 excitation conditions. At the bow apex, the strong (dissociative) shocks destroys all molecules and ionizes atoms and ions creating a dissociative cap. In the bow tail, the H_2 emission becomes weaker due to the smaller normal velocity component. Between the dissociative cap and the bow tail, there is a varying degree of excitation conditions determined by the angle between the shock velocity vector and the normal at the surface. These type of shocks have been proposed to explain the H_2 line profile properties in DR21 (Davis and Smith, 1996).

However, we have shown that the observed scatter in the column density diagram (CDR) for the H_2 line emission cannot be reproduced with any type of shock model alone (Fernandes and Brand, 1995). Shock heating will always thermalize the gas and thus generate an upward collisional distribution of the rotational-vibrational energy levels. This "local Boltzmann distribution" produces a single straight line on the CDR connecting all the H_2 line measurements that is inconsistent with the observations.

The observed scatter on the CDR diagram then requires an extra source of H_2 excitation which must be of non-collisional nature. We have shown for HH7 and DR21 that H_2 fluorescence is able to generate just enough extra emission (and thus the scatter) on the higher-excitation energy H_2 lines and therefore explain the CDR. In this scenario, both shocks and fluorescence are competing mechanisms for the excitation of H_2 in these outflows.

When the surface of a molecular cloud is permanently exposed to intense ultraviolet radiation fields, a photo-dissociation region (PDR) is created (Black and van Dishoeck, 1987). These authors have shown that in PDRs the ultraviolet photons with energies < 13.6 eV can be absorbed by H_2 in the Lyman and Werner band systems. These absorptions are then followed by fluorescence to the vibrational continuum of the ground state (90%) or the molecule dissociates (10%). The infrared H_2 emission is produced by the cascade down the rotational and vibrational energy ladder of the ground electronic state. In a PDR, the H_2 molecules can be excited by strong recombination emission lines such as H I Lyα and Lyβ with which there are accidental resonances with the H_2 lines (Black and van Dishoeck, 1987).

2.1. DR21

The high-mass star-formation region known as DR21 (Downes and Rinehart, 1966), is a complex molecular outflow with several compact H II regions at the center and with several near-infrared sources hiding deep in the cloud. Earlier observations of 2 μm line emission from the vibrationally excited H_2 in this bipolar outflow suggested shock excitation at temperatures of $\sim$ 2000 K and gas densities of $\sim$ 10^5–10^6 cm^{-3} (Fischer et al., 1985). In this complex, both high-velocity flows and large UV photon fluxes originating from the central expanding H II regions are present.

A high angular resolution image of DR21 (Davis and Smith, 1996) shows that the H_2 line emission is intrinsically inhomogeneous and clumpy at both bipolar lobes sown to scales $\leq 5''$. The 1–0 S(1) intensity structure observed within each lobe may correspond to variations in density within the outflowing gas, either pre-existing or the result of periodic outbursts in the acceleration mechanism. The dense clumps of shocked gas may in fact represent a larger scale variation of the HH phenomenon common to the high-velocity winds of less luminous pre-main-sequence stars (Schwartz, 1983).

As described in Fernandes *et al.* (1997), we have assumed a paraboloidal shape for the bow and a shock velocity of $v_s = 120\,\mathrm{km\,s^{-1}}$. The best fit "shock + fluorescence" model run in DR21 gives a pre-shock density of $n_0 = 10^6\mathrm{cm}^{-3}$ and $G_0 = 10^3$ (using BHT fluorescent model) and $n_0 \sim 3 \times 10^3\mathrm{cm}^{-3}$ and $10^2 \leq G_0 \leq 10^3$ (using BvD fluorescent model). For these parameters, the C-shock model yields a maximum excitation temperature of $\sim$ 2700 K in the eastern lobe and $\sim$ 2000 K in the western lobe.

2.2. HERBIG-HARO 7

In HH7, the absence of a strong nearby continuum source suggests that the fluorescent H_2 emission originates from the vicinity of the shock that is propagating into the molecular cloud. Since the shock speed is high enough to dissociate partially the H_2 molecules in the post-shock gas, the same shock can also explain the low-excitation optical lines. The stronger shocks that develop near the bow apex produce intense radiation affecting not only the pre-shock gas (magnetic precursor) but also the post-shock gas. In this manner, the rotational-vibrational H_2 line emission is produced through X-ray, ultraviolet heating and Lyman-α pumping. It is likely that the non-ionizing far-UV radiation (6–13.6 eV) generated near the bow apex will excite the ambient molecular cloud and produce the H_2 fluorescence. In this picture, the density of the pre-shock gas is about 10^3–10^4 cm^{-3}, which is consistent with PDR model densities (Black and van Dishoeck, 1987; Burton *et al.*, 1990). On the other hand, an H_2 reformation scenario would predict higher densities, 10^5–10^6 cm^{-3} and thus we favour the lower density fluorescent component scenario in HH7.

3. Conclusions

The many young stellar outflows that have been studied so far provide an extensive set of data on the physical conditions of the interstellar gas, and on the emission lines of the abundant hydrogen molecule especially suited to allow the study of the optically obscured star-forming regions. The intensities of the near-infrared H_2 lines and respective column densities obtained for the Herbig-Haro 7 object and the DR21 bipolar outflow have been used to construct diagnostic CDRs. We then show that the H_2 emission from the shocked gas in the HH7 outflow can be modeled by employing a mixed "shock + fluorescent" model, with a C-type bow-shock producing the warm H_2 gas measured in the low vibrationally excited levels. The emission from the higher excitation levels ($v > 2$) is explained by the Lyα pumping of the Lyman band system of H_2 which produces the H_2 fluorescent emission component.

Acknowledgments

This work was supported in part through project PESO/C/PRO/1033 (PRAXIS XXI).

References

Allen, D. and Burton, M.: 1993, *Nature* **363**, 54
Black, J. and van Dishoeck, E.: 1987, *Astrophys. J.* **322**, 412, (BvD)
Burton, M., Hollenbach, D., and Tielens, A.: 1990, *Astrophys. J.* **365**, 620, (BHT)
Burton, M., Hollenbach, D., and Tielens, A.: 1992, *Astrophys. J.* **399**, 563
Davis, C. and Smith, M.: 1996, *Astron. Astrophys.* **310**, 961
Downes, D. and Rinehart, R.: 1966, *Astrophys. J.* **144**, 937
Fernandes, A. and Brand, P.: 1995, *Mon. Not. R. astr. Soc.* **274**, 639
Fernandes, A., Brand, P., and Burton, M.: 1997, *Mon. Not. R. astr. Soc.* **290**, 216
Fischer, J., Sanders, D., Simon, M., and Solomon, P.: 1985, *Astrophys. J.* **293**, 508
Schwartz, R.: 1983, *Astrophys. J.* **268**, L37
Wolfire, C. and Königl, A.: 1991, *Astrophys. J.* **383**, 205

MOLECULAR OUTFLOWS FROM HIGH-MASS PROTOSTARS

MICHAEL D. SMITH
Astronomisches Institut der Universität Würzburg,
Am Hubland, 97074 Würzburg, Germany

Abstract. The youngest protostars are obscured from direct view by a high column of molecular gas. Nevertheless, their presence is betrayed through spectacular infrared outflows. I demonstrate here that infrared spectroscopy has the potential to reveal a remarkable variety of details concerning the underlying physics.

Near-infrared spectroscopic analyses of the OMC-1, DR 21 & Cepheus A outflows are discussed here. Molecular hydrogen is vibrationally excited by collisions in shock waves. In OMC-1, the ortho-para ratio has been mapped. The ratio is close to 3, suggesting efficient shock thermalisation. In DR 21, shocked (up to the first vibrational level) and fluorescent (higher v-levels) components have been successfully separated. In Cepheus A, non-LTE effects imply low densities.

1. Interpretation

Molecular hydrogen exists in two forms, according to the nuclear spin directions: ortho and para. Smith, Davis & Lioure (1997) have mapped the ortho-para ratio for the first time. OMC 1 was chosen, and three rotationally-consecutive lines in the first vibrational level were mapped. Fortunately, errors arising due to extinction are effectively eliminated by this choice. Smith et al displayed the ratio along east-west cuts. Here, three ratio images of the northern lobe including the 'bullets' are presented (Fig 1). These are harder to intepret since the errors are large at the edges of the outflow where the intensities are weak. However, no evidence for large-scale systematic variations are found.

In DR 21, shock wave excitation is not the mechanism by which gas is excited to high vibrational levels. Fernandes, Brand & Burton (1997) uncovered a fluorescent component by modelling K-band data. Smith, Eislöffel &

Astrophysics and Space Science is the original source of publication of his article. It is recommended that this article is cited as: *Astrophysics and Space Science* **261**: 183–184, 1999.

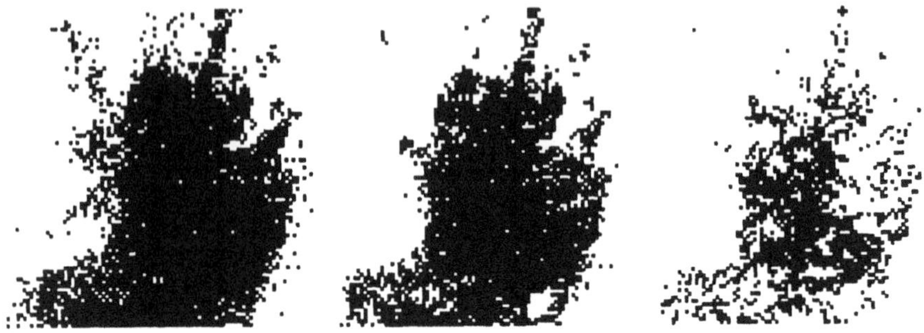

Figure 1. The ortho-para ratio, ϕ, in the northern lobe of OMC 1, as derived from the 1-0 S(0), S(1) & S(2) lines. Left: $0.3 < \phi < 5.3$; center: $1.3 < \phi < 4.3$; right: $2.3 < \phi < 3.3$. Resolution: 1.04 arcsec (see Smith et al (1997)).

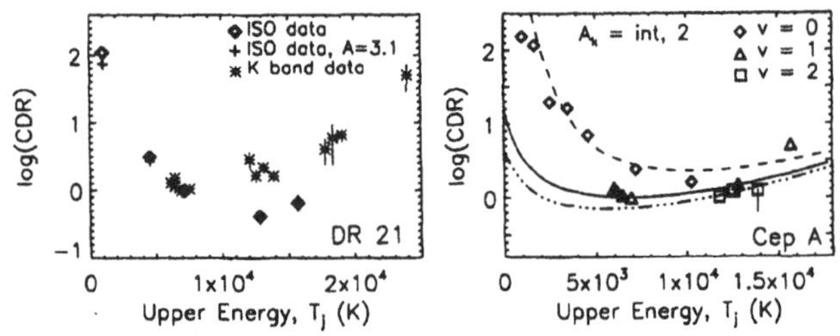

Figure 2. The log(CDR) diagrams for DR 21 & Cep A, including the DR 21 K-band data of Fernandes et al (1997), normalised to the 1-0 S(1) line. Note that ISO has observed the weak lines from pure shock-excited gas even for lines from T_j above 10,000 K, while the high-T_j K-band data is the consequence of a fluorescent cascade. The Cep A model shows that the levels splits up for another reason: low densities mean LTE is not achieved. The superimposed model is a C-type (s=1.4) bow shock. Dashed, full and dot-dashed lines correspond to the ground, first and second vibrational levels (Smith, 1998, MNRAS submitted). The original data were presented by Wright et al (1996, A&A, 315, L301).

Davis (1998) have now used ISO data in the 5-7μm region to show that the shocked component is indeed present, but with a low-excitation spectrum. Smith et al developed a method which employs pure rotation transitions lying within vibrationally excited levels. The columns of H_2 from the ISO and K-band lines are presented in Fig. 2.

References

Fernandes, A.J.L., Brand, P.W.J.L., Burton, M.G., 1997, MNRAS, 290, 216
Smith, M.D., Davis, C.J., Lioure, A., 1997, A&A, 327, 1206.
Smith, M.D., Eislöffel, J., Davis, C.J., 1998, MNRAS, in press.

A MICROWAVE JET INSIDE THE ROSETTE NEBULA?

A.P. TSIVILEV
P.N.Lebedev Physical Institute, Academy of Sciences of Russia, Moscow, Russia

S. CORTIGLIONI, M.R. CIONI, M. ORSINI AND E. PALAZZI
I.Te.S.R.E./C.N.R.,Bologna, Italy

AND

G.G.C. PALUMBO
Universita' degli Studi, Dipartimento di Astronomia, Bologna, Italy

1. Observations and results

New spectral measurements of the well known HII region Rosette Nebula have been performed with the 32m Medicina (Italy) VLBI single dish [Beamwidth: $5'$, beam efficiency: 83%; $T_{sys} \sim 45K$; 1024 channels autocorrelator, bandwith: $12.5MHz$, velocity resolution: $0.88Kms^{-1}$]. Hydrogen Radio Recombination Lines (RRLs) observations at 8.3 MHz (92α) are useful in attempting to build a non-LTE model for this ring like HII region in order to explain the observed behaviour with frequency of the line to continuum ratio Tl/Tc. Measurements were performed at Celnik's (1985) positions. Standard spectral fitting and data analysis have led to the values reported in the Table.

Most available observations were made towards Celnik's (1985) position 1 ($\alpha = 6^h28^m15.7^s$, $\delta = +5°21'14"$, delta=). RRL radial velocity (V_{LSR}) versus transition number n in the direction of the above position 1 is plotted in the figure for all the available measumements (for references see caption).

The observed sharp decrease of V_{LSR} from $n \sim 110$ downwards is not observed in the other two positions of the nebula.

If real, this decrease may mean that some dense inner part of the shell is moving towards the observer. The following scenario may be proposed: a stellar wind comes in contact with the outer surrounding dense material; the corresponding shock forms a jet. The following scheme can be envisaged. A

Astrophysics and Space Science is the original source of publication of his article. It is recommended that this article is cited as: *Astrophysics and Space Science* **261**: 185–186, 1999.

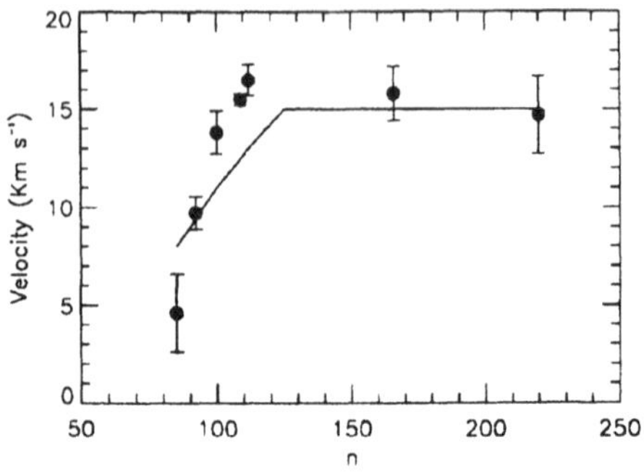

Figure 1. The hydrogen lines V_{LSR} versus n in the direction of Position 1. 92α; A weighted average was performed for data with same n. Solid line: model calculations.

Results from $H92\alpha$ line and continuum observation.

Pos.	T_L (mK)	ΔV (Km s^{-1})	V_{LSR} (Km s^{-1})	T_c (mK)	T_e(LTE) (K)
1	16.1 ± 3.3	25.1 ± 1.9	9.7 ± 0.8	140.0 ± 5.6	6000 ± 1160
2	21.5 ± 3.2	24.6 ± 1.1	16.8 ± 0.5	182.1 ± 5.5	5980 ± 820
3	9.4 ± 3.3	30.1 ± 2.9	15.0 ± 1.3	133.2 ± 5.5	7850 ± 2500

dense clump ($N_e = 5 \cdot 10^4 cm^{-3}$, radius 1 pc) is moving at $20 km/s$ relative to the stationary shell ($N_e = 10 cm^{-3}$, $R_{in} = 2$ pc and $R_{out} = 22$ pc) in the direction of the observer. The filling factor of the clump is such that both clump and shell have the same EM$\sim 4 \cdot 10^3 cm^{-6} pc$. Its Stark broadening become strong for RRLs with n ¿ 100. The whole system $V_{o,LSR} = 15$ km/s is assumed. From the figure one can see how, in principle, such scheme may describe the observed data. Such hypotheses may also be supported by recent work of Schneider et al. (1996) who, to explain the line wing emission found in their spectra, suggested the existence of high velocity clumps or outflows. One of such outflows could have been detected more recently by Schneider et al. (1996).

References

Celnik, W.E. (1985) The Rosette Nebula, *Astronomy & Astrophysics* **Vol. no. 144**, pp. 171–178.

Schneider, N., Stutzki, J., Winnewisser, G., Blitz, L. (1996) The Nature of the Molecular Line Wing Emission in the Rosette Molecular Complex, *Astrophysical Journal* **Vol. no. 468**, pp. L119–L122.

ISO-LWS OBSERVATIONS OF THE NGC7129 STAR FORMING REGION

ELISABETTA TOMMASI
ISO Science Operation Centre, Villafranca del Castillo, P.O.Box 50727, 28080 Madrid, Spain
DARIO LORENZETTI
Osservatorio Astronomico di Roma, Italy
IFSI-CNR, Roma, Italy
TERESA GIANNINI AND BRUNELLA NISINI
IFSI-CNR, Roma, Italy
AND
FRANCESCO PALLA
Osservatorio Astrofisico di Arcetri, Firenze, Italy

Abstract. We present far-infrared observations of the surroundings of the Herbig Be star LkHα234 in NGC7129, obtained with the Long Wavelength Spectrometer on board of the Infrared Space Observatory. [CII]158μm, [OI]63μm and [OI]146μm lines are detected everywhere in the mapped region and their intensity ratios are consistent with the model predictions of a photodissociation region. The spatial distribution of the emission features indicates the presence of a peak which does not coincide with the position of LkHα234, but is more likely associated with another nearby Herbig star, namely BD+65°1637; the far-ultraviolet field intensity and the density derived from our observations are fully consistent with this hypothesis.

1. Introduction

NGC7129 is a complex star forming region at a distance of about 1kpc; two Herbig Be stars, LkHα234 and BD+65°1637, are located in the region, along with a far-infrared (FIR) source (Bechis et al. 1978), two molecular outflows (Edwards and Snell 1983), several Herbig-Haro objects (Ray et al. 1990) and water masers (Sandell and Olofsson 1981). Given all these observational evidences we have included a map of this region in the ISO-LWS observative programme on Herbig Ae/Be stars (Lorenzetti et al. 1997) aiming to recognize the prevailing excitation mechanism(s) in the enviroments of these sources. The results of the programme suggest that photodissociation is responsable for FIR line excitation in most cases, and this raster

Astrophysics and Space Science is the original source of publication of his article. It is recommended that this article is cited as: *Astrophysics and Space Science* **261**: 187–190, 1999.

TABLE 1. Line intensities and derived parameters.

pos. Fig. 1	[OI]63μm	[OI]146μm	[CII]158μm	log G_o obs.+model	log G_o BD+65°1637	log n_o
	F (ΔF^*) $10^{-19} W/cm^2$					
1-1	3.4(.5)	<0.6	4.4(.5)	1.58(.04)	1.8(.15)	3.7(.2)
1-2	19(1)	1.8(.5)	27.1(.6)	2.37(.01)	2.5(.15)	2.6(.2)
1-3	58(2)	5.0(.5)	49.7(.8)	2.63(.01)	2.8(.15)	2.9(.2)
1-4	10(1)	1.0(.5)	12.8(.3)	2.05(.01)	2.0(.15)	3.4(.2)
1-5	3(1)	<0.6	4.9(.4)	1.62(.03)	1.6(.15)	3.4(.2)
1-6	1.6(.9)	<0.3	2.4(.2)	1.32(.04)	1.3(.15)	>3.7
2-1	2.8(.7)	0.2(.1)	3.1(.2)	1.43(.03)	1.6(.15)	>4
2-2	14(1)	0.7(.2)	8.2(.5)	1.85(.03)	1.9(.15)	4.1(.2)
2-3	9.7(.2)	1.4(.5)	13(2)	2.05(.07)	1.9(.15)	3.2(.2)
2-4	3.5(.5)	0.5(.1)	6.5(.9)	1.75(.06)	1.7(.15)	3.2(.2)
2-5	2.5(.5)	<0.5	3.1(.2)	1.43(.04)	1.4(.15)	>4
2-6	1.3(.6)	<0.5	2.8(.3)	1.39(.05)	1.2(.15)	>3.2
LkHα234	54(1)	5(1)	32.0(.7)	2.44(.01)	2.3(.15)	3.6(.2)

* 1σ statistical error due to the residuals from a gaussian fit

map offers the opportunity to study at a larger scale the morphology and the spatial distribution of such excitation.

2. Observations

The Herbig Be star LkHα234 ($\alpha_{1950} = 21^h41^m57.1^s$, $\delta_{1950} = +65°53'09''$) has been observed with the Long Wavelength Spectrometer (LWS, Clegg et al. 1996) on board of the Infrared Space Observatory (ISO, Kessler et al. 1996) on February 15, 1997 and a 2×6 raster map in the surroundings of the source (100″ grid, center in $\alpha_{1950} = 21^h41^m56.7^s$, $\delta_{1950} = +65°51'30''$) was performed on November 17, 1996. All the observations were carried out with the LWS AOT01 full grating scan mode (*i.e.* range=43-196.7μm, $\lambda/\Delta\lambda \sim 200$, FOV$\sim$80″), with an oversampling of 4. Total integration time per spectral sample is 6 sec for the star and 1.6 sec at each map pointing.

Raw data have been processed with an internal version of the reduction pipeline, which improves the quality of the data in comparison with previous published results (Tommasi et al. 1997). Flux calibration is based on observations of Uranus and the absolute accurancy is better than 30%.

3. Results and Discussion

[CII]158μm, [OI]63μm and [OI]146μm lines are detected everywhere in the mapped region; in columns 2 to 4 of Table 1 the line intensities for each pointing of the map (column 1, see Fig.1*right*) and for the observation cen-

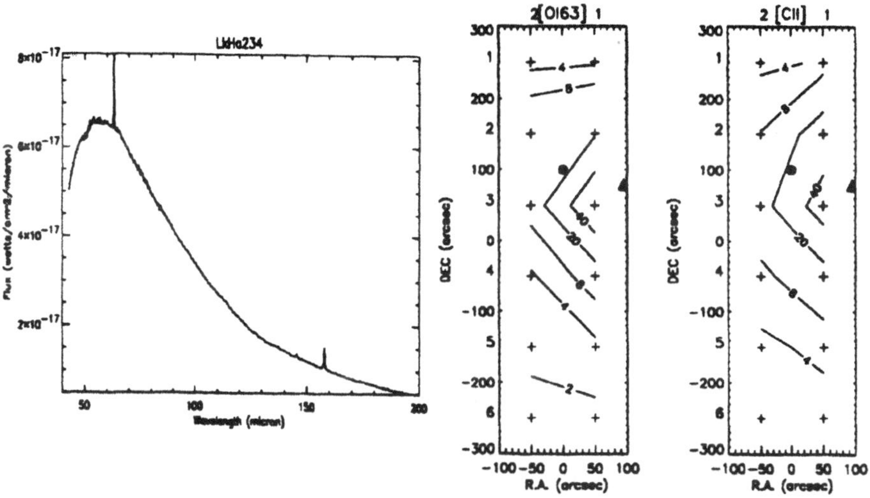

Figure 1. (*left*) LWS spectrum of LkHα234. (*right*) Maps of FIR line emission; positions of LkHα234 and BD+65°1637 are marked by a circle and a triangle respectively and contour unit is $10^{-19} W/cm^2$.

tered on LkHα234 are reported. The ratio of [OI]63μm and [CII]158μm fluxes is always less than 2, as expected if the lines are excited in a photodissociation region (PDR, Luhman et al. 1997).

Figure 1 shows the whole LWS spectrum of LkHα234 (*left*) and the contour plots for [OI]63μm and [CII]158μm lines obtained using the raster map data, along with the 12 pointing positions and the location of the two Herbig stars (*right*). The morphology of the emission suggests that the PDR is more likely associated with BD+65°1637 rather than LkHα234. This hypothesis is strongly supported by molecular emission distributions on a large scale: the ^{13}CO map obtained by Fuente et al. (1997) shows clearly the cavity created by BD+65°1637, coincident with the peak of FIR atomic line emission, and the ridge of nearby molecular cloud, in which LkHα234 is still partially embedded.

To quantitatively assess this possibility, we have used the [CII]158μm line intensities observed in each position to derive the far-ultraviolet (FUV) field G$_o$, using the relationship expected for low density PDRs: $I_{[CII]158\mu m} \sim 10^{-6}$ G$_o$ ergs cm^{-2} s^{-1} sr^{-1}, where G$_o$ is measured in units of $1.6 \cdot 10^{-3}$ ergs cm^{-2} s^{-1} (Hollenbach et al. 1991). The derived values of G$_o$ are reported in Table I along with those expected for BD+65°1637, adopting a luminosity of 2400L$_\odot$ (Hillenbrand et al. 1992); the error on the latter values is estimated assuming uncertainties of 20% on both flux and distance. We note that the agreement between the two values is very good.

We have also calculated the density n_o in each observed position (column 7 of Table 1) using the derived value of G_o and the ratio $\frac{[OI]63\mu m}{[CII]158\mu m}$ (Luhman et al. 1997). In this way we verify *a posteriori* that our assumption of a low density PDR is correct, being the density always $<10^4 cm^{-3}$; lower limits are imposed by the limited range of G_o values in the model ($\log G_o > 1.5$). We also note that our n_o values are consistent with the density gradient shown by molecular emission (Fuente et al. 1997, Güsten and Marcaide 1986).

A further evidence in favour of the presence of a PDR has been recently provided by Schultz et al. 1997: they obtained IR images of the edge structure in the filters H_2 1-0 S(1) and 2-1 S(1) deriving for the ratio 1-0/2-1 S(1) the value expected for fluorescent H_2. They ascribe the PDR origin as due to the FUV field from BD+65°1638, whereas we suggest the Herbig star BD+65°1637 because of both its location more consistent with the edge morphology traced by ^{13}CO emission and the very good agreement found in the derived parameters (G_o, n_o).

In conclusion, results of ISO-LWS observations in this region are compatible with a scenario in which the effects of the PDR generated by the Herbig Be BD+65°1637 dominate the FIR line emission in a quite wide area, despite all the other phenomena which are present in the region.

Acknowledgements

ISO is an ESA project funded by ESA Member States and with the participation of ISAS and NASA. The observations presented here are part of the Guaranteed Time Programme and as such are the results of a joint effort of all the ISO-LWS team.

References

Bechis, K.P., Harvey, P.M., Campbell, M.F., Hoffmann, W.F. 1978, *ApJ*, **226**, 439
Clegg, P.E. et al. 1996, *A&A Letters*, **315**, L38
Edwards, S., Snell, R.L. 1983, *ApJ*, **270**, 605
Fuente, A. et al. 1997, submitted to *A&A*
Güsten, R., Marcaide, J.M.: 1986, *A&A*, **164**, 342
Hillenbrand, L.A., Strom, S.E., Vrba, F.J., Keene, J. 1992, *ApJ*, **397**, 613
Hollenbach, D.J., Takahashi, T., Tielens, A.G.G.M. 1991, *ApJ*, **377**, 192
Kessler, M.F. et al. 1996, *A&A Letters*, **315**, L27
Lorenzetti, D. et al. 1997, *Proc. of "Star formation with ISO"*, in press
Luhman,M.L., Jaffe,D.T., Stenberg,A., Hermann,F., Poglitsch,A. 1997, *ApJ*, **482**, 298
Ray, T.P., Poetzel, R., Solf, J., Mundt, R. 1990, *ApJ Letters*, **357**, L45
Sandell, G., Olofsson, H. 1981, *A&A*, **99**, 80
Schultz, A.S.B., Rank, D.M., Temi, P., Holbrook, J.C. 1997, Poster proc. of IAU Symp. No.182, eds. F. Malbet & A. Castets, p.39
Tommasi, E. et al. 1997, *Proc. of "Star formation with ISO"*, in press

SUPERSONIC AMBIPOLAR DIFFUSION:

Estimating the Magnetic Field Strength in Protostellar Outflows

MICHAEL D. SMITH
*Astronomisches Institut der Universität Würzburg,
Am Hubland, 97074 Würzburg, Germany*

AND

MORDECAI-MARK MAC LOW
*Max-Planck-Institut für Astronomie, Königstuhl 17,
D-69117 Heidelberg, Germany*

Abstract.
The magnetic field plays a crucial role in star formation. It is involved in rotational braking, collapse braking, outflow formation and jet collimation. Direct observations of the field are difficult. However, the field can be indirectly estimated through the field-cushioned C-shocks which produce strong infrared molecular emission lines. In particular, a high field in the outflows will generate the 'shock absorber' signature: very broad H_2 lines. Such lines are indeed observed.

Here we summarise recent progress in C-shock formation and stability. We demonstrate numerically that the Shock Absorbers are evolutionary and stable. The widths of H_2 lines then limit the magnetic field strength. A field of 6 mG is suggested for HH 212.

1. Do C-Shocks exist?

The evolution away from jump shocks toward the numerous steady C-shock sub-types has been investigated by Smith & Mac Low (1997). The evolution passes through four stages, which possess distinctive observational properties. The time scales and length scales cover broad ranges. Specific results were included for shock types including switch, absorber, neutralised, oblique, transverse and intermediate. Only intermediate Type II shocks and 'slow shocks', including switch-off shocks, remain as J-type under the low ion levels assumed. All other shocks transform to C-type via a steadily growing neutral precursor to a diminishing jump.

Astrophysics and Space Science is the original source of publication of his article. It is recommended that this article is cited as: *Astrophysics and Space Science* **261**: 191–192, 1999.

Stone (1997) and Mac Low & Smith (1997) also computed the nonlinear development of the instabilities in C-shocks first described by Wardle. The latter used a version of the ZEUS code modified to include a semi-implicit treatment of ambipolar diffusion. We found that, in three dimensions, thin sheets parallel to the shock velocity and perpendicular to the magnetic field lines form. High resolution, two-dimensional models show that the sheets are confined by the Brandenburg & Zweibel (1994) ambipolar diffusion singularity, forcing them to numerically unresolvable thinness.

We compared steady-state to unstable C-shocks, showing excitation diagrams, line ratios, and line profiles for molecular hydrogen lines visible in the K-band, with the Infrared Space Observatory and with NICMOS on the Hubble Space Telescope.

How can we deduce the magnetic field strength? High magnetic fields imply low Alfven Number shocks. Such C-shocks are stable. Also, the molecules remain quite cool and shock-resistant. The main path for their destruction is by ion impact dissociation: when the ion-neutral streaming speed reaches $v_d \sim 42\,km\,s^{-1}$, runaway dissociation occurs. For a shock speed v_s, Alfven speed v_a (Alfven number $A = v_s/v_a$), it was estimated that hydrogen molecules in dense regions can survive fast shocks since the streaming speed in a steady shock with transverse field reaches the maximum value of just

$$v_{impact} = v_s \left[1 + \frac{1}{2A^2} - \frac{3}{2A^{2/3}}\right] \tag{1}$$

which can be a small fraction of v_s (Smith et al 1991). This analysis has been confirmed in the numerical experiments. One cannot use this equation directly to predict the H_2 line widths: one must assume a shock geometry and consider the modification required for an oblique field (Smith 1992). Nevertheless, a velocity jump of $80\,km\,s^{-1}$ can be produced by a shock of speed $\sim 120\,km\,s^{-1}$ and $A = 3.1$, without destroying the molecules. The low A and high Alfven speed ($39\,km\,s^{-1}$) are consistent with expectations from recent models of protostellar jets, in which A increases slowly beyond the Alfven radius (where $A = 1$) in a collimated wind. Hence the broad lines observed in HH 212 (Zinnecker et al, in prep.) may well be indicating a strong field: for a density of $10^5\,cm^{-3}$, a field of 6 mG is predicted.

References

Brandenburg & Zweibel, 1994, ApJ, 427, L91
Mac Low, M.-M. & Smith, M.D., 1997, ApJ, 491, 596
Smith, M.D. & Mac Low, M.-M., 1997, A&A, 326, 801
Smith, M.D., Brand, P.W.J.L., Moorhouse, A. 1991, MNRAS, 248, 730
Smith, M.D., ApJ, 406, 520
Stone, J.M., 1997, ApJ, 487, 271.

PROPER MOTIONS OF THE H_2O MASERS IN W49N: THE VELOCITY VARIANCE/COVARIANCE MATRIX

E. E. BLOEMHOF
California Institute of Technology
Pasadena, CA USA 91125

1. Introduction

The proper motions of the H_2O masers in W49N have been measured with multi-epoch VLBI by Gwinn, Moran, and Reid (1992). The number of motions is large (105), and the coverage is excellent, making this source a prime candidate for investigation with model-independent analytic tools involving diagonalization of the velocity variance/covariance matrix (VVCM),

$$\sigma_{ij} = \frac{1}{N-1} \sum_{n=1}^{N} (v_{i,n} - \bar{v}_i)(v_{j,n} - \bar{v}_j),$$

as described by Bloemhof (1993). Here i, j denote RA, Dec, and Z (radial) coordinates, the sum is over N maser spots, and the bar indicates averaging over spots. The VVCM is invariant with respect to choice of the unknown velocity of the reference maser spot. In any VLBI proper motion experiment, the absence of an absolute interferometric phase reference implies that absolute velocities are not known, so the velocity dispersions of the VVCM analysis are the natural, objective quantities with which to work.

2. The VVCM for W49N H_2O Masers

Using the velocities given in Table 4 of Gwinn *et al.* (1992), the VVCM, before and after diagonalization, is (in units of km^2 s^{-2}):

$$\sigma = \begin{pmatrix} 2369.6 & 709.3 & 1940.4 \\ 709.3 & 1133.3 & 663.9 \\ 1940.4 & 663.9 & 3825.9 \end{pmatrix}; \quad \tilde{\sigma} = \begin{pmatrix} 1179.5 & 0 & 0 \\ 0 & 766.3 & 0 \\ 0 & 0 & 5383.0 \end{pmatrix}.$$

Astrophysics and Space Science is the original source of publication of his article. It is recommended that this article is cited as: *Astrophysics and Space Science* **261**: 193–194, 1999.

The eigenvalues, displayed on the main diagonal of the second matrix, describe the velocity dispersions along three new "principal" axes that are orthogonal to each other but rotated from the original coordinate system. One is substantially larger than the other two, as expected for a bipolar outflow. The axis of the outflow is computed in the diagonalization as the eigenvector corresponding to the largest eigenvalue. This "VVCM axis" is [0.564,0.219,0.796], expressed in the original spatial coordinates...right ascension, declination, and a radial coordinate increasing toward the source. In two-dimensional projection, this axis is very well aligned, as expected, with the spatial elongation of the maser distribution seen in Figure 2 of Gwinn *et al.* (1992).

3. Azimuthal Symmetry of the Outflow

Although there is gross equality of the two smaller eigenvalues, the degree to which they differ may be taken as a measure of the azimuthal asymmetry of the flow. In a plane perpendicular to the VVCM axis, one may identify the direction along which velocity dispersions are maximal; for W49N, this "major transverse axis" is [0.569,0.595,-0.567].

This azimuthal symmetry has not previously been emphasized in standard modelling of maser flows, but it may give vital information about the collimating structures. It could not be determined, of course, in the 2D VVCM treatment of the OH masers of W3(OH) (Bloemhof 1993).

4. Significance of the Outflow Axis

It will be important to seek spatial correlations between the kinematic axes of the H_2O maser outflow, as extracted by the VVCM diagonalization procedure presented here, and other kinematic or morphological structures in the W49N environment, such as the remarkable ring of HII regions observed by Welch *et al.* (1987). Correlations with the orientation of the maser outflow may help to explain how the outflow is collimated, and perhaps even how the young star that created the outflow is formed in the first place.

5. References

Bloemhof, E. E. (1993) Diagonalization of the Velocity Variance/Covariance Matrix as a Diagnostic of Maser Proper Motions, *Ap.J.*, **406**, L75–L78

Gwinn, C. R., Moran, J. M., and Reid, M. J. (1992) Distance and Kinematics of the W49N H_2O Maser Outflow, *Ap.J.*, **393**, 149–164

Welch, W. J., Dreher, J. W., Jackson, J. M., Terebey, S., and Vogel, S. N. (1987), Star Formation in W49A: Gravitational Collapse of the Molecular Cloud Core Toward a Ring of Massive Stars, *Science*, **238**, 1550-1555

THE DECAY OF SUPERSONIC AND SUPER-ALFVÉNIC TURBULENCE IN STAR-FORMING CLOUDS

The Spatial Distribution of Shocks

MORDECAI-MARK MAC LOW
Max-Planck-Institut für Astronomie, Königstuhl 17, D-69117 Heidelberg, Germany

MICHAEL D. SMITH
Astronomisches Institut der Universität Würzburg, Am Hubland, 97074 Würzburg, Germany

AND

RALF S. KLESSEN AND ANDREAS BURKERT
Max-Planck-Institut für Astronomie, Königstuhl 17, D-69117 Heidelberg, Germany

Abstract.
Supersonic and super-Alfvénic turbulent motions generate an intricate pattern of shock waves through which the turbulence decays. We here illustrate the spatial structure during shock formation and decay.

1. A Display of Decadence

We have recently presented numerical studies of compressible, decaying turbulence, with and without magnetic fields, with initial rms Alfvén and Mach numbers ranging up to five, and applied the results to the question of the support of star-forming interstellar clouds of molecular gas (Mac Low et al 1998). In all the regimes that we have studied 3D turbulence—super-Alfvénic, supersonic, sub-Alfvénic, and subsonic—the kinetic energy decays as $t^{-\eta}$, with $0.85 < \eta < 1.2$. We compared results from a grid and an SPH algorithm in the unmagnetized case, and have performed extensive resolution studies in all cases, reaching resolutions of 256^3 zones or 350,000 particles. We conclude that the observed long lifetimes and supersonic motions in molecular clouds must be due to external driving, as undriven turbulence decays far too fast to explain the observations.

Astrophysics and Space Science is the original source of publication of his article. It is recommended that this article is cited as: *Astrophysics and Space Science* **261**: 195–196, 1999.

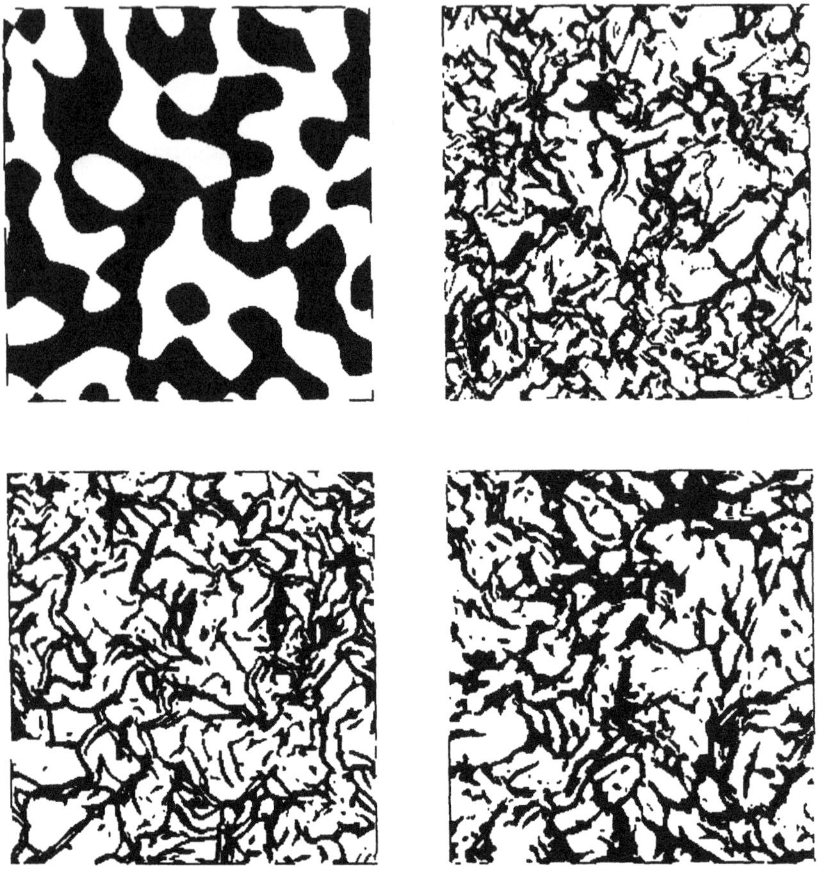

Figure 1. The evolution of the shock fields as measured by regions of negative velocity divergence. The initial wave field (top left, t = 0, rms Mach number is 50) generates an intricate shock pattern (top right, t = 0.2) which maintains its character (bottom left, t = 2) while slowly deteriorating (bottom right, t = 10). The resolution here is 256^3 zones.

Here, we show the decay of supersonic hydrodynamic turbulence. We present, not density slices, but convergence slices (div v < 0). These demonstrate that a complex shock pattern rapidly develops and is maintained as the shocks weaken and converging regions spread out. Simulated observations of these cubes are in preparation.

References

Mac Low, M.-M., Klessen, R.S., Burkert, A. & Smith, M.D., 1998, PRL, submitted.

THE "SEYFERT-LIKE" SUPERNOVA SN 1997AB.

I. SALAMANCA
Royal Greenwich Observatory (RGO), Cambridge, UK
& Leiden Observatory, P.O. Box 9513, 2300 RA Leiden,
The Netherlands

Abstract. New spectroscopic data of the peculiar type II supernova 1997ab is presented. The most remarkable features are the presence of broad ($\lesssim$3000 $\mathrm{km\,s^{-1}}$) Balmer emission lines and a featureless blue continuum. Prominent FeII lines are also observed. When observed with very high spectral resolution the narrow component of the Hα and Hβ lines is resolved into a P-Cygni profile of 180 $\mathrm{km\,s^{-1}}$ of Full Width at Zero Intensity (FWZI). All of these features can be explained by the interaction of the supernova ejecta with a dense (n $\sim 10^{7-8}$ $\mathrm{cm^{-3}}$) circumstellar medium (CSM), probably created by a wind of the progenitor star.

1. Introduction

One of the most spectacular events is a subclass of type II supernovae (SN), called "Seyfert 1 impostors" (Filippenko 1989) or "peculiar type IIn SN" (Schlegel 1990). The most remarkable observational features known up to now are: (i) Their optical spectrum is dominated by intense emission lines. Balmer lines are often observed to have composite profiles: a narrow component on top of a broad (few thousands of $\mathrm{km\,s^{-1}}$ of FWHM). Broad bands of FeII lines are also often present, as well as HeI 5876 emission line and the infrared CaII triplet. (ii) They have slow luminosity and spectral evolution compared to normal type II SN: ~0.007 mag/day as opposed to 0.01 mag/day. Besides, some of them (but not all) are among the most powerful radio supernovae and/or X-ray emitters (Van Dyk *et al.* 1993, Fabian & Terlevich 1996).

The slow evolution means that there is an additional source of energy (besides the radioactive decay) effective up to few years after the explosion, and whose origin is most probably strongly radiative shocks produced dur-

Astrophysics and Space Science is the original source of publication of his article. It is recommended that this article is cited as: *Astrophysics and Space Science* **261**: 197–200, 1999.

ing the interaction of the fast SN ejecta (few$\times$1000 km s^{-1}) with a dense circumstellar medium, CSM (n $\sim 10^{5-7}$ cm^{-3}).

1.1. SN 1997AB

We present here new spectroscopic data of SN 1997ab, one example of these peculiar supernova. Its discovery was reported in the IAU Circular No. 6589 and in Hagen *et al.* , 1997. It was first observed on 1996 April 11 (at B=14.7 $\pm$05), but only noticed one year later, by Hagen & Reimers. The magnitude then was $B = 16.2 \pm 0.5$, i.e. SN 1997ab had only varied 1.5 magnitudes in about one year. The exact date of explosion is not known, but due to its high luminosity in 1996, Hagen *et al.* assume that the SN must have been close to its maximum at that epoch (for more details, refer to the letter of Hagen *et al.* , 1997). From the narrow emission lines in the spectra of SN 1997ab, a redshift of 0.012 is deduced. This gives a distance of 47.2 Mpc, and an absolute B magnitude of -18.7 at discovery ($H_o = 75$ km s^{-1} Mpc^{-1}). Finally, SN 1997ab has not yet been detected in X-rays by ROSAT (Fabian, private communication), although it is expected to become a bright X-ray source.

2. The Data

The data presented here consist of spectra in the Hα and/or Hβ region, taken between 1997 March 26 and May 30 at "Guillermo-Haro" observatory in Cananea, Mexico and at "El Roque de los Muchachos" in La Palma, Spain. The spectral resolution varies from 800 to 8 km s^{-1} . The spectra, as well as the most relevant information is given in figure 1.

If we first look at the lowest resolution spectra (fig 1a), we notice that the emission is dominated by the intense Hα emission line, followed by Hβ , FeII bumps at $\sim$5250 Å and a strong CaII triplet at λ8542 Å. All these lines are on top of a blue continuum. When examining the intermediate resolution spectra (see fig 1b) we can identify the rest of the Balmer series, Hδ and Hγ , as well as the H-K Ca lines, HeI 4922 Å, [OIII]4959,5007, [NII]5754 and HeI 5875 Å. Also, the FeII becomes more evident: one can easily identify the multiplets no. 27, 28, 37, 38, 42, 48 and 49.

Regarding the profile of the Balmer lines, they are quite broad: their core is about 2760 km s^{-1} . The Hα line also have wings with a width of about 7000 km s^{-1} . Besides that, there is a very extended (up to almost -20000 km s^{-1}) but very weak blue wing. The red part of Hβ line is contaminated mainly by FeII emission and therefore we cannot measured the full extension of its profile.

The high resolution spectra (see fig 1c and 1d) allow us to determine the precise shape of the profile of the Hα and Hβ lines: they appear "flat-

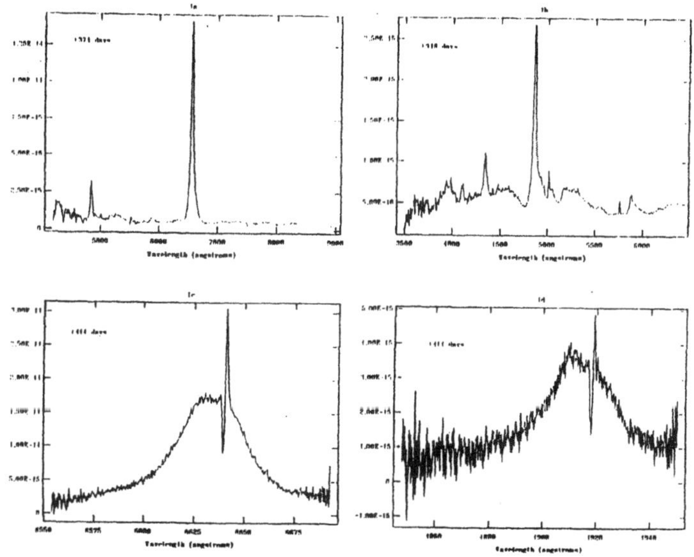

Figure 1. Spectra of the peculiar SN 1997ab. The flux is in units of erg s^{-1} cm^{-2} $Å^{-1}$ (a) Average of the low resolution spectra (18 Å) taken on 19, 20 and 21 of April 1997 at Guillermo-Haro Observatory (Cananea,Mexico). (b) Medium resolution spectrum (12 Å) taken on the 26 March 1997 at INT/IDS (La Palma, Spain). (c) and (d) High resolution spectra (0.18 and 0.13 Årespec.) taken on the 30 May 1997 at WHT/UES (La Palma, Spain). A "smooth" has been applied to them.

topped", with a very narrow P-Cygni profile on the flat top, displaced from the center to the red part of the broad component. After a close examination, we notice that the broad component is not symmetric and that its red part is missing. This explains the red displacement of the narrow P-Cygni profile, which is probably centered on the total broad component. The FWZI of the P-Cygny profile is 180 km s^{-1} , which gives us an expansion velocity of 90 km s^{-1} . We must notice that the width of the emission and absorption components is very similar, and that in general there is a remarkable symmetry in the P-Cygni profile.

3. Interpretation and Discussion

The above mentioned features strongly suggest the presence of a radiative shock produced by the interaction of the fast SN ejecta with a dense (n $\sim 10^{7-8}$ cm^{-3}) CSM. When a supernova explosion occurs, two shocks are produced: the leading one that heats the CSM, and a reverse shock, that thermalizes the ejecta. In the standard case, when the SN explosion occurs in a medium of density $\sim$1 cm^{-3}, both shocks are adiabatic, and remain so for thousands of years. However, for very high densities (n $\sim 10^7$) the adiabatic phase is practically nonexistent, and the shocks become radiative a few months after the explosion, when the ejecta is still moving at high speeds, and when the size of the remnant is less than 0.01 pc. This is accompanied by the formation of a very dense (n $\sim 10^{9-12}$ cm^{-3}) cold (T $\sim 10^4$ K) and fast (v $\sim$ thousands of km s^{-1}) shell which is formed by the so-called "catastrophic cooling". Theoretical calculations show that if the CSM has a constant density, both the leading and reverse shocks become radiative (see Terlevich et al 1992, 1995). However, if the density of the surrounding media decreases as r^{-2} as is expected from a wind, the leading shock may stay adiabatic, and only the reverse shock becomes radiative (Chevalier 1994). This second case is successful in explaining those supernovae (like SN 1980K, SN 1979C) that having evidence for ejecta-CSM interaction and the presence of radiative shocks, are less luminous than SN 1997ab. The first case (density of the CSM constant) is more successful in explaining the more extreme cases, like SN 1988Z or SN 1997ab, whose luminosities in Hα are of the order of 10^{41-42} erg s^{-1}. In any case, when either both or one shock becomes radiative, they will emit copious amounts of UV and X-ray photons which will ionize the freely expanding ejecta, the dense and fast thin shell and the surrounding CSM. The ejecta and the thin shell will emit this newly acquired energy in the form of broad permitted lines. The low speed and lower density CSM will produce narrow forbidden and permitted lines. In the case of SN 1997ab, the fact that the Hα and Hβ profiles are flat-topped is consistent with them being emitted in a thin expanding shell. Moreover, the narrow P-Cygni profile seen on top of the Balmer lines is direct proof of the existence of the dense CSM into which the ejecta is expanding.

Acknowledgements

I would like to acknowledge my collaborators, G. Tenorio-Tagle, R. Terlevich, E. Carrasco, E. Telles and A. Díaz, for their collaboration in the obtention and interpretation of these data. I also would like to thank Matt Lehnert for helping me to improve the English of this manuscript.

References

Chevalier R. A. & Fransson C. 1994, *Astrophys. Journal* **Vol. no. 420** 268
Fabian A.C. & Terlevich R., 1996 *Mont. Not. R. Astr. Soc* **Vol. no. 280** L5
Filippenko A., 1989, *Astron. Journal* **Vol. no. 97** 726
Hagen H.-J., Engels D., Reimers D. (1997), *Astron. & Atrophys.* **Vol. no. 324** L29
Terlevich R.J. *et al.* , 1992, *Mont. Not. R. Astr. Soc* **Vol. no. 255** 713
Terlevich R.J. *et al.* , 1995, *Mont. Not. R. Astr. Soc* **Vol. no. 272** 198
Schlegel E., 1990, *Mont. Not. R. Astr. Soc* **Vol. no. 244** 269
Van Dyk S.D. *et al.* , 1993, *Astrophys. Journal* **Vol. no. 419** L69

CHIMNEYS AND WORMS IN THE GALAXY

3D Numerical Modelling

M. A. DE AVILLEZ
Dept. Mathematics, University of Évora
R. Romão Ramalho 59, 7000 Évora, Portugal

Abstract. A three-dimensional model for the evolution of the disk gas is presented here. This model reproduces many of the features that have been observed in the Galaxy, by means of 21 cm line emission surveys. The cold structures obtained in the model are associated with supernova shells and chimney walls.

1. Introduction

HI emission surveys of the Galactic disk have revealed the presence of holes with sizes varying between 150 pc and 300 pc, surrounded by wiggly structures that appear to crawl out of the disk (Heiles, 1990). The holes are associated with isolated as well clustered supernovae. As the clustered supernovae occur they release the hot inner parts of the remnants into the halo in well collimated cylindrical-like structures that resemble chimneys (Tomisaka & Ikeuchi, 1986).

Here, the results of a three-dimensional simulation of the Galactic disk powered by supernovae and populated with chimneys and worms is reported.

Section 2, deals with the numerical method. Section 3 presents the initial and boundary conditions used in the simulations. Section 4 presents a description of the evolution of the simulations. Finally, section 5 deals with a discussion and summary of the main results.

2. Numerical Scheme

A section of a Galactic arm is simulated with the three-dimensional hydrodynamical scheme described in Avillez (1997). This is a Godunov class scheme that solves the Riemann problem at the cell interfaces, by approx-

Astrophysics and Space Science is the original source of publication of his article. It is recommended that this article is cited as: *Astrophysics and Space Science* **261**: 201–204, 1999.

imating the profiles of density, pressure and velocity through a third order interpolation polynomial algorithm. The scheme uses a local uniform reggriding method (Berger & Colella, 1989), that allows for an automatic refinement of the grid cells where steep variations of pressure and density occur.

The scheme does not include thermal conduction but, cooling is allowed by approximating the cooling curve of Raymond *et al* (1976) except in the range of temperatures of $5 \times 10^5 - 5 \times 10^7$ K where Kahn (1976) approximation is used.

Hydrodynamical tests (Avillez, 1998) performed in the scheme have shown that it is able to resolve the complex features that arise in the flow without introducing spurious effects. The largest relative error between numerical and analytical calculations is 1%.

3. Initial and Boundary Conditions

The simulations were carried out on a cartesian grid centred on the Galactic plane with an area of 1 kpc^2 and a vertical extent from -500 pc to 500 pc. Initially, a coarse grid with a resolution of 5 pc is set up. this grid can be refined automatically until the finest resolution of 1.25 pc is achieved. The boundary conditions along the vertical axis are periodic; outflow boundary conditions are used in the upper and lower parts of the grid parallel to the Galactic plane.

The computational domain is filled with hydrogen in neutral and ionized forms having a vertical distribution given by Dickey & Lockman (1990) and inferred from 21 cm line emission surveys of the Galaxy.

The disk is filled with type Ib and II supernovae having a vertical distribution compatible with that of Population I stars and given by (de Avillez, 1997)

$$\rho_\star = \rho_{\star,o} sech^2 \left[(2\pi G \beta_\star \rho_\star)^{1/2} z \right] \tag{1}$$

where $\rho_\star$ is the vertical density distribution, $\rho_{\star,o} = 3.0 \times 10^{-24}$ g cm^{-3} is the mass density contributed by Population I stars near the Galactic plane (Allen, 1991) and the constant $\beta_\star = 1.9 \times 10^{-13}$ cm^{-2} s^2. 40% of all supernovae occur randomly in the disk, whereas the remaining supernovae are clustered in OB associations (Cowie *et al*, 1979). Their rates of occurrence are 2.2 per century for the isolated supernovae (van den Bergh, 1990) and 0.65 per century in the associations (Evans, 1989).

4. Evolution of Simulations

The simulations were carried out for a simulated period of 500 Myr. They start from a highly unstable configuration where initially there is no equilibrium between cooling and heating. As the supernovae occur they eject high velocity gas into the surroundings, expanding both in the vertical and horizontal directions establishing a redistribution of mass and energy through out the computational domain. The system comes to a quasi-equilibrium between the descending and ascending gases during the rest of the simulations. The major fraction of the disk is filled with warm and hot gas. Most

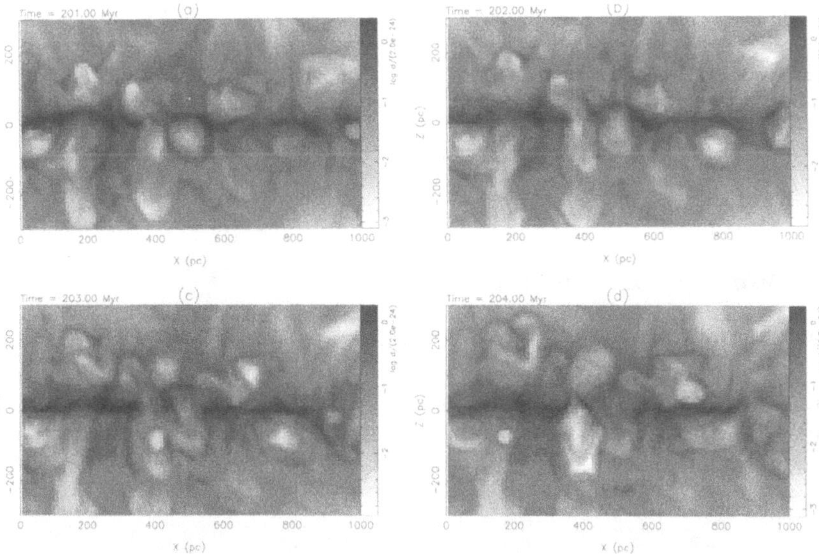

Figure 1. Gray scale maps of the gas density for y=400 pc at: (a) 201 Myr, (b) 202 Myr, (c) 203 Myr, (d) 204 Myr after start of calculations.

of it is related to supernovae that are spread over the disk, and warm it up as they occur. Embedded in this medium there are cold sheets, filaments and clouds. The cold gas is mainly confined to the thin disk which has a thickness varying between 60 pc and 100 pc and shows a wiggly structure (see Figure 1).

As the supernovae occur they change the local structure of the interstellar medium, but are unable to change the global structure. Individual and clustered supernovae dominate the local environment depending on their spatial location. The isolated supernovae dominate over the clustered ones in the inner, dense regions of the Galactic disk, whereas the clustered supernovae dominate the outer regions, which have a lower density and thus, effectively, more direct path to the rarefied space above the Galactic plane.

Supernovae located at $z \sim 100$ pc have shells with elongated appearance, that result partly because the medium is not smooth on the small scale (5 pc) and partly because the steep variation of the density on the large scale in directions perpendicular to the Galactic plane. As a consequence, and due to hydrodynamical instabilities, the shells break in the directions where there is a larger density gradient. Thus, the hot inner parts of the remnant are released into the interstellar medium in collimated channels. Effects like these are observed in Figure 1, which shows sequential evolution of the disk gas in a region where a chimney evolves. Figure 1 (c) shows the presence of a supernova with a radius of some 50 pc (the supernova is well inside phase III of its evolution) located at $x = 400$ pc, $z = -100$ pc. As the shell breaks, the hot gas in the remnant escapes into the halo through a tunnel with a width of 110 pc (Figure 1 (d)).

The vertical structure of the holes is observed in most of the images. Some are partially surrounded by thin sheets of cold gas which result from old shells. The sheets have a vertical orientation and appear to "crawl" out of the disk resembling "worms".

5. Discussion and Summary

The simulations show qualitative resemblances to the results of HI emission surveys of the Galaxy by Heiles (1990) and Koo *et al* (1992). HI gas in the Galaxy is mainly confined to a thin layer around the Galactic plane. This layer has a thickness that varies from 100 pc to 200 pc. HI emission maps also show a series of low density regions surrounded by thin sheets of cold gas having a wiggly vertical orientation (Koo *et al*, 1992).

Acknowledgments

The author would like to acknowledge the ESO/JNICT for funding this research and Wolfgang Dreyer of Quant-X, Austria, for providing access to one the machines where most of the calculations were carried out.

References

Allen C. W., 1991, *Astrophysical Quantities*, the Athalon Press, London.
Avillez M.A., 1997, *Ph.D Thesis*, University of Évora.
Avillez M.A., 1998, *Test Suit for Hydrodynamical Schemes*, *J. Fluid Mech.*, submitted.
Berger M.J. & Colella P., 1989, *J.Comput.Phys.*, **82**, 64.
Cowie L.L., Songaila A. & York D., 1979, *ApJ* **230**, 469.
Dickey J. M. & Lockman F. J., 1990, *ARA&A* **28**, 215.
Evans R., van den Bergh S. & McClure R. D., 1989, *ApJ* **345**, 752.
Heiles C., 1990, *ApJ* **354**, 483.
Kahn F.D., 1976, *A&A* **50**, 145.
Koo B.-C., Heiles C. & Reach W.T., 1992, *ApJ* **390**, 108.
Raymond J.C., Cox D.P. & Smith B.W., 1976, *ApJ* **204**, 290.
Tomisaka K. & Ikeuchi S., 1986, *PASJ*, **38**, 697.
van den Bergh S., 1990, *AJ* **99**, 843.

THE INFLUENCE OF MASS LOSS ON THE ECCENTRICITY OF DOUBLE STAR ORBITS

J. A. DOCOBO[1] , C. PRIETO[2], J. F. LING[1]
Observatorio Astronomico "Ramon Maria Aller"
Universidade de Santiago de Compostela
P.O. Box 197. Santiago de Compostela. Spain
[1]*Dpto. Matemática Aplicada. Universidade de Santiago. Spain.*
[2]*Dpto. Matemática Aplicada. Universidade de Vigo. Spain.*

Abstract. In this comunication we study the behaviour of the eccentricity of double star orbits (visual and wide spectroscopic binaries) according to simplified laws of mass loss. Applications to the systems WDS 05245S0224 - HD 35411, WDS 05387S0236 - HD 37468 and WDS 06154S0902 - HD 43362 are included.

1. Introduction

The study of the mass loss in stars is a difficult task, especially if we take into account the involvement of several parameters: mass, luminosity, spectral type, terminal velocity, radius, rotation, etc. Nevertheless, different authors have made important contributions in this field in recent years.

It is clear that the problem becomes more complex when we study a binary or perhaps multiple star.

Putting aside close binaries which may have a tidal deformation or even mass change, it is interesting to investigate how mass loss influences stellar system dynamics.

The main goal of this comunication is to emphasize different behaviours of eccentricity and to obtain mathematical data in real orbits which can be performed using distinct laws of mass loss. In some cases it can produce a secular increase, in other cases only a periodic variation.

For this reason we are reminded that in the two-body problem with slowly decreasing mass (Hadjidemetriou, 1963) the semi-major axis increases secularly.

This fact has been interpreted by various authors as a mechanism capable of producing wide binaries from close binaries (e.g. Valls-Gabaud, 1988). However eccentricity behaviour is quite different with regards to the function $m(t)$.

In 1924, J. H. Jeans (Jeans, 1924) established a law of mass loss of the type

$$\dot{m} = -\alpha m^n \tag{1}$$

Astrophysics and Space Science is the original source of publication of his article. It is recommended that this article is cited as: *Astrophysics and Space Science* **261**: 205–207, 1999.

based on the theory of A. S. Eddington, where n and α are two real positive numbers, the first with values between 1,4 and 4,4 and the second close to zero. Since then, this law has been known as the Eddington-Jeans law.

In the case $n = 2$ we are presented with one of Mestscherskys cases in which an exact solution is available (Prieto and Docobo, 1996).

As regards Eddingtong-Jeans type laws, eccentricity shows a lack of secular perturbation, which always fluctuates around its mean value.

However, if the mass loss of the double star is inversely proportional to a distance power between components, eccentricity behaviour is radically different: that is, it increases secularly.

E. L. Martin (Martin, 1934) had reached this conclusion first on proposing the law

$$\dot{m} = -\frac{\alpha m^n}{r^2} \tag{2}$$

We also demonstrated that a similar behaviour of eccentricity is produced when the power of r in (2) is any positive number although the increase in eccentricity is faster when its power is bigger.

Since 1963, J. Dommanget (Dommanget, 1963,1964,1981,1982 and 1997) has published some papers in which he states that a correlation exists between eccentricities and orbital periods (P), that is to say, on average, a bigger eccentricity corresponds to a bigger period.

J. Dommanget has also suggested that this correlation is related to a substantial mass loss in binary star components.

As m decreases P always increases, therefore it seems reasonable that eccentricity must also increase in order for the empirical correlation to be achieved. If such a correlation really exists, the binaries must lose mass according to laws in which the distance between their components plays an important role.

2. Applications

In order to put the previous comments into practice we shall consider as examples the systems WDS 05245S0224, WDS 05387S0236 and WDS 06154S0902 all of which belong to spectral types (O and B) with a big mass loss ($\sim 10^{-6}$ solar masses/year). We have also chosen them with different eccentricities: 0.43, 0.051 and 0.666 respectively.

Stellar data together with current orbital elements and their variations for these stars are given in Table 1.

In the three cases we calculate the variation of the orbital elements over a period of 500.000 years taking into account two laws with different mass loss: $\dot{m} = -10^{-6}m^2$, (Eddington-Jeans' law, with $n = 2$) and $\dot{m} = -\frac{10^{-6}m}{r^2}$ (Martin's law, with $n = 1$). In the case of the B9 spectral type star (WDS 06154S0902) we took $\alpha = 10^{-7}$ instead of 10^{-6}.

As was expected and already foreseen by J. Dommanget, variations in orbital elements and mass (for the same value of α) are much smaller when eccentricity is smaller.

In the same way for a concrete double star, variations are greater according to Martin's law. Nevertheless, when eccentricity is small there is practically no difference between the results obtained from both laws.

Table 1

Star (WDS)	a (AU)	P (Years)	e	m (Solar masses)	Law	Δa	ΔP	Δe	Δm
05245S0224	9.945	9.219	0.430	11.572	E-J	+1.732	+3.472	~ 0.0	-1.709
					M	+8.690	+17.121	+0.248	-2.245
05387S0236	93.028	155.3	0.051	33.381	E-J	+0.926	+3.106	~ 0.0	-0.329
					M	+0.932	+3.116	+0.001	-0.327
06154S0902	47.143	107.2	0.666	9.117	E-J	+0.070	+0.320	~ 0.0	-0.014
					M	+3.891	+14.636	+0.024	-0.163

For each star the line E-J represents the results obtained with Eddington-Jeans' law and M for those obtained with Martin's law.

The integration of movement equations has been carried out with Mestschersky's exact solution in the case of the Eddington-Jeans law with $n = 2$, while with Martin's law, a numerical integration was achieved using the Runge-Kutta method of eighth-order for systems of first-order ordinary differential equations.

Finally, we should point out that the system WDS 05245S0224 (ADS 4002 - HD 35411) is actually known as a triple star although the third component forms a spectroscopic system along with the primary star having aproximately an 8 day period. For this reason it has been considered to be a double star (ab, c) in this study.

References

Dommanget, J. (1963) *Ann. Obs. Royal de Belguique*, 3eme Ser.**IX**, Fas.c. 5.

Dommanget, J. (1964) *Comm. Obs. Royal de Belguique*, No. 232.

Dommanget, J. (1981) Effects of Mass Loss on Stellar Evolution *Eds. C. Chiosi and R. Stalio* 507-513.

Dommanget, J. (1982) Binary and Multiple Stars as Tracers of Stellar Evolution *Eds Z. Kopal and J. Rahe* 3eme Ser.**IX**, Fas.c. 5

Dommanget, J. (1997) Visual Double Stars: Formation, Dynamics and Evolutionary Tracks. *Eds J. A. Docobo, A. Elipe and H. A. McAlister* ASSL, vol 223, 403. KAP.

Martin, E. L. (1934) *Reale Stazione Astron e Geof. di Carloforte (Cagliari)* No. 30.

Prieto, C. and Docobo, J. A. (1996) *Astron. Astrophys. Suppl. Ser.* **318**, 657-661

Prieto, C. and Docobo, J. A. (1997) Visual Binary Stars: Formation, Dynamics and Evolutionary Tracks *Eds. J. A. Docobo, A. Elipe and H. A. McAlister*, ASSL, Vol 223, 331. KAP.

Valls-Gabaud, D. (1988) *Astrophys and Space Science* **142**, 289-304.

CO COCKTAILS: STIRRED... BUT NOT SHAKEN

T.C. TEIXEIRA
Queen Mary & Westfield College, Mile End Road
London E1 4NS, England

Institute of Physics and Astronomy, Aarhus University,
Ny Munkegade, 8000 Aarhus C, Denmark

J.P. EMERSON
Queen Mary & Westfield College, Mile End Road
London E1 4NS, England

AND

M.E. PALUMBO
Osservatorio Astrofisico di Catania, V.le A. Doria 6
I-95125 Catania, Italy

Dust grains in molecular clouds are covered by ice mantles composed of complex "cocktail mixtures" of molecules frozen onto the grains (e.g. Schutte 1996). Solid CO in the ice mantles is identified by an absorption feature at 4.67 μm, corresponding to the stretch of the CO molecule. The observed features generally show two components: a narrow dominant one corresponding to CO in a nonpolar or weakly polar mixture, and a broader one appearing as a long-wavelength wing corresponding to CO in a polar mixture (Tielens et al. 1991, Kerr et al. 1993, Chiar et al. 1995, Teixeira et al. 1998). The comparison of the observed CO–ice bands with laboratory astrophysical ice analogs might be used as a probe of the astrophysical conditions in the coolest regions of the cloud.

In this work, observations of the solid-CO feature at 4.67μm (2140 cm^{-1}) were presented towards low mass YSOs in nearby molecular clouds, an example of which can be found in Figure 1. A detailed study of the feature by comparison of the observed spectra with the spectra of laboratory astrophysical ice analogs shows that ion irradiated ices produce excellent fits to the observations. In particular the nonpolar component of the band seems to be well constrained within the (limited) laboratory sample of CO-ices used in this work (Teixeira et al. 1998 and references therein), and cor-

Astrophysics and Space Science is the original source of publication of his article. It is recommended that this article is cited as: *Astrophysics and Space Science* **261**: 209–210, 1999.

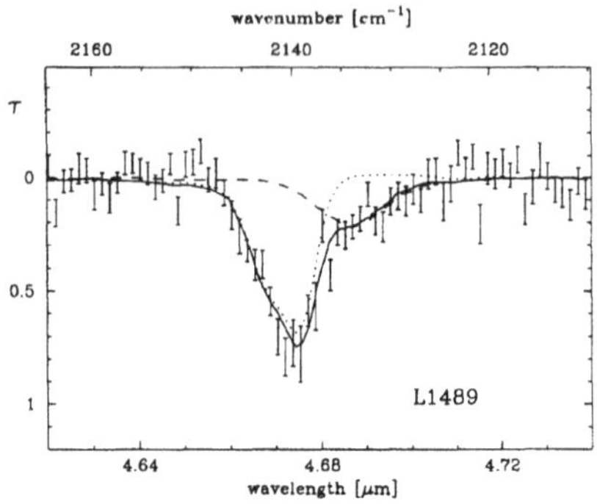

Figure 1. Solid CO spectrum of the Class I source L1489 in Taurus, in optical depth units. The solid line is the best-matching fit using laboratory spectra, which is the sum of the nonpolar component (dotted line) and the polar component (dashed line).

responds to irradiated pure CO for all the objects with a significant CO feature. From the statistical point of view, it is not possible to constrain the composition of the polar component of the band, although irradiated and warmed-up ices seem to be preferred.

The origin of the ion irradiation is not clear. One possibility is that it may be due to high energy particles accelerated in flares, CME events and winds from the embedded YSO. If the clouds are clumpy enough to allow the penetration of a sufficient flux of low energy cosmic-rays, X-rays or UV radiation deep into the cloud, then the structural changes of the icy mantles may be due to a combination of cosmic-ray bombardment and UV and X-ray irradiation.

In view of the goodness of the fits obtained in this work, the results are meant to be a *possibility* for the composition of the CO ices in the grains, although other possibilities should *not* be excluded. The problem of the composition of the ices will be very hard to settle definitely.

TCT supported by grants CIENCIA/BD/2095/92-RM and PRAXIS XXI/ BD/5622/95 from JNICT, Portugal.

References

Chiar J.E., Adamson A.J., Kerr T.H., Whittet, D.C.B., (1995), *ApJ*, **455**, 234

Kerr T.H., Adamson A.J., Whittet D.C.B., (1993), *MNRAS*, **262**, 1047

Schutte W.A., (1996), in *The Cosmic Dust Connection*, ed. J.M. Greenberg, Kluwer Academic Publishers, 1

Teixeira T.C., Emerson J.P., Palumbo M.E., (1998), *A&A*, **330**, 711

Tielens A.G.G.M., Tokunaga A.T., Geballe T.R., Baas F., (1991), *ApJ*, **381**, 181

THE UNIVERSITY OF PORTO RADIO SPECTROGRAPH

ANTÓNIO MAGALHÃES AND JORGE CARNEIRO
Observatório Astronómico Prof. Manuel de Barros
Monte da Virgem - 4430 V.N.Gaia - Portugal

The microcomputerized Solar Radio Spectrograph (RSM) of the University of Porto is situated 70 km away from Porto by road, at Espiunca ($\phi = 40^o$ 59' 57" N, $\lambda = 8^o$ 13' 52" W), a site which was carefully selected for general electrical inactivity. Since May 1997 there have been daily observations interrupted only by power failures due to lightning strokes.

As in any radio telescope the RSM comprises three main parts: the antenna, the receiver and the data acquisition and recording system. The antenna is a single equatorially mounted paraboloid dish having 5 meter diameter and a 0.43 f/D ratio. A single polarization log-periodic dipole array, designed to cover 100 to 1000 MHz illuminates the dish. Polarisation information is not of main concern since priority is given to fast events, namely type III radio bursts, which originate in weak, open magnetic field–line structures. Unfortunately it is necessary to provide a low pass filter between the antenna and the low noise pre-amplifier to avoid overloading by strong interfering signals. Fifteen meters of coaxial cable bring the signal into the shelter where a commercial spectrum analyser is used as a receiver.

This instrument sweeps 150 to 650 MHz, eighteen times per second, with one megahertz resolution. The choice of this band is a natural consequence of several needs and constraints. In plasma hypothesis and quiet corona, 650 MHz corresponds roughly to an altitude of 0.85, and 150 MHz to 1.05 solar radii above the photosphere, which is an interesting region on the solar atmosphere. Furthermore, since the antenna aperture is small, its gain would be very poor at lower frequencies, a region strongly plagued by radio frequency interference.

The data at the completion of each sweep of the operating band is an almost instantaneous power spectrum of the solar radiation. By gathering consecutive spectra a dynamic spectrum is built up. Broadcasting and TV transmitters, which work on single frequencies, appear on the spectra as sharply defined horizontal lines.

Astrophysics and Space Science is the original source of publication of his article. It is recommended that this article is cited as: *Astrophysics and Space Science* **261**: 211–212, 1999.

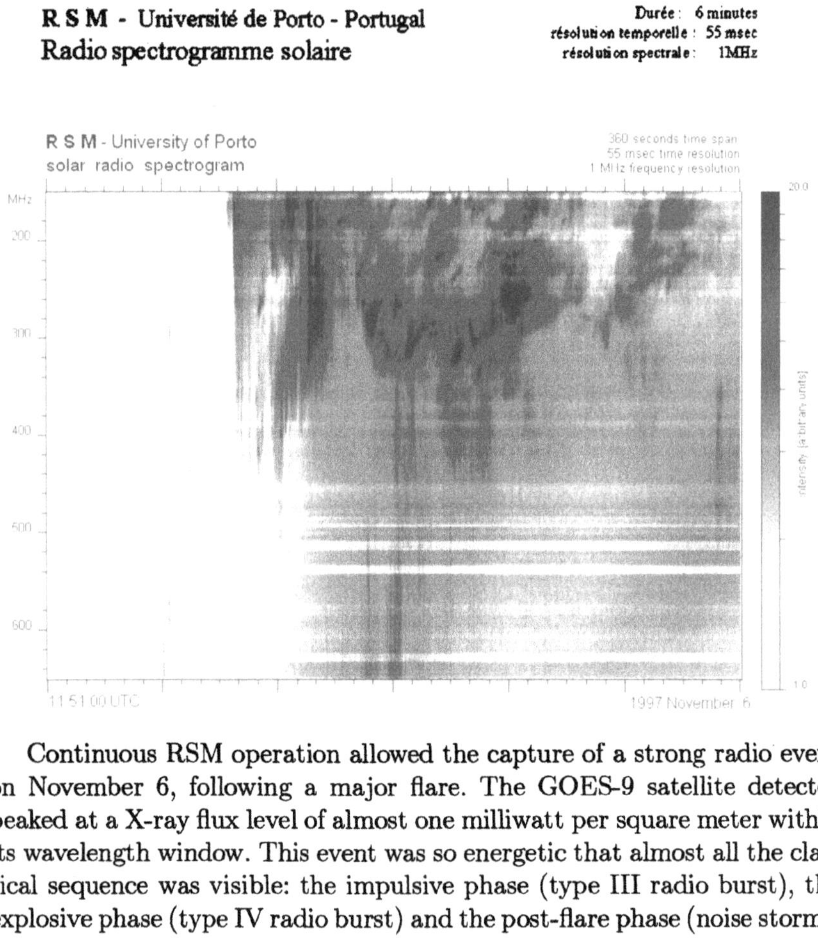

Continuous RSM operation allowed the capture of a strong radio event on November 6, following a major flare. The GOES-9 satellite detector peaked at a X-ray flux level of almost one milliwatt per square meter within its wavelength window. This event was so energetic that almost all the classical sequence was visible: the impulsive phase (type III radio burst), the explosive phase (type IV radio burst) and the post-flare phase (noise storm).

The University of Porto is privileged in having their Solar Radio Spectrograph located in the most western and quiet solar radio observing site in Europe, which is now beginning to prove its reliability in filling the Atlantic gap and naturally contributing to the world wide solar observation network. This justifies the high time resolution solar monitoring which provides valuable data not only to the direct collaborators but also to the scientific community in general.

Some improvements are programmed for the near future, in particular an increase in sensitivity.

AFTER DINNER POSTER PAPER

IAN W ROXBURGH
Astronomy Unit, Queen Mary and Westfield College,
University of London, London E1 4NS, UK.

Neste lugar essencialmente Inglês, penso que é aceitável falar Inglês em vez de Português

I wish on this occasion to make three points:
1. To congratulate Teresa Lago on the new Institute and Planetarium
2. To thank all who have contributed to the success of the meeting
3. To give yet one more poster paper!

As a child I had learned of Portugal as that country whose explorers and navigators always seemed to get there before the Brits!. Then I learned of it as the place from which port came (we call it port in English not port wine). Port had a special place in the drinking habits of ordinary working English people at a time when wine was unknown. It was what you had at Christmas and offered to visiting Aunts and Uncles, and it was drunk by "ladies" in pubs as "Port and Lemon", a mixture of port with English lemonade. The Portuguese may wince at this, but I suspect that the decline in the British consumption of Port that we heard of earlier, could probably be explained by the decline in popularity of this mixture in the 1960's, when the ordinary Brits first started to go on holidays "abroad", mostly to Spain, and learned about wine.

Another feature connecting Britain and Portugal (our oldest ally), is the perfection of bureaucratic structures in their old empires. This bureaucratic "excellence", which seems to have been a feature of all the old European empires, now seems have found a modern outlet in the European Union administration in Brussels. Teresa Lago is famed for her ability to get money out of the EU, so she probably has more experience than most of us of dealing with the uncontrolled Brussel's bureaucracy, and might therefore appreciate my after dinner poster paper - which is a purported letter from the Duke of Wellington to the bureacrats in London, during the time he

Astrophysics and Space Science is the original source of publication of his article. It is recommended that this article is cited as: *Astrophysics and Space Science* **261**: 213–214, 1999.

was in Portugal and Spain campaigning against the armies of Napoleon. With due apologies to my French colleagues, the letter is as follows:

Gentlemen:

Whilst marching to Portugal to a position which commands the approach to Madrid and the French forces, my officers have been diligently complying with your requests which have been sent by H.M.ship from London to Lisbon and then by dispatch rider to our headquarters.

We have enumerated our saddles, bridles, tents and tent poles, and all manner of sundry items for which His Majesty's Government holds me accountable. I have dispatched reports on the character, wit and spleen of every officer. Each item and every farthing has been accounted for, with two regrettable exceptions for which I beg your indulgence.

Unfortunately the sum of one shilling and ninepence remains unaccounted for in one infantry battalion's petty cash and there has been a hideous confusion as to the number of jars of raspberry jam issued to one cavalry regiment during a sandstorm in western Spain. This reprehensive carelessness may be related to the pressure of circumstance since we are at war with France, a fact which may come as a bit of a surprise to you gentlemen in Whitehall.

This brings me to my present purpose, which is to request elucidation of my instructions from His Majesty's Government, so that I may better understand why I am dragging an army over these barren plains. I construe that perforce it must be one of two alternative duties, as given below. I shall pursue either one with the best of my ability but I cannot do both:

1. To train an army of uniformed British clerks in Spain for the benefit of the accountant and copy boys in London, or perchance,

2. To see to it that the forces of Napoleon are driven out of Spain.

Conclusion

Finally I would like again to convey our congratulations to Teresa on the establishment of the new Institute and Planetarium. I would dearly like to have a similar facility in the East End of London! On behalf of all the participants I convey our thanks to Teresa, the local organising committee, the local staff, and to those whose generous support made the meeting possible.

GAMMA-RAY BURSTS: WHERE ARE WE NOW?

CHARLES MEEGAN
NASA/Marshall Space Flight Center
Huntsville AL 35812

Abstract. There has been significant progress recently in our understanding of gamma-ray bursts. The long-sought counterparts at other wavelengths have finally been found for a few bursts. This breakthrough is the result of coordinated observations involving several satellites and ground-based optical and radio observatories. In one case, GRB970508, redshifted absorption lines have been detected, finally settling the debate about the distance scale. The consensus is that the burst sources lie at cosmological distances, requiring at least $\sim 10^{51}$ ergs to be emitted in gamma rays in just a few seconds. The gamma radiation is thought to be produced by shocks in a highly relativistic fireball. Many mysteries remain. There is no consensus on the nature of the sources, although coalescing neutron stars are the leading candidate. There is evidence that the sources of the faintest bursts may be at redshifts above 2. If so, gamma-ray bursts may ultimately tell us something about the early Universe.

1. Introduction

Since their discovery by Klebesadel, Strong, & Olson (1973), gamma-ray bursts (GRBs) have remained one of the most puzzling phenomena in astrophysics. We have only recently made progress in determining the distance scale, and the source objects and physical mechanisms are still poorly understood. This review will describe our current understanding of gamma-ray bursts, including some discussion of the interesting history of the evolution of our thinking.

I distinguish three eras in the history of gamma-ray burst studies. In the "Disk Era", GRBs were generally considered to arise from nearby galactic neutron stars. Many theoretical models were developed, but none worked well enough to become the accepted paradigm. The state of the field at

Astrophysics and Space Science is the original source of publication of his article. It is recommended that this article is cited as: *Astrophysics and Space Science* **261**: 215–224, 1999.

this stage is reviewed by Higdon & Lingenfelter (1990). The "Great Debate Era" began in late 1991 when the Burst and Transient Source Experiment (BATSE) established that the burst sources are isotropic but inhomogeneous, effectively falsifying the current paradigm of a galactic disk population of sources. Consequently, two possibilities for the distance scale emerged: either bursts arose from sources in an extended Galactic halo (EGH model) or at cosmological distances. In spite of the huge difference in these distances, corresponding to about nine orders of magnitude for the required energy, evidence could be marshalled for both scenarios. The status of research at this stage is reviewed by Fishman & Meegan (1995). The current era, which I somewhat optimistically call the "Counterparts Era", began with the detection of an optical transient associated with the burst of 28 February, 1997. With the detection of counterparts, a wealth of new information has became available.

In Sections 2 and 3, the temporal and spectral properties of gamma-ray bursts are summarized. In Section 4, the BATSE results that led to the Great Debate Era are reviewed. In Section 5, the Counterparts Era is covered, with particular emphasis on GRB970228 and GRB970508. Section 6 presents our current theoretical understanding of GRBs, including areas of controversy. In Section 7, future observations are considered.

2. Temporal Structure

One of the most striking characteristics of bursts is the wide range of temporal behavior. Bursts can last from less than 0.1 seconds to over 1000 seconds. Their time histories may be relatively smooth or extremely complex and chaotic. Figure 1 shows some examples from the first BATSE catalog (Fishman *et al.*, 1994).

The distribution of burst durations is bimodal, with peaks at ~ 0.5 s and ~ 30 s (Kouveliotou *et al.* 1993). The shorter bursts have, on average, harder spectra. There appear to be no differences, however, in the spatial distributions of the two classes.

3. Energy Spectra

The spectral properties of bursts are not nearly as varied as the temporal properties. A typical burst spectrum shows a peak in νF_ν at a few hundred keV. Most burst spectra are well described by a four-parameter function (Band *et al.* 1993), consisting of two power laws smoothly joined by an exponential at a break energy. The indices of the power laws are typically 0 to -1.5 below the break energy and -2 to -2.5 above the break energy. Break energies typically range from 100 keV to 1 MeV.

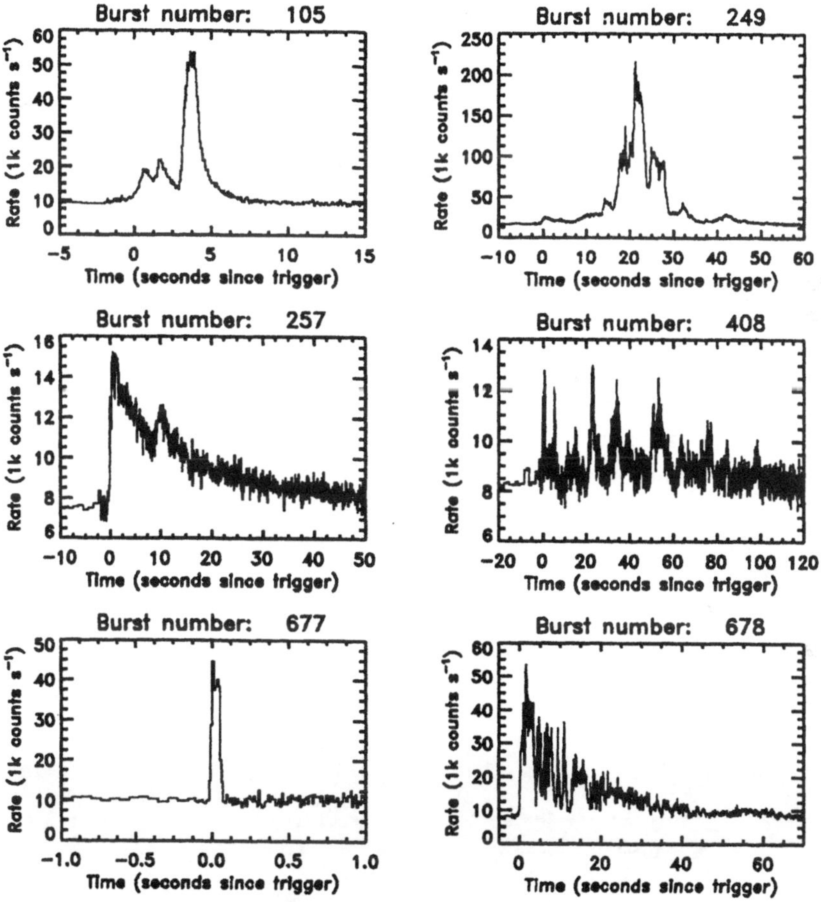

Figure 1. A sample of BATSE bursts showing a wide range of temporal behavior

Spectral features have been seen in data from several instruments, most notably the absorption features at 20 and 40 keV from *Ginga* (Murakami *et al.* 1988). These have been interpreted as cyclotron absorption lines in 1.7×10^{12} G fields, typical of neutron stars (e.g. Fenimore *et al.* 1988). The cyclotron lines were widely considered to be conclusive evidence that bursts are produced by neutron stars. Briggs *et al.* (1998) have searched the BATSE data for similar features. Out of 133,000 spectra from 117 strong bursts, 12 interesting spectral features have been found, but none can be unambiguously identified as a narrow line. Band *et al.* (1997) investigated the BATSE sensitivity to lines similar to those seen by *Ginga*. The conclusion is that there is some discrepancy between the *Ginga* and BATSE results, but that the statistical significance is not high enough to cast doubt on either instrument.

4. The Great Debate

The accepted paradigm of galactic neutron stars as sources of GRBs was dealt a serious blow with the early data from BATSE (Meegan *et al.* 1992). The intensity distribution showed a deficit of weak bursts, indicating a decrease in source density with distance. However, the angular distribution was isotropic to within the statistical uncertainty. The combination of inhomogeneity and isotropy ruled out disk geometries.

Figure 2 shows the angular distribution, in galactic coordinates, of 1637 bursts observed with BATSE. The galactic dipole and quadrupole moments, corrected for anisotropy of the sky exposure, are -0.014 ± 0.014 and 0.004 ± 0.007 respectively, consistent with isotropy.

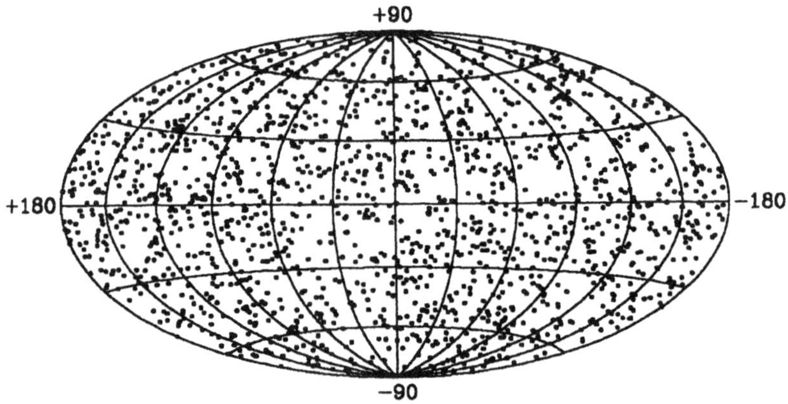

Figure 2. The angular distribution in galactic coordinates of 1637 BATSE bursts

The GRB intensity distribution is shown in Figure 3. The integral number of bursts is plotted against their peak flux (averaged over 1.024 s), and is corrected for detection efficiency, but not for atmospheric scattering. The dashed line represents the $-3/2$ power law expected for a spatially homogeneous distribution in Euclidean space. The data show a statistically convincing deviation from the $-3/2$ law. The V/V_{max} statistic (Schmidt, Higdon & Heuter 1988) is 0.33 ± 0.01, indicating that the inhomogeneity is significant and not an artifact of trigger threshold variations.

The combination of isotropy and inhomogeneity observed with BATSE is not compatible with a distribution of sources in a flattened disk of any

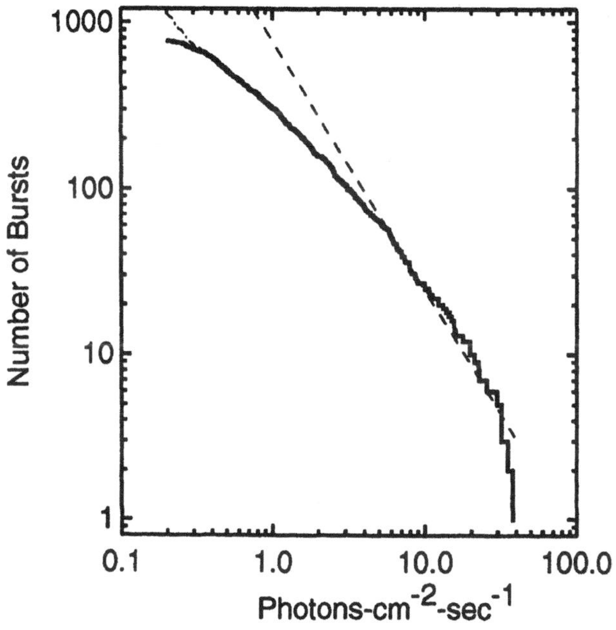

Figure 3. The integral intensity distribution for BATSE bursts that triggered on the 1024 ms time scale.

size. It appears that we are near the center of a spherically symmetric bounded distribution of sources, but with an undetermined distance scale. Thus, the BATSE observations inaugurated the era of the Great Debate between the "Galactophiles" and the "Cosmophiles".

In the Cosmological model, the isotropy arises naturally, and the deviation from the $-3/2$ power law is due to redshift effects. Several authors (e.g., Wickramasinghe *et al.* 1993, Fenimore *et al.* 1993) have fitted cosmological models to the brightness distribution, typically obtaining redshifts of about 1 for the faintest observed bursts, and peak luminosities of $\sim 10^{51}$ ergs/s. The best evidence for the cosmological viewpoint was provided by the isotropy, which became more compelling as the statistical sample increased. Supporting evidence was provided by indications of time dilation in the durations of bursts (e.g., Norris *et al.* 1994) and redshift in the distribution of spectral peak energies (Mallozzi *et al.* 1995).

Those who held for a galactic neutron star origin cited the earlier observations of cyclotron absorption lines, and hints of multiple bursts from a single source (Quashnock & Lamb 1993; Wang & Lingenfelter 1993). In the

EGH model, high-velocity neutron stars were considered to be the sources of GRBs. To reproduce the isotropy, the size of the halo must be significantly larger than the distance from the Sun to the Galactic Center.

The defining event of this era was the debate in 1995 between Bohdan Paczyński, representing the cosmological camp, and Don Lamb, representing the galactic camp, commemorating the 75th anniversary of the Shapley-Curtis debate (Nemiroff 1995).

5. Counterparts

Much effort has been devoted to finding burst counterparts at other wavelengths. Possible counterparts are classified according to the time since the gamma-ray event. Flaring counterparts are those that occur simultaneously with the gamma rays, fading counterparts would be detectable for some limited time after the event, and quiescent counterparts would be continuously detectable. The searches for counterparts have ranged across the electromagnetic spectrum. The successes have been few, but are extremely significant. In particular, fading X-ray and optical counterparts have been associated with GRB970228 and GRB970508, described in the next two subsections.

5.1. GRB970228

GRB970228 provided the first convincing case for a fading counterpart to a GRB. This burst triggered the BeppoSAX Gamma-Ray Burst Monitor and was detected by the Wide Field Camera (Costa *et al.* 1997a). Within eight hours, BeppoSax was reoriented to observe with the narrow field instruments. A fading X-ray source was detected (Costa *et al.* 1997b). Optical observations of the field revealed a source at R $\sim$ 21 on Feb. 28 that faded to R $>$ 23.3 by March 4 (van Paradijs *et al.* 1997). An extended source was later found at the location of the transient. Two subsequent observations were made with the Hubble Space Telescope (Sahu *et al.* 1997; Fruchter *et al.* 1998). The most recent HST observation was made on 4 September 1997, and found the point source at magnitude $V = 28.0$, and fading approximately as $t^{-1.14}$. It lies near the edge of a nebula of magnitude $V = 25.7$ that appears to be a moderate redshift galaxy.

5.2. GRB970508

Quick response by the BeppoSAX team (Costa *et al.* 1997c) resulted in the identification of X-ray and optical counterparts for GRB970508. Bond (1997) found a brightening optical source within the 3 arcminute error box over the next day. Subsequent observations by a number of groups established a peak at about R=19.8 two days after the GRB, followed by a decay of about 2 magnitudes over the next 10 days. The most noteworthy observation of this transient was the discovery by Metzger *et al.* (1997) of absorption lines at a redshift z=0.835, which represents a minimum distance to the source. The absence of a Lyman decrement provides a maximum redshift of ~ 2.2.

GRB970508 has the distinction of being the only GRB detected in radio. Frail *et al.* (1997) found a flaring radio source coincident with the optical emission. The radio flux exhibited early rapid fluctuations that dampened after a few weeks. This was interpreted as interstellar scintillation from an expanding fireball.

6. Current Models

This section presents a review of our current theoretical understanding of gamma-ray bursts, and most directly addresses the question presented in the title of this paper.

6.1. THE DISTANCE SCALE

The observation of a host galaxy for GRB970228 and redshifted absorption lines in the spectrum of GRB970508 are now widely considered to have settled the issue of the distance scale. Although this conclusion strictly applies to only two events, it is clear that Nature has found a way to make gamma-ray bursts at cosmological distances, thus removing the theoretical objections for all the other bursts. While multiple-population models may inevitably be entertained, the isotropy of over 2000 events indicates that the vast majority, and perhaps all, of the gamma-ray bursts are cosmological.

6.2. FIREBALLS AND THE AFTERGLOW

A successful model of cosmological GRBs must satisfy the contradictory requirements of short time scale, large energy release, and non-thermal gamma-ray spectrum. If 10^{51} ergs is released in a region of a few light-seconds radius, by whatever means, a thermal pair plasma called a "fireball" is created. The fireball is initially opaque to gamma rays. It expands and cools, eventually radiating a thermal spectrum. The problem has been how to get a hard non-thermal spectrum, with rapid time variability, out of such a fireball. The current thinking is that the fireball is highly relativistic, with Lorentz factors of order 10^3, and that kinetic energy is converted to non-thermal gamma radiation via shocks (Mészáros & Rees 1993). It is thought that numerous internal shocks in the expanding fireball could produce the complex time histories (Rees & Mészáros 1994, Sari & Piran

1997). External shocks with an interstellar medium could produce simple bursts with one or a few peaks, or might be responsible for only the fading counterparts – the so-called "afterglow". A recurring difficulty with this scenario is that there must be a very low contamination of the fireball by baryons, or the expansion will not attain the required high Lorentz factors. A number of authors (e.g. Wijers, Rees & Mészáros 1997; Waxman 1997; Vietri 1997) have pointed out that the temporal behavior of the fading counterparts observed from GRB970228 and GRB970508 are in agreement with the predictions of the fireball model.

6.3. PROGENITORS

The nature of the ultimate source of the energy is still unknown. Many possibilities, some quite exotic, have been suggested as sources for cosmological GRBs. Only a few of the top contenders will be mentioned. Most attention has focused on coalescing neutron stars (Narayan, Paczyński & Piran, 1992). Such events must occur, they provide about the right amount of energy, and they have about the right rate if the faintest bursts are at z ~ 1. A similar model is a neutron star – black hole merger (Paczyński 1991). Usov (1992) proposed accretion-induced collapse of a highly magnetic white dwarf. Woosley (1993) suggested the collapse to a black hole of a rapidly rotating Wolf-Rayet star, which he terms a "failed supernovae". Paczyński (1998) proposed a similar model, which he refers to as a "hypernova".

All of the models rely on one of two possibilities for driving the fireball, neutrino annihilation and Poynting flux. High production of neutrinos and antineutrons is expected in some models, such as neutron star mergers. The neutrino pairs must annihilate in a region of low density to prevent "quenching" of the fireball by baryon contamination. The efficiency of this annihilation process is problematic. Magnetically driven fireballs are postulated for other models, including the white dwarf collapse model of Usov and the hypernova model of Paczyński. The main hurdle is that very high magnetic fields ($\sim 10^{15}$ G) are required.

Hartmann (1996) provides a recent, but pre-counterparts era, review of theoretical models of gamma-ray bursts.

6.4. HOST GALAXIES

There has been a long-standing controversy as to whether there are suitable candidate galaxies in the error boxes of GRBs – the "no host galaxy" problem (e.g. Schaefer *et al.* 1997). In the case of the optically identified bursts, GRB970228 is superimposed on a very faint galaxy (van Paradijs *et al.* 1997), while no galaxy is seen at the position of GRB970508 (Pian *et al.* 1998). It appears that GRBs do not trace the distribution of stellar mass. One interesting solution to this conundrum is that GRBs may be considerably more distant even than redshifts of ~ 1. It is reasonable to expect that GRBs arise from compact objects and thus trace the star formation rate. Recent measurements indicate that the star formation rate increases rapidly back to $z \sim 1$ (Madau *et al.* 1996). When such evolution is taken into account in fitting the brightness distribution, the distance to the faintest bursts is significantly greater (Totani 1997, Wijers 1998). The energy requirement then grows accordingly, perhaps ruling out coalescing neutron stars. If bursts are actually at such great distances, they may provide new information on the early star formation rate.

7. Future Work

The discovery of optical counterparts to GRB970228 and GRB970508 has re-energized efforts to obtain burst counterparts. BeppoSax will continue to provide fast and accurate burst locations. The launch of HETE-2 in 1999 will produce even more. BATSE burst locations are rapidly disseminated to the astronomical community. The prospects for optical detections continue to improve, as a number of rapidly responding telescope systems are in operation or under development. There is also optimism that gamma-ray burst instrumentation will be selected for a future SMEX or MIDEX mission. After many years of frustration and false leads, we are becoming more optimistic that the gamma-ray burst riddle will finally be solved.

References

Band, D. *et al.*: 1993, *ApJ* **413**, 281.
Band, D. *et al.*: 1997, *ApJ* **485**, 747.
Bond, H.: 1997, IAU Circular 6654.
Briggs, M. *et al.*: 1998, in *Gamma-Ray Bursts: 4th Huntsville Symposium*, C. Meegan, R. Preece & T. Koshut, eds., (New York: AIP), in press.
Costa, E. *et al.*: 1997a, IAU Circular 6572.
Costa, E. *et al.*: 1997b, IAU Circular 6576.
Costa, E. *et al.*: 1997c, IAU Circular 6649.
Fenimore, E. *et al.*: 1988, *ApJ* **335**, L71.
Fenimore, E. *et al.*: 1993, *Nature* **366**, 40.
Fishman, G. *et al.*: 1994, *ApJS* **92**, 229.
Fishman, G. and Meegan, C.: 1995, *Ann. Rev. Astron. Astrophys.* **33**, 415.
Frail, D. *et al.*: 1997, *Nature* **389**, 261.
Fruchter, A. *et al.*: 1998, in *Gamma-Ray Bursts: 4th Huntsville Symposium*, C. Meegan, R. Preece & T. Koshut, eds., (New York: AIP), in press.
Hartmann, D.: 1996, *Astron. Astrophys. Suppl. Ser.* **120**, 31.
Higdon, J. and Lingenfelter, R.: 1990, *Ann. Rev. Astron. Astrophys.* **28**, 401.
Klebesadel, R., Strong, I. and Olson, R.: 1973, *ApJ* **182**, L85.
Kouveliotou, C. *et al.*: 1993, *ApJ* **413**, L101.
Madau, P. *et al.*: 1996, *MNRAS* **283**, 1388.
Mallozzi, R. *et al.*: 1995, *ApJ* **454**, 597.
Meegan, C. *et al.*: 1992, *Nature* **355**, 143.

Mészáros, P. and Rees, M.: 1993, *ApJ* **405**, 278.
Metzger, M. *et al.*: 1997, *Nature* **387**, 879.
Murakami, T. *et al.*: 1988, *Nature* **335**, 234.
Narayan, R., Paczynski, B. and Piran, T.: 1992, *ApJ* **395**, L85.
Nemiroff, R.: 1995, in *Gamma-Ray Bursts: 3rd Huntsville Symposium*, C. Kouveliotou, M. Briggs & G. Fishman, eds., AIP Conference Proceedings **384** (New York: AIP), p. 973.
Norris, J. *et al.*: 1994, *ApJ* **424**, 540.
Paczyński, B.: 1991, *Acta Astr.* **41**, 257
Paczyński, B.: 1998, in *Gamma-Ray Bursts: 4th Huntsville Symposium*, C. Meegan, R. Preece & T. Koshut, eds., (New York: AIP), in press.
Pian, E. *et al.*: 1998, *ApJ* **492**, L103.
Quashnock, J. and Lamb, D.: 1993, *MNRAS* **265**, L59.
Rees, M. and Mészáros, P.: 1994, *ApJ* **430**, L93.
Sahu, K. *et al.*: 1997, *Nature* **387**, 476.
Sari, R. and Piran, T.: 1997, *ApJ* **485**, 270.
Schaefer, B., Cline, T., Hurley K. and Laros, J.: 1997, *ApJ* **489**, 693.
Schmidt, M., Higdon, J. and Heuter, G.: 1988, *ApJ* **329**, L85.
Totani, T.: 1997, *ApJ* **486**, L71.
Usov, V.: 1992, *Nature* **357**, 472.
van Paradijs, J. *et al.*: 1997, *Nature* **386**, 686.
Vietri, M.: 1997, *ApJ* **488**, L105.
Wang, V. and Lingenfelter, R.: 1993, *ApJ* **416**, L13.
Waxman, E.: 1997, *ApJ* **485**, L5.
Wickramasinghe, W. *et al.*: 1993, *ApJ* **411**, L55.
Wijers, R.A.M.J.: 1998, in *Gamma-Ray Bursts: 4th Huntsville Symposium*, C. Meegan, R. Preece & T. Koshut, eds., (New York: AIP), in press.
Wijers, R.A.M.J., Rees, M. and Mészáros: 1997, *MNRAS* **288**, L51.
Woosley, S.: 1993, *ApJ* **405**, 273.

AXION - DILATON COUPLING AND GAMMA - RAY BURSTS

O. BERTOLAMI
Instituto Superior Técnico, Departamento de Física,
Av. Rovisco Pais, 1096 Lisboa Codex, Portugal

Abstract

Axions emitted in supernovae are interesting candidates to account for Gamma-Ray Bursts provided their energy can be effectively converted into electromagnetic energy elsewhere. We point out that this convertion may occur more efficiently if one considers the coupling between intermediate scale axions and the string theory dilaton along with the inclusion of string loops.

1. Introduction

Gamma-Ray Bursts are among the most striking astrophysical events of the century. Ever since their discovery in the late sixties [1] these bursts have been the subject of great interest and debate. There has been more than 2600 bursts recorded, about 1900 of which by the Burst and Transient Source Experiment (BATSE) at Nasa's Compton Gamma-Ray Observatory [2] and till February 1997 no observation of bursts counterparts in any other wavelength had ever been recorded. Data from BATSE clearly indicate that the distribution of bursts is isotropic around us although not homogeneous in a 3-dimensional Euclidean space [3]. This last feature, which arises from the lack of low intensity sources is, of course, a strong evidence for the cosmological nature of the bursts. This suspicion has been recently confirmed by the observations of the so-called burst afterglows in the X-Ray region by BeppoSAX [4], a dutch-italian satellite, and in the optical [5] for the burst of February 28th 1997 (GRB 970228), and also in radio for the burst GRB 970508 [6] as they allowed the determination of the afterglow's red-shift ($z = 0.835$ for GRB 970508 [7]). Models for explaining observations require either quite drastic phenomena involving compact objects and ultra relativistic motion of the source or new physics. A prototype example of the former is Paczyński's hypernovae model which considers a 10 $M_{\odot}$

Astrophysics and Space Science is the original source of publication of his article. It is recommended that this article is cited as: *Astrophysics and Space Science* **261**: 225–228, 1999.

Kerr black hole and the mechanism of Blandford-Znajek in order to obtain the large magnetic fieds, $B \sim 10^{15}\ G$, required to match the observed energy of the bursts [8]. An interesting suggestion involving new physics assumes energy scapes from supernovae via the emission of axions and is converted into electromagnetic energy elsewhere [9]. This proposal solves the so-called compactness problem related with the quick thermalization of the electromagnetic energy one should expect in optically thick media due to Compton scattering with $e^{+}e-$ pairs [10]. This is the typical situation one encounters in merging scenarios to explain gamma-ray bursts. To this problem one should add that, in case the electromagnetic energy is contaminated by more than $10^{-5}\ M_{\odot}$ of baryons, it is shown that the photon energy tends to degrade to lower energies implying that the burst duration will be stretched beyond the observation limits [11]. Neutron star merging scenarios involve, for instance, about $10^{-2}\ M_{\odot}$ of baryonic debris [12]. Of course, these difficulties can be, at least in principle, overcome assuming the source of the bursts is accelerated to the ultra relativistic regime as is achieved in fireball models (see [13] for a recent review). In here, however, we consider the proposal of Ref. [9] in the context of string theory. We stress that it is quite natural to consider astrophysical and cosmological phenomena involving axions as these particles stand out as the most prominent candidates for the dark matter of the Universe in the much observationally favoured Cold Dark Matter Models.

2. Axion-Dilaton Coupling

Axions arise in particle physics to explain the strong CP problem via the Peccei-Quinn mechanism [14] (see eg. Ref. [15] for a review). Axions correspond to pseudogoldstone bosons associated with the spontaneous breaking of an anomalous chiral $U(1)_{PQ}$ symmetry with vacuum expectation value (vev) f_{PQ}. Axions acquire a mass as QCD instantons break $U(1)_{PQ}$ and $m_a = \frac{f_\pi\ m_\pi}{f_{PQ}}$, where $f_\pi = 93\ MeV$ and $m_\pi = 134.5\ MeV$. For standard axions, bounds on the vev, f_{PQ}, can be obtained from astrophysical and cosmological considerations [15]:

$$10^9\ GeV < f_{PQ} < 10^{12}\ GeV \quad , \tag{1}$$

where the lower bound arises from the demand that axions do not lead to a quick cooling of stars while the upper bound comes from requiring that the classical coherent oscillations of axions do not dominate the Universe dynamics at present [16]. In the context of string theory, however, the latter bound may not hold as moduli fields which are assumed to have masses comparable to the gravitino are therefore substantially heavier than the axions, meaning that their coherent oscillations may dominate the Universe dynamics quite early on [18]. This may imply the axion can be much more "invisible" than usually assumed and that axions arising from Grand Unified Theories (GUTs) or supersymmetric GUTs via the breaking of R-type symmetries are feasible. The possibility that f_{PQ} is greater than the intermediate scale $10^{12}\ GeV$ implies, of course, that axions are much less interesting candidates for solving the compactness problem of gamma-ray bursts. However, this conclusion may be altered if one considers the effect of string loops and assumes these can stabilize the dilaton in the absence of a potential [19]. This is relevant as moduli fields, including the dilaton, are shown to remain massles in all orders in string perturbation theory. Indeed, after assuming that the dilaton coupling is universal, the dynamics of the axion-dilaton-gravity system is described by the following lowest order bosonic action [19, 20]:

$$S_B = \int d^4x\sqrt{-g}\left\{-\frac{R}{2k^2} + 2(\partial\phi)^2 + \frac{c}{4}\ B(\phi)^3\ \frac{a}{f_{PQ}} Tr\left(F_{\mu\nu}\tilde{F}^{\mu\nu}\right) + ...\right\} \quad , \tag{2}$$

where $k^2 = 8\pi M_P^{-2}$, M_P is the Planck mass, c is a constant, the field strength $F^a_{\mu\nu}$ corresponds to the one of a Yang-Mills theory with a gauge group G which must be a subgroup of $E_8 \times E_8$ or $SO(32)$ and $\tilde{F}^{\mu\nu}$ is the dual of the gauge field strength; we assume that there is no potential for the dilaton, ϕ, and following Ref. [19] we introduce the universal function:

$$B(\phi) = e^{-2k\phi} + c_0 + c_1 e^{2k\phi} + c_2 e^{4k\phi} + ... \quad , \tag{3}$$

which expresses the fact that string-loop interactions have an expansion in powers of the dilaton, that is $g_S \equiv e^{2k\phi}$; the coefficients c_0, c_1, c_2, ... are presently unknown. As discussed in Ref. [19], when accounting for string loops the dynamics of the dilaton is such that the function $B(\phi)$ must reach a maximum at present, when say $\phi = \phi_0$, as fermion masses are shown to be proportional to inverse powers of $B(\phi)$. The gauge coupling constants are extracted from the dilaton-gauge field coupling $\frac{k_i}{4}\ B(\phi)\ Tr\,(F_{\mu\nu}F^{\mu\nu})$ in the bosonic action, from which follows that $g_i^{-2} = k_i\ B(\phi_0)$, where k_i is an order one constant and the index i refers to the appropriate gauge subgroup of G. Of course, the coupling between the axion and the electromagnetic field, i.e. $-\frac{c_{a\gamma\gamma}}{2}\alpha_{EM}B(\phi)^4\frac{a}{f_{PQ}}\mathbf{E}.\mathbf{B}$, is contained in the third term in (2) for the $U_{EM}(1)$ subgroup of G with $c_{a\gamma\gamma} = 8\pi^2 k_1 c$. This coupling ensures the photon conversion of axions due to magnetic fields surrounding the progenitor neutron star or interstellar medium. For $m_a << 10^{-2}\ eV$ axions free stream through neutron stars with temperature T_{NS} leading for standard axions to a total luminosity given by [21]:

$$L_a \approx 2\ \times 10^{50}\ \times 10^{\pm 1.5}\left(\frac{m_a}{10^{-4}\ eV}\right)^2\left(\frac{T_{NS}}{30\ MeV}\right)^{3.5}\ erg\ s^{-1} \tag{4}$$

which is consistent with the estimated output of a typical gamma-ray burst, $L_{GRB} \sim 10^{50}\ erg\ s^{-1}$, provided $m_a \approx 10^{-4}\ GeV$ [9]. In string theory however, the presence of heavy moduli implies, as mentioned, that there is no upper bound for f_{PQ} and that axions are lighter and hence that L_a is lower, but this may be compensated by the dilaton insertion as one should multiply the luminosity (eq. (4)) by a factor $B(\phi_0)^8$ and $k_i B(\phi_0) \sim 10$ [20].

Another way the dilaton may intervene is in the fireball expansion itself. Supposing that this expansion and the ensued cooling is adiabatic, then the process can be described through the Milne cosmological model [22]. If one further assumes the radiation corresponds, for simplicity, to an $SO(3)$ gauge field, it follows that one can build an homogeneous and isotropic model and show that the dilaton energy can be transferred into radiation as far as $B(\phi) > 0$ and its derivative with respect to ϕ is negative [17, 20].

References

1. R. Klebesabel, I. Strong and R. Olson, Ap. J. **182** (1973) L85.
2. C. Meegan et al., Nature **355** (1992) 143.
3. G. Fishman, Astron. Soc. Pacific **107** (1995) 1145.
4. E. Costa et al., I.A.U. Cir. **6572** (1997).
5. J. van Paradijs et al., Nature **386** (1997) 686.
6. D. Frail et al., I.A.U. Cir. **66y62** (1997).
7. J. Metzger et al., I.A.U. Cir. **6655** (1997).
8. B. Paczyński, astro-ph/9706232; astro-ph/9710086.
9. A. Loeb, Phys. Rev. **D48** (1993) R3419.
10. J. Goodman, Ap. J. Lett. **307** (1986) L47;
    B. Paczyński, Ap. J. Lett. **308** (1986) L43.
11. D. Eichler, M. Livio, T. Piran and D.N. Schramm, Nature **340** (1989) 126;
    R. Narayan, B. Paczyński and T. Piran, Ap. J. **395** (1992) L83.
12. S.E. Woosley, Ap. J. **405** (1993) 273;
13. P. Mészáros, "Theoretical Models of Gamma-Ray Bursts", astro-ph/9711354.
14. R.D. Peccei and H.R. Quinn, Phys. Rev. **D16** (1977) 1791.
15. J.E. Kim, Phys. Rep. **150** (1987) 1.
16. J. Preskill, M. Wise and F. Wilczek, Phys. Lett. **B120** (1983) 127;
    L. Abbot and P. Sikivie, Phys. Lett. **B120** (1983) 133;
    M. Dine and W. Fischler, Phys. Lett. **B120** (1983) 137.
17. M.C. Bento and O. Bertolami, Phys. Lett. **B336** (1994) 6;
    Gen. Relativity and Gravitation **28** (1996) 565.
18. G. Dvali, "Removing the Cosmological Bound on the Axion Scale", hep-ph/9505253;
    T. Banks and M. Dine, Nucl. Phys. **B505** (1997) 445.
19. T. Damour and A.M. Polyakov, Nucl. Phys. **B423** (1994) 532.
20. O. Bertolami, work in preparation.
21. G. Raffelt and D. Seckel, Phys. Rev. Lett. **60** (1988) 1793; **67** (1991) 2605; **68** (1992) 3116;
    A. Burrows, M.S Turner and R.P. Brinkmann, Phys. Rev. **D39** (1989) 1020.
22. T. Piran, Gen. Relativity and Gravitation **28** (1996) 1421.

INITIAL DATA SET FOR COSMOLOGY

H. ZIAEEPOUR
ESO, Schwarzschildstrasse, 2,
85748, Garching b. München, Germany

Abstract. The method of conformal metric is used to find an analytic set of consistent initial data for the linearized Einstein field equations. We discuss the relation between independent degrees of freedom and perturbations.

In cosmology the task of defining a set of initial conditions for solving the Einstein-Boltzmann field equations is not trivial (Arnowitt 62; Ó Murchadha and York 74). A priori, if we knew the processes responsible for the creation of fluctuations in an initially homogeneous universe, and if we could calculate the exact back-reaction of these processes on the space-time metric, the definition of initial conditions would be easy. However, even for heavily studied models like inflation and topological defects, our present knowledge and our calculation power is far from being enough for this purpose, and usually one has to use simplified solutions based on physical arguments to define a reasonable relation between the initial conditions of the matter and space-time fields.

For the metric, this problem has another dimension. Due to diffeomophism gauge symmetry, the initial value of the metric and its time derivative can not be specified in an arbitrary way (York 71, 72, 74). The configuration of matter fields partially determines them. In a more mathematical language (Misner et al. 73), one must introduce a time slicing in the space-time by defining a diffeomorphism between the space-time and a 3+1 manifold. The initial conditions are defined on the 3-space. This results on a restriction of the Einstein field equations on this 3-space called constraints.

The general form of the metric in 3+1 space is:

$$ds^2 = -\alpha^2 dt^2 + g_{ij}(dx^i + \beta^i dt)(dx^j + \beta^j dt) \quad (1)$$

The Einstein field equations and energy-momentum tensor conservation can be expressed by a set of first-order differential equations that depend

Astrophysics and Space Science is the original source of publication of his article. It is recommended that this article is cited as: *Astrophysics and Space Science* **261**: 229–230, 1999.

on the field set $\{\alpha, \beta^i, g_{ij}, K_{ij}\}$ where K_{ij} is the extrinsic curvature. With respect to these fields, constraints take the following form:

$$^{(3)}R - K_i^{\ j}K_j^{\ i} + K^2 = 16\pi T^*_*. \qquad K_{i\,|j}^{\ j} - K_{|i} = -8\pi T^*_i. \tag{2}$$

T^*_i and T^*_* are projection of energy-momentum tensor on the 3-space. J.W. York (York 71, 72, 74) showed that the independent degrees of freedom in a gravitational system on a 3-space are conformally equivalent. This permits one to separate dependent and independent degrees of freedom of g_{ij} and K_{ij}. The conformal transformation of a 3-metric g_{ij} is defined as $\bar{g}_{ij} = \phi^4 g_{ij}$. We call g_{ij} the base metric. The traceless-transverse part S_{ij} is chosen to transform as $\bar{S}^{ij} \equiv \phi^{-10} S^{ij}$. It is an independent degree of freedom. Constraint equations become dependent only on 4 fields ϕ and W^i. Taking an initial 3-space with constant K, trace of K_{ij}, and the flat Freedman-Lemaître metric as the base metric, for a flat universe with small perturbations, ϕ and W^i can be determined analytically (Ziaeepour 98):

$$W^i(k) = \frac{2\pi}{k^4}[k^2 \bar{T}^{*i}(k) - k^i k_j \bar{T}^{*j}(k)] \tag{3}$$

$$\phi = \delta(k) + \frac{2\pi \delta T^*_*}{k^2 + \frac{5}{12}K^2 + 6\pi T^*_{*homo}}. \tag{4}$$

The transverse-traceless component of the extrinsic curvature is an independent quantity. Its initial value can be separately related to the kinematics of matter and relic gravitational waves: $S^{ij} = S^{ij}_{relic} + S^{ij}_{matter}$,

$$\begin{aligned} S^{ij}_{matter} &= \int \sqrt{g} d^3x' \frac{3}{2|\vec{r} - \vec{r}'|^2} (p^i(r')u^j + p^j(r')u^i - (g^{ij} - u^i u^j)p^c(r')u_c) + \\ & \int \sqrt{g} d^3x' \frac{3}{|\vec{r} - \vec{r}'|^3} (\epsilon^{imn} s_m(r') u_n u^j + \epsilon^{jmn} s_m(r') u_n u^i). \end{aligned} \tag{5}$$

$p^i(r)$ and $s^i(r)$ are linear and angular momentum field densities. This method is easily applicable to higher-order perturbative calculations and, with more calculation complexity, to non-flat spaces (Ziaeepour 98). Apart from its mathematical robustness, the physical interpretation of relations and results is easy and clear.

References

R. Arnowitt, S. Deser, and C.W. Misner in *Gravitation: An Introduction to Current Research*, ed. L. Witten, John Wiley, New York (1962).

N. Ó Murchadha and J.W. York, *Phys. Rev.* **D10**, 428 (1974); J.W. York, *Phys. Rev. Letter* **26**, 1656 (1971); J.W. York, *Phys. Rev. Letter* **28**, 1082 (1972); J.M. Bowen and J.W. York, *Phys. Rev.* **D21**, 2047 (1980);

C. Misner, K. Thorne, J.A. Wheeler, *Gravitation*, Freeman and Co. (1973).

H. Ziaeepour, submitted (1998).

ON THE HYPOTHESES OF A NEW MODEL OF GALAXY FORMATION

J.-F. PASCUAL-SÁNCHEZ
Fundamental Applied Math Dept., Section of the Sciences Faculty, University of Valladolid, 47011, Valladolid, SPAIN

1. Introduction

Recently, Carvalho & Macedo have studied in some works, the behaviour of small, adiabatic, density perturbations in a two component (baryonic and dark matter) self-gravitating system in the non-relativistic aproximation, i.e., continuity plus Euler-Poisson equations. These works have generalized the classical model of Jeans for one component, which was proposed as a model of structure formation under Newtonian self-gravity, but have some problems in the case of galaxy formation.

In these papers, a new model of galaxy formation was studied. They assume a self-gravitating two component fluid made by baryons and hot dark matter embedded in a flat ($\Omega = 1$) Universe.

They start from the usual set of non-relativistic hydrodinamical system of equations for the mixture of two components.

Using a first-order perturbative analysis of these equations, one obtains a system of two 2nd-order ODE of coupled oscillators for the density contrast Δ_i of component i, which is defined by

$$\Delta_i = \frac{\delta\rho_i}{\rho_{io}} = \frac{\rho_i - \rho_{io}}{\rho_{io}}$$

ρ_{io} is the initial unperturbed density of mass of component i, and hereafter, k_i is the wave number of the perturbation, $W_i = 4\pi G\ \rho_{io}$ and c_i is the adiabatic sound speed.

In the first referenced paper, was proved the existence of a resonance in the amplitude of the growing mode of the baryonic component for a wave

Astrophysics and Space Science is the original source of publication of his article. It is recommended that this article is cited as: *Astrophysics and Space Science* **261**: 231–232, 1999.

number near to k_c, exactly at k_R given by

$$k_R = \left(k_c^2 + \frac{2W_B^2}{W_D\,(c_D^2 - c_B^2)} \right)^{1/2}$$

Where

$$k_c = \left(\frac{W_D - W_B}{c_D^2 - c_B^2} \right)^{1/2}$$

which must be real and positive.

Carvalho & Macedo have supposed the following sufficient condition for the resonance, $W_D > W_B$ and $c_D > c_B$, i.e., a hot dark matter (HDM) dominated, flat ($\Omega = 1$) Einstein-de Sitter Universe, which implies $k_c \in \mathbb{R}^+$.

The result was that for a mass of the hot dark matter component of 30 eV, the resonant scale at which galaxy formation occurs, decreases from a value of $10^{13}\ M_\odot$ at recombination to a few $10^8\ M_\odot$ at recent epochs, just in the range of the observed mass of different kinds of galaxies.

2. New Hypotheses

In this paper, I would like to point out that the appearance of a resonance that only affect the baryonic component, also may occur by taking the opposite physical hypotheses.

That is, if we first assume that one component is cold dark matter CDM, and hence its sound velocity c_D is less that the sound velocity of baryonic matter c_B and second, that the fluids are embedded in a open ($\Omega < 1$) baryon dominated Universe, i.e., the mass density of cold dark matter CDM, W_D, is less that the density of baryonic matter, W_B. These two assumptions together, $c_D < c_B$ and $W_D < W_B$, give a sufficient condition for the existence of a baryonic resonance different for the Carvalho & Macedo one, but with similar results with respect to the formation of galaxies.

From the observational point of view the Universe is open and hence we consider that this sufficient condition, i.e., baryon dominated open Universe with some CDM, is more realistic that the hot dark matter dominated flat Universe used in the commented works.

References

J.P.M. de Carvalho and P.G. Macedo, *Astron. Astrophys.* **299**, 326-338 (1995), and astro-ph/9605133.

J.P.M. de Carvalho and P.G. Macedo, in "Mapping, measuring and modeling the Universe", Astron. Soc. of Pacific Conf. Series, **vol. 94**, 69-74, (1996).

J.P.M. de Carvalho and P.G. Macedo, "The effective critical scale for gravitational instability in a two-component expanding fluid", Preprint GC-CAUP, 02/95.

RECENT RESULTS ON THE HUBBLE DEEP FIELD

R. D. BLANDFORD

Theoretical Astrophysics, 130-33 Caltech, Pasadena, CA 91125 USA

Abstract. A brief summary of some recent research on the Hubble Deep Field is presented. Particular attention is paid to the incidence of strong gravitational lensing and it is argued that the optical depth is nearly ten times smaller than average. Preliminary estimates of the incidence of galaxy-galaxy lensing are consistent with the presence of a large population of faint galaxies at large redshift. An estimate of the galaxy luminosity function is consistent with the local version undergoing passive evolution out to $z \sim 1$. The OII luminosity function shows more rapid evolution to this redshift.

1. INTRODUCTION

The Hubble Deep Field, henceforth HDF, (Williams et al 1996) is an unique data set for studies of faint galaxies. A small area was chosen well away from known local density enhancements and sources of obscuration. It was imaged for 30 complete orbits in each of four filters centered on wavelengths $\lambda 300, 450, 606, 814$ nm. There are three, contiguous WFPC2 frames and a solid angle 15,500 arcsec^2 is usable. This tiny region is teeming with galaxies. At least 2500 significant features can be identified down to $R = 30$, though it is not clear that these are all entire galaxies.

This data archive has, as intended, spawned a variety of follow up studies, including infrared and radio surveys and a redshift survey using the Keck telescope. This last program has garnered over 500 redshifts on the HDF and neighboring "flanking fields" (Cohen *et al* 1996, Phillips *et al* 1997, Hogg, 1997, unpublished thesis, Caltech, Cohen *et al* 1998, in preparation), of which over 100 are on the HDF itself and therefore imaged with unprecedented depth. The redshift sample is essentially complete to $R \sim 23$.

Astrophysics and Space Science is the original source of publication of his article. It is recommended that this article is cited as: *Astrophysics and Space Science* **261**: 233–240, 1999.

2. SOME KEY RESULTS

For completeness, let me list some developments that have come from analysis of the HDF. Some of these are original discoveries; others are verifications and elaborations of prior results. All of them are discussed in a forthcoming proceedings of a conference held at STScI ("The Hubble Deep Field", ed. Livio & Ealley).

- There appear to be $\sim$ 100 billion faint sources observable over the whole sky. These may be galaxies, but are surprisingly blue in color and they **appear** to be very compact with characteristic sizes $\sim 3h^{-1}$ kpc.
- The slope of the galaxy counts flattens fainter than $B \sim 25$ and most of the galaxian contribution to the brightness of the night sky comes from around this magnitude. (Again some caution is necessary because HST may not be sensitive to extended low surface brightness regions. It would be of particular interest to study the same field at depth using observations from Keck.)
- Perhaps the most significant discovery is of a substantial population of normal galaxies at redshifts $z \sim 3$. These are about 3 percent of the $\sim 25^m$ galaxies and represent a major breakthrough in the study of galaxy formation (Steidel *et al* 1996, Lowenthal *et al* 1997).
- It has also been found that just under half of galaxies with $z > 0.5$ are irregular morphologically, a far higher proportion than is the case at lower redshift.
- The redshift surveys have shown that galaxies with $z \sim 0.5$ are less clustered than locally. In addition, about 40 percent of these are concentrated within narrow features in redshift space (Cohen *et al 1996*). (As we only have a pencil beam survey, with linear width much smaller than the two point correlation function, this finding must be interpreted judiciously.) The redshift survey has also allowed a convincing demonstration of the reliability of the color-redshift method, at least for $z < 1$, $R < 24$ (Hogg *et al* 1997 preprint).

3. STRONG GRAVITATIONAL LENSING

These prefatory remarks provide a context for an ongoing investigation of strong gravitational lensing on the HDF. This is also motivated by radio and other surveys which are showing that a fraction ~ 0.002 of high redshift sources are multiply-imaged by intervening galaxies. On this basis, Hogg *et al* (1996) argued that there should be $\sim 5 - 10$ strong lenses on the HDF and identified the most promising candidates on morphological grounds. Unfortunately, it has proved hard to follow this up as the candidates are very faint for spectroscopic study. There are still no convincing examples

of multiply-imaged galaxies on the HDF. I have therefore tried another approach to attempt to locate these gravitational lenses or argue that they are not present, (which would immediately imply that the faint galaxy sources are predominantly relatively close by).
What I have tried to do is to combine the exquisite imaging of the potential lens galaxies on the HDF with their measured redshifts to identify the galaxies most likely to be lenses and then to use the faint galaxy counts to quantify the lensing probability. In addition I have performed simulations to see if it is possible to detect multiple imaging of a background against the light from the lens galaxy. As this research is ongoing, I can only report preliminary conclusions.

3.1. LENSES

As these are field galaxies with no discernible cluster along the line of sight, multiple imaging will be confined to the central regions of the most concentrated galaxies and least affected by dark matter in the halo. Now for every galaxy in the redshift sample, we can use the redshift to estimate the surface brightness in the frame of the source in the emitted B band. To accomplish this, we use the $F814W$ filter, (as this image will be least susceptible to the effects of dust), and make a spectral correction based upon the integrated color. (This is preferable to performing the spectral correction on a pixel by pixel basis because we are interested in the global mass distribution.)
We next convert this surface brightness into a surface mass density by assuming a constant mass to light ratio in the B-band. Local measurements of hM/L_B suggest a value of 5 for spirals and 10 for ellipticals. Measurements of known gravitational lenses (which may be a biased sample) suggest twice this value (Keeton, Kochanek & Falco, 1997 preprint). (I included irregulars and unclassified galaxies in with the spirals.) Both observations and the expectations from passive stellar evolution indicate that galaxies were about twice as luminous at $z \sim 1$ and a simple, empirical evolutionary correction was made.
We must next make some cosmographic assumptions. For illustration, I shall adopt an open universe with $\Omega_0 = 0.3$. (The outcome is independent of h.) I shall also place all the faint sources at $z = 3$.
The next step is to adopt the surface brightness maps of each of the 114 galaxies with known redshifts $z < 1.5$ and solve Poisson's equation using a Fourier method to obtain the equivalent surface potential. This can, in turn, be converted into a deflection vector field and a magnification field, obtained from the gradient of the deflection field. Images with negative magnification are multiple and so we can compute the cross section for

multiple imaging by projection back to the source plane.

There were many surprises. The first was that most galaxies do not multiple image, at least at the resolution of HST. (Of course if every galaxy contains a massive black hole in its nucleus, they will each multiple image every other galaxy an infinite number of times (Rauch & Blandford 1994). However, this is not a concern for observational astronomy!) Those that do are almost all elliptical galaxies that have higher central surface brightness and are assumed to have higher mass to light. ratio. In fact, over half of the cross section is contributed by three galaxies.

Secondly, the propensity for multiple imaging is strongly dependent upon the mass to light ratio. For example, increasing the central mass to light ratio for local galaxies from 10 solar units to 20 solar units increases the cross section of the strongest lens from ~ 1.0 arcsec^2 to ~ 3.0 arcsec^2. As always, finding gravitationally lensed sources of known redshift is a very sensitive measurement of the central mass to light ratio of the lensing galaxy.

Thirdly, the four strongest lenses all have $z \sim 1$ and $L \sim L^*$ and are therefore quite ineffectual if the sources have redshifts much less than $z \sim 2$. Finding multiple images systematically from a complete sample of lens galaxies is a potentially powerful probe of the faint source redshift distribution.

The integrated cross section over the whole HDF is surprisingly low, only ~ 1 arcsec^2 for the "local" choice of M/L increasing to ~ 3 arcsec^2 for the lens-normalized choice.

3.2. SOURCES

The sources were classified by size rather than integrated flux. Size was defined operationally as the number of contiguous 0.04 arcsec pixels with counts above 2.5σ. There are $\sim$ 1600 galaxies with sizes between 5 and 50 pixels (about two-thirds of the total number of galaxies). The median galaxy in this sample has a size of 13 pixels. As we extrapolate to smaller images, we do not expect to find a large increase in the potential source density unless there is a sharp upturn in the magnitude source counts below $\sim 30^m$. The total density of source galaxies is therefore ~ 0.1 arcsec^{-2} and the total probability of finding one of them multiple-imaged if they are at redshift $z \sim 3$ is ~ 0.4 assuming the local M/L and ~ 1 for the lens-normalized M/L. For the individual strongest galaxies, the probabilities per lens are therefore ~ 0.1 and ~ 0.3 respectively.

3.3. DETECTABILITY

What is clear from the foregoing discussion is that it may pay to seek multiple images very carefully in judiciously chosen locations, rather than by inspecting a large sample of galaxies. Now, it goes without saying that this is a daunting task. What we are looking for are faint, inclined arc-like features within ~ 5 kpc. They are likely to be patchy and blue. In other words, they look just like spiral arms! Now, I believe that the prospects are not so bleak if one has images with a resolution and depth comparable with those on the HDF in several filters. In particular, ellipticals are clearly preferred and can be identified on spectral and not just morphological grounds. They have relatively smooth light profiles against which arcs can be brought out by image processing techniques. Furthermore the image pattern that is expected, and can be easily simulated, is quite distinctive. The magnified images should have similar colors although variable reddening by the lens galaxy can confound this expectation.

A quick inspection of the images of all of the galaxies in the redshift survey reveals several features that could be formed by galaxy lensing. However, none of these is convincing, based upon the criteria listed above. Looking at the four best candidates, only the strongest of these exhibits a feature that is worth considering further and this is a large, tangential arc at a radius of ~ 1 arcsec. It has a similar color to the rest of the galaxy and would require $hM/L \sim 40$ solar units. It is far more likely to be a shell produced by tidal stripping of a merging galaxy.

Using the observed surface brightnesses of galaxies on the HDF, we estimated a strong lensing frequency about an order of magnitude lower than that based on radio surveys and advertised in Hogg *et al* 1996. If this is the correct approach to estimating the frequency of multiple imaging, then it should not be a surprise that we have no good examples to offer as yet. Put another way, the best multiple image candidates selected from the HDF on morphological grounds, required M/L ratios far in excess of those measured in local, normal galaxies. Perhaps, in retrospect, this is quite reasonable. Many of the secure examples of gravitational lensing are unusual because they belong to a rich compact group or cluster which magnifies the focusing power of an individual galaxy and the HDF was selected, as far as possible, to avoid such concentrations. Furthermore, the source population in the radio surveys is known to be at high redshift, whereas not all of the faint optical sources may be distributed in this manner.

4. GALAXY-GALAXY LENSING

As first discussed by Webster (1983) and Tyson et al (1984), individual lens galaxies at modest redshift, ($z_d \sim 0.5$), are expected to produce a weak tangential stretching of the images of background source galaxies of significantly greater redshift, ($z_s \sim 1-3$). If we make a very simple model of the potential of the lens galaxies, (presumably dominated by dark matter), as a truncated isothermal sphere then, in round numbers, the Einstein ring subtends an angular radius $\theta_c \sim 1''$ whereas the halo is expected to extend out to $\theta_H \sim 10''$. The cross section for weak lensing by the halo is therefore roughly a hundred times that for strong lensing by the central regions of the galaxy. If we measure the halo density by its (roughly constant) circular velocity, then the expected ellipticity of a background galaxy separated by an angle θ from the galaxy center is

$$\epsilon \sim \frac{2\pi V_c^2}{\theta c^2}; \qquad \theta < \theta_H \tag{1}$$

$$\sim \frac{2\pi\theta_H V_c^2}{\theta^2 c^2}. \tag{2}$$

If we now concentrate on sources with impact parameters close to the halo radii of intervening galaxies, we find that the expected tangential ellipticity will be $\epsilon \sim 0.1$. this will be contributed by all the lens galaxies within $\theta \sim \theta_H$ and, to a pretty good approximation, the signals will add linearly. The noise in this measurement is the intrinsic galaxy ellipticity which has a value ~ 0.3. Now the number of lens galaxies within angle θ of a given source increases $\propto \theta^2$ whereas the signal decreases $\propto \theta^{-1}$ for $\theta < \theta_H$ and $\propto \theta^{-2}$ for $\theta > \theta_H$. Therefore the signal to noise ratio is maximized if we include all lenses within a few θ_H of the source and ignore those beyond θ_H.

A positive measurement of this effect has previously been reported by Brainerd, Blandford & Smail(1996) who analyzed a $9.6' \times 9.6'$ red image with average seeing $\sim 0.8''$, taken with the COSMIC imager on the 5m Hale reflector. The detected galaxies were divided into 511 sources with $23 < r < 24$ and 439 deflectors with $r < 23$. Source-deflector separations in the range $5'' < \theta < 34''$ were included. A polarization signal $\epsilon \sim 0.01$ was measured which is consistent with a simple model of the deflector galaxies in which the circular velocity of a L^* galaxy is ~ 220 km s^{-1} and the physical radius of the halo of an L^* galaxy is $\sim 50h^{-1}$ kpc. (This last quantity is not well constrained by the measurement.) In addition, Griffiths *etal* (1996) reported detecting this field using HST images drawn from the MDS survey. They were able to distinguish ellipticals and spirals and found that the effect was much stronger for the former. Turning to the HDF itself. Dell Antonio & Tyson(1996) reported a detection of galaxy-galaxy lensing for pairs of galaxies with $\theta < 5''$, consistent with an L^* circular velocity of ~ 260 km s^{-1}.

Each of these studies considered sources and deflectors in a blind, pairwise manner. In practices this dilutes the signal by weighting all pairs equally including those where the "source" redshift is less than that of the deflector. It is apparent that if we have additional information, in particular the redshift distribution of the deflector galaxies, then we should be able to measure a much larger signal to noise ratio using from a given set of imaging data. In addition, these studies assumed a simple form of the source redshift distribution and focused on describing the average mass distribution in the deflector galaxies. In the present study, we restrict our deflector population to those galaxies with known redshifts, and, consequently, well-determined luminosities.

The deflector field comprises over 200 galaxies of which 70 are on the HDF and the remainder are on the flanking area. This ensures that the average distribution of deflector galaxies about the source galaxies is isotropic on the sky. This removes a potentially serious source of systematic error. A subset of the galaxies have nine wavelength photometry and most of the remainder have six wavelength photometry. We can therefore estimate their rest frame B magnitudes and thereby obtain estimate of L_B/L_B^*. We then adopt the deflector galaxy model of Brainerd *etal* (1996) in which the circular velocity and halo radius are given by

$$V_c = 220(L/L^*)^{1/4}\text{kms}^{-1} \tag{3}$$

$$r_H = 50h^{-1}(L/L^*)^{1/2}\text{kpc}. \tag{4}$$

Using this model, it is possible to compute the polarization field across the whole of the HDF as a function of source redshift z_s, $p(X, Y, z_s)$, where X, Y are angular coordinates on the sky.

There are two potentially serious sources of systematic error in galaxy-galaxy lensing. The first is that when the source is too close to the deflector, the isophotes can overlap so that galaxies will appear to be elongated radially. The second is that some designated source galaxies will actually be companions of the deflector and be subject to Newtonian tidal forces that may stretch them out in a tail stretched out along the orbit. Both of these effects can be minimized by excluding source-deflector pairs with $\theta < 5''$, which we do.

The total source sample that we use comprises ~ 1000 galaxies each covering 16 or more pixels. The magnitude distribution extends down to $V \sim 29$ and the typical galaxy ellipticity if $\epsilon \sim 0.2$ For each galaxy, we compute the complex orientation, χ and then correlate with the predicted polarization field for different assumptions about the source redshift distribution. The complex correlation coefficient $C =< p\chi^* >$ can be formed, averaging over the whole source sample. Its expectation is real, the imaginary part measuring the correlation of the source images with the polarization field after rotation through $\pi/4$. For the limiting case when all the source galaxies are assumed to lie at large redshift, we measure $C = 0.0008 + 0.0000i$. The expected error in C is ~ 0.0003 and so this is a $\sim 3\sigma$ detection of the polarization signal. However it only a third of the predicted value. On this basis, we suggest that perhaps half of the source galaxies have redshifts $z_s < 1$.

A fuller discussion of this analysis and related matters will be presented in Blandford *etal* (1998, in preparation).

5. GALAXY LUMINOSITY FUNCTION

The HDF can also be used to derive a galaxy luminosity function using the measured redshifts of its bright galaxies and also those on the flanking fields. As might be expected this is quite an involved exercise in practice as there are terrible selection effects with which to contend. The worst of these is establishing the probability that the spectroscopic procedure that is followed will actually furnish a redshift for a particular galaxy. This has been investigated in detail by Hogg(1997, unpublished thesis, Caltech) and applied to the redshift sample. The result is that the B magnitude luminosity function appears to be no quite consistent with a passively evolving version (one magnitude in L^* out to $z \sim 1$). It is not yet clear whether or not this is inconsistent with the more rapid evolution reported by some other surveys.

By contrast, a measurement of the OII luminosity function (Hogg et al 1998, in preparation), does show fairly rapid evolution to the extent that the star formation rate at $z \sim 1$ is roughly 5 times the local star formation rate. This is consistent with determinations of the Balmer line luminosity function.

Acknowledgments
I thank Pedro Avelino, Teresa Lago and their colleagues for the invitation to attend this meeting and financial support. I also thank my colleagues Tereasa Brainerd, David Hogg and Tomislav Kundić for collaboration on various parts of this investigation. Special thanks are due to Judith Cohen who has led the Caltech-Keck redshift survey and colleagues at UCSC and UH who have also contributed many of the redshifts. Acknowledgment must also be made of Bob Williams who had the foresight to see the long term value of the HDF project. Support under NSF grant AST 95-29170 and NASA grant AR-06337.IS-94A is gratefully noted.

References

Brainerd, T. G., Blandford, R. D. & Smail, I. R. 1996 ApJ 466 623
Cohen, J. G. *et al* 1996 ApJ 471 L5
Dell Antonio, I. P. & Tyson, J. A. 1996 ApJ 443 L17
Griffiths, R. E. *et al* 1996 MNRAS 282 1159
Hogg, D. W. *et al* 1996 ApJ 467 L73
Lowenthal, J. D. 1997 Ap J 481 673
Phillips, A. C. *et al* 1997 ApJ 489 543
Rauch, K. P. & Blandford, R. D. 1994 ApJ 421 46
Steidel, C. C. *et al* 1997 AJ 112 352
Tyson, J. A., Valdes, F., Jarvis, J. F. & Mill, J. R. 1984 ApJ 281 L59
Webster, R. L. 1983 Unpublished thesis, University of Cambridge
Williams, R. E. *et al* 1996 AJ 112 1335

RELATIVISTIC MOTION ON KILO-PARSEC SCALES IN POWERFUL RADIO SOURCES

J. DENNETT-THORPE
University of Lisbon
Observatório Astronómico, Tapada da Ajuda, Lisboa, Portugal

1. Introduction

Strong evidence already exists for relativistic motion on parsec scales (superluminal motion and variability, and also relations between radio core dominance and optical characteristics). Evidence for bulk relativistic motion on kilo-parsec scales is somewhat less secure, and it has been variously argued that the supposed evidence may be attributable to other, intrinsic, effects. Evidence for such motion is the existence of one-sided jets in quasars, which is easily understood as a Doppler flux-boosting effect of the approaching jet. Other evidence is the depolarisation asymmetries correlated to jet side (Laing 1988, Garrington et al. 1988), which is interpreted as an effect of viewing the counterjet side of the source through more Faraday medium. It is a fundamental tenet of the relativistically-based unification schemes of radio galaxies and radio-loud quasars that the jet is an orientation indicator. A possible problem for the standard interpretation had emerged which I outline below. I then describe work undertaken to investigate this directly. I show that there is no conflict with the standard models, and further, that relativistic bulk motion extends all the way to the hotspots in powerful classical double radio sources.

There have been a number of correlations reported in the literature which related to the asymmetric appearance of powerful radio sources. These correlations were in themselves all extremely strong. The following chain emerged from a combination of a number of studies:
jet side $\leftrightarrow$ least depolarised lobe $\leftrightarrow$ flatter spectrum hotspot and lobe $\leftrightarrow$ less extended emission line gas $\leftrightarrow$ long side. (Garrington et al 1991, McCarthy et al 1991, Pedelty et al 1989). Although the samples used were different in all cases, none found evidence for any correlations in the opposite sense, and Liu & Pooley (1991) and Garrington et al (1991) found explicit correlations between more than two of the above. The problem is

Astrophysics and Space Science is the original source of publication of his article. It is recommended that this article is cited as: *Astrophysics and Space Science* **261**: 241–244, 1999.

that the unbroken chain of reasoning leads from those attributed by unified schemes to orientation effects, to supposed intrinsic or environmental asymmetries.

The objective of this work was to focus on the expected 'weak link' in the chain: the spectral asymmetries. Two samples of powerful radio sources were selected: one of which was chosen to be dominated by orientation effects if unified schemes were correct (i.e. a sample of quasars) and one which would be dominated by intrinsic /environmental effects (i.e. radio galaxies). For the quasar sample we chose 10 of the largest brightest quasars (Bridle et al. 1994), due to the availability of high quality data. In all sources we required a detected jet: as radio jets are notoriously difficult to detect in radio galaxies, we were restricted to selecting our radio galaxies from a sample of nearby radio galaxies which had been subjects of a high-sensitivity, high-resolution imaging campaign (Black et al. 1992, Leahy et al. 1997). This results in a power/redshift difference between the two samples.

2. Observations and Results

Observations were taken with the VLA for most sources, and with MERLIN for the two smallest quasars. This new data combined with archival data was used to create high quality images at two frequencies, using similar uv-plane coverage at both frequencies. The data was reduced in the AIPS package in the standard way and imaged with the CLEAN algorithms. The analysis proceeded by cutting up the images on the lower frequency map into $\sim$ ten logarithmically equally spaced intensity contours. These contours were then used as templates to select areas on the higher frequency maps. Spectral indices were then calculated from the fluxes in each surface brightness region. The advantage of this method over other possibilities is that it is well suited to the reality of sources which are generally not axisymmetric, and does not impose subjective decisions on what constitutes the 'hotspot'.

The spectra of the high surface brightness regions showed:

- A strong correlation with jetside for the quasars.
- No correlation with jet side in the radio galaxies.
- No correlation with lobe length, lobe brightness, or hotspot flux densities in either sample.

There are asymmetries in the low surface brightness regions which are *not* dependent on jet side, which I do not discuss further here.

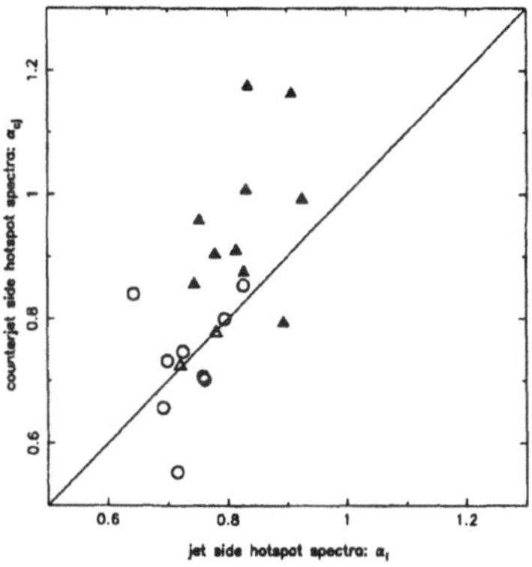

Figure 1. Spectral asymmetries of highest brightness regions of quasars (filled triangles), broad and narrow-line radio galaxies (open triangles and circles respectively).

3. Interpretation

3.1. SPECTRAL ASYMMETRIES AS A RELATIVISTIC EFFECT

This result is predicted by relativistically-based unified schemes if the spectral asymmetry can be explained as a Doppler effect. Of the three possible ways to explain the asymmetry as such an effect, only two could explain the results.

The simplest model which just adds an extra flat-spectrum component on the approaching side due to Doppler flux boosting (e.g. Tribble 1994), cannot explain the observations because (i) this predicts no asymmetry when the analysis is in terms of surface brightness, as it is here; (ii) there is no tendency in these samples for the jet side hotspot to be brighter.

The apparent frequency shift caused by relativistic motion could account for the result, if the hotspot spectra was curved downwards, as the emission from the material moving towards the observer would have been emitted at a lower frequency.

The final possibility is that Doppler suppression of the material on the counterjet side might lead us to identify the 'hotspot' as a place other than that of the termination of the jet; thus it would contain more aged electrons. Doppler suppression could also play a role, even if the site of jet termination had been correctly identified as the high surface brightness regions, as the observer might see different parts of the flow from the hotspot (possibly preferentially backflowing, and therefore or aged, material in the case of the counterjet-side hotspot)

3.2. THE CASE OF THE BROAD-LINED RADIO GALAXIES

In simple models of unified schemes the broad-lined radio galaxies (BLRG) are often assumed to be 'nearby quasars', or objects at an intermediate angle to the line-of-sight. It is therefore noteworthy that neither of the BLRG investigated show the spectral asymmetry seen in the quasars.

In view of the power/redshift difference between the two samples, however, we are unable to draw any conclusions. It is possible that this difference could account for the lack of asymmetry in the BLRG by a number of means, as lower radio power may be related to lower bulk speeds, or to less spectral curvature in the hotspots, or the lower redshift might not shift the observed spectrum into a sufficiently curved region to enable the detection of the asymmetries at these observing frequencies.

However, in view of the findings of Hes et al. (1995) this result may also be suggesting that BLRG are not 'quasar-like'. Studies of the source asymmetries, with properly matched samples, provide an ideal way to investigate this possibility, and such a study is presently underway.

4. Conclusions

The spectral asymmetry in the hotspots of quasars is consistent with unified schemes and is easily explicable as an effect of relativistic motion. For such explanations to work we require either that the hotspot spectra are curved, or that we do not see the site of jet termination on the counterjet side within the highest surface brightness regions. It seems likely that both of these play a part.

The unbroken chain of correlations outlined in the introduction, is now broken into orientation and other effects, where the spectral asymmetries are dependent on orientation in the high surface brightness regions, but not in the extended material.

The lack of asymmetries in the BLRG is possibly explicable as a power/ redshift effect, but is worthy of further investigation.

References

Bridle A.H., Hough D.H., Lonsdale C.J., Burns J.O., Laing R.A., 1994, AJ, 108, 766
Dennett-Thorpe J., Bridle A., Scheuer P., Laing R., Leahy J., 1997, MN, 289, 753
Garrington S. T., Leahy J. P., Conway R. G., Laing R. A., 1988. Nat, 331, 147
Garrington S.T., Conway R.G., Leahy J.P. 1991, MNRAS, 250, 171
Hes R., Barthel P.D., Hoekstra H., 1995, A&A, 303, 8
Laing R. A., 1988, Nat, 331, 149
Liu R., Pooley G. G., 1991, MNRAS, 249, 343
McCarthy P.J., van Breugel W.J.M., Kapahi V.K., 1991, ApJ, 371, 478
Pedelty J.A., Rudnick L., McCarthy P.J., Spinrad H., 1989, AJ, 97, 647
Tribble P., 1992, MNRAS, 256, 281

RECENT RESULTS ON ACTIVE GALACTIC NUCLEI

R. D. BLANDFORD
Theoretical Astrophysics, 130-33 Caltech, Pasadena, CA 91125 USA

Abstract. There is now little doubt that the nuclei of most galaxies contain massive black holes with 10^6 $M_\odot < M < 10^{10}$ $M_\odot$ and that these are ultimately responsible for most nuclear activity when they are supplied with gas. Parallel observations of Galactic X-ray sources are providing small scale counterparts that have similar properties except that they vary far more rapidly. As we can now determine the masses of these black holes in several instances, we can ask much more quantitative questions about the mode of accretion and the origin and collimation of outflows. In this contribution, I will emphasize the importance of magnetic field for extracting the angular momentum from the flow and explain how the radiative efficiency is controlled by the mass accretion rate and the actual properties of magnetized, plasma.

1. INTRODUCTION

As much of this material has appeared elsewhere, eg Blandford (1998ab) and references therein, I will only present the briefest of summaries and the most cursory of references.

2. THE BLACK HOLE MODEL

The discovery of quasars (Hazard et al 1963, Schmidt 1963) led quickly to the realization that the nuclei of galaxies could easily outshine their hosts. It was also soon realized that one possible, physical explanation for this phenomenon was a massive black hole (Zel'dovich & Novikov 1964, Salpeter 1964). Since then, the arguments for favoring this generic interpretation of the most prodigious displays of nuclear activity have steadily multiplied and strengthened so that observations are now routinely interpreted, at

Astrophysics and Space Science is the original source of publication of his article. It is recommended that this article is cited as: *Astrophysics and Space Science* **261**: 245–252, 1999.

least implicitly, in the context of a black hole model. Of course, this does not imply that the case is rigorously made; neither does it require that all nuclear activity is of this form. However, the association of massive black holes with nuclear activity is much more secure than most of what follows.

With the important exception of the Balbus-Hawley (1998) instability, the major developments in recent years have been observational or at least strongly motivated by observations. Masses of the "central dark object" have been estimated in about forty cases, using stellar dynamics, emission lines of orbiting gas and, most accurately, using water masers. They range from $\sim 2 \times 10^6$ $M_\odot$ to $\sim 3 \times 10^9$ $M_\odot$ (Magorrian et al 1998, preprint) and, in many cases, the compactness is sufficient to rule out star clusters with confidence (Maoz 1995). We have learned that although the Eddington limit remains a fiducial power for an accreting black hole, it is not a typical power. Some sources appear to be more powerful than this while others appear to have very low radiative efficiency and may be advecting most of the thermal energy generated during the accretion into the hole.

3. GENERAL RELATIVISTIC EFFECTS

A black hole of mass $10^6 M_6$ $M_\odot$ has an equivalent length- and time-scales $m = 1.5 \times 10^{11} M_6$ cm $\equiv 5M_6$ s. Apart from this scale, astrophysical black holes form a one parameter family. This parameter is most conveniently chosen to be the spin angular velocity $\Omega < 1/2m$. Now, the mass of a spinning black hole, has an irreducible part $m_0 = m(1 + 4\Omega^2 m^2)^{-1/2} \geq 2^{-1/2} m$. It is convenient to define an effective radius for the hole, $r_0 = 2m_0$, so that the surface area of the event horizon (proportional to the entropy) is $4\pi r_0^2$, the spin angular momentum per unit mass is $a = r_0^2 \Omega < m$ and the effective rotational speed is $\beta = \Omega r_0 < 2^{-1/2}$. (The surface gravity, which is proportional to the temperature, is $(1 - 2\beta^2)m/r_0^2$.) In terms of these quantitities, the mass m is related to the irreducible mass m_0 through the memorable relation

$$m = \frac{m_0}{(1 - \beta^2)^{1/2}}. \tag{1}$$

The difference between the total mass m and the irreducible mass m_0, up to $0.29m$, is extractable in principle. It is also extractable in practice though large scale electromagnetic torques associated with magnetic field lines that thread the black hole event horizon and are supported by external electrical currents.

The geometry of the spacetime around an astrophysical black hole can be described by the Kerr metric, whose two parameters are conventionally chosen to be m, a. When $\Omega \neq 0$, the geometry is not spherically symmetric and there is a well-defined equatorial plane into which accreting gas can settle. Circular orbits in this plane are dynamically stable down to a minimum radius $r_{\min}$ that decreases from $6m$ to m as Ω increases from 0 to its maximum allowed value $1/2m$. The corresponding binding energy per unit mass associated with this marginally stable orbits increases from 0.06 to 0.42. There is an additional effect, called the dragging of inertial frames, or Lense-Thirring precession . One manifestation of this phenomenon is an orbital precession with an angular frequency $\sim 2am/r^3$. All of these effects are confident predictions of strong field, general relativity. At present it is generally best to assume their truth and try to interpret the observations in terms of them. However, ultimately we hope to use black holes to perform quantitative tests of the theory.

Most observations of accretion phenomena can be interpreted using a Newtonian analysis, crudely truncated at $r \sim r_{\min}$. However, some important exceptions are starting to emerge. The most immediate of these is the measurement of Fe lines from nearby Seyfert galaxies (Tanaka et al 1995) with profiles that require relativistic corrections to the gas kinematics and ray propagation. Indeed, it has been argued on these grounds that holes are spinning rapidly. (Although some think that this conclusion is premature, it is remarkable that we are even arguing about it. All of this is very encouraging for the next generation of X-ray satellites which should provide line profiles with exquisite detail.)

Another development is the widely publicized report that Lense-Thirring precession has been detected in GRS 1915+105 and other Galactic black hole candidates (Morgan et al 1997). This comes from associating the observed QPOs with the precession frequency as evaluated at $\sim 1.5r_{\min}$, where, it is supposed, that the emissivity is concentrated. (The relatively high quality factor of the QPOs requires that the emission extend over a limited range of radius.) Of course, this is not the only explanation that has been advanced for QPOs and it is possible that we could be observing the fundamental orbital frequency at some radius (Zhang et al 1997) or a trapped g-mode (Nowak et al 1997).

4. MODES OF ACCRETION

"Standard" accretion disks are Keplerian, coplanar, steady, conservative and driven by local viscous torques that extend to $r_{\min}$ and release the associated binding energy as thermal radiation. There is now reason to suspect that each of these assumptions can be violated. Accretion at low rates is argued to be advection-dominated and quasi-radial (Narayan & Yi 1994). We observe (as well as expect) warped disks (Herrnstein et al 1996). Dynamical, thermal and viscous instability, long a feature of specific disk models, is surely being observed directly, most interestingly in

GRS 1915+105. Significant mass loss is seen directly in the protostellar jets and inferred in other examples (eg Livio 1997). Finally release of large parcels of mechanical or electromagnetic energy, by its very nature, needs coordinated action over a large volume and may be required to produce non-thermal X-ray and γ-ray outbursts in these objects.

INTERNAL ANGULAR MOMENTUM TRANSPORT

It now appears to be fairly certain that magnetic fields are responsible for the torque in accretion disks. Unmagnetized, Keplerian disks are stably stratified. However, when a small, vertical magnetic field is introduced, the slow magnetosonic mode is destabilized by the rotation. The magnetic tension restoring force is overcome by centrifugal force, in analogy to what happens when two orbiting satellites are connected by an elastic rope. A gas element at radius $r+\delta r$ will fall behind a second element at radius r to which it is connected by magnetic field. The magnetic stress will exert a positive torque on the outer element, increasing its angular momentum ℓ. However, for circular orbits, $r \propto \ell^2$. Therefore the radius of the outer element will increase with time. The instability grows on a dynamical timescale $\sim \Omega^{-1}$ for lengthscales $\sim a/\Omega$, where $a = (4\pi\rho)^{-1/2}B$ is the Alfvén speed. The instability is more powerful than conventional dynamo modes.

DISSIPATION

This leads us to consider the nature of the dissipation. A magnetic torque G acting within an accretion disk with angular velocity ω, does work on the exterior disk at a rate $G\omega$. This corresponds to an outwardly directed flux of angular momentum, which may be offset by the inflow. In addition, there is an inevitable energy release at a rate $\int d\omega G$. Normally this is thought of as viscous dissipation that appears as a direct heating of the gas analogous to the nuclear heating that occurs inside stars. There are two ways to bring about this dissipation, magnetic reconnection and driving a turbulent cascade. In magnetic reconnection, an "anomalous", (ie enhanced), resistivity develops around a neutral point and as gas flows through this region, its entropy increases though Ohmic heating at a rate that corresponds to the rate of decrease of magnetic energy. Note that reconnection proceeds in a piecemeal fashion in an accretion disk as it can only take place over a length scale $< 0.1a\omega^{-1}$ in a shearing timescale. In a cascade, a purely hydromagnetic development of the Balbus-Hawley instability may create disturbances on some "outer" scale that is far too large for microscopic dissipation to operate, though still small compared with the thickness of the disk. The energy deposited at this scale then forms a hydromagnetic turbulence spectrum extending down to small length scales on which ion-cyclotron or transit time damping can operate again heating the gas directly. Both of these processes preferentially heat ions.

The issue of magnetic dissipation is particularly important for advection-dominated accretion flows, "ADAFs". The observational stimulus for reconsidering these solutions is the low apparent radiative efficiency of some accreting black holes. A good example is the Galactic center where it is argued that the mass supply rate could be as high as $\sim 10^{22}$ g s^{-1}, while the bolometric luminosity could be as small as $\sim 10^{36}$ erg s^{-1} (Falcke & Melia 1996). The radiative efficiency might then be as low as $\epsilon \sim 10^{14}$ erg g$^{-1} \equiv 10^{-7}c^2$ and is unlikely to exceed $\sim 10^{-4}c^2$.

ADAFs address this challenge by making two assumptions. The first is that the torque and the dissipation are both measured by a value of the parameter α that is large enough for the gas to remain relatively tenuous. The second assumption is that the associated dissipation just heats the ions not the electrons. The electrons maintain a Maxwellian distribution with temperature much less than that of the ions through Coulomb scattering and synchrotron self-absorption. For sufficiently low mass accretion rate, $\dot{M} < \dot{M}_{\rm ADAF} \equiv 10^{24}\alpha^2(M/10^6{\rm M}_\odot)$ g s^{-1}, the disk will thicken within some transition radius and follow a quasi-spherical inflow of low radiative efficiency, a feature which distinguishes black hole accretion from accretion onto neutron stars. It has proved possible to model the observed spectra of several black holes sources in this manner. Now, it is crucial that the electrons not be accelerated to suprathermal energies where they can become very efficient radiators through the synchrotron and inverse Compton processes. At present it is controversial whether or not reconnection and cascade development can avoid direct acceleration to the extent required by the most extreme cases. In any case, all of this is in marked contrast to what is assumed in models of X-ray Seyfert galaxies, where it is supposed that electrons in a hot corona above a cooled accretion disk (as is supposed to occur when $\dot{M} >> \dot{M}_{\rm ADAF}$), are highly efficient sources of a non-thermal power law, hard X-ray spectrum that illuminates the disk.

However, if most of this energy release occurs in the corona, the energy need not directly increase the entropy of the gas. Loops of tangential magnetic field may be created on scales that are sufficiently large compared with the local scale height that they can rise buoyantly in the effective gravitational field of the disk and escape from the corona via Parker and allied instabilities. If global disk evolution is to be encapsulated in the α prescription, it may be necessary to distinguish two values of this parameter, one defined, as usual, by the shear stress and the second by the local entropy production. Both of these quantities could be measured in numerical simulations of sufficient scale that allow magnetic flux to escape. If it turns out that $\alpha_{\rm dissipation} << \alpha_{\rm shearstress}$, then this may help account for underluminous, black hole accretors.

ELECTROMAGNETIC ENERGY EXTRACTION

As mentioned above, and discussed extensively elsewhere, jet cores may be powered by Poynting flux extracted directly from the spin of the central black hole. In particular, fields of strength $\sim 10^4$ G threading a $\sim 10^9$ $M_\odot$, hole are adequate to account for the most energetic radio galaxies. (Electromagnetic energy extraction is quite familiar. For example, the rotational energy, that is extracted from the neutron star in the Crab Nebula, is transported initially as a Poynting flux before being transformed into a particle flux in the outflowing relativistic wind.)

However this is not the only way to extract energy from the hole. Another possibility is that magnetic flux is convected radially inward by the accreting gas until it threads the event horizon. Now, the field lines will be frozen into the inner accretion disk and move with its angular velocity. Therefore, as long as the hole's angular velocity exceeds that of the disk gas (and this will be true for Keplerian orbits of radius $r > r_{\rm min}$ if $\Omega > 0.093/m$), there will be a positive, magnetic torque G acting on the disk increasing its angular momentum at the expense of the spin of the hole. (This is like the "propellor" mechanism that takes place when there is accretion onto a rapidly spinning neutron star except that now the field lines are attached to the disk.) Normally, it is supposed that the net angular momentum flow through the disk, including the angular momentum advected by the inflowing gas, vanishes. However, if enough gas accumulates close in the inner disk the magnetic field strength can be comparatively large so that the hole can exert a large torque $G >> \dot{M}\ell$, where ℓ is the specific angular momentum, and that this torque is communicated radially outward through the disk, creating power through the differential rotation. Indeed, there may be so large an outward transport of angular momentum that the accretion is almost completely arrested with much of the mass inflow at large radius being driven away as a wind. Many accreting sources are associated with jets which, in some cases,(eg SS433), appear to be more powerful than the underlying disks. The jets in these sources may also be important channels for removing energy and angular momentum.

5. FORMATION OF JETS

The observational progress on jets has been almost as impressive as that on holes and disks and, consequently, our understanding of the relationship between disks and jets is developing. We are starting to observe relativistic jets on a very small scale. The best case so far is M87 where it appears from VLBI observations, that the eastern jet is at least partially collimated on a transverse length scale $\sim 3 \times 10^{16}$ cm $\sim 60m$. In addition, observations of the jet in the LINER galaxy NGC 4258 (Herrnstein et al 1996) have verified that the flat spectrum radio core is really the synchrotron self-absorbed inner region of the approaching jet instead of, for example, a spherical isotropic halo surrounding the central black hole. Indeed, in this source, the absence of radio emission from the dynamical location of the black hole sets an important constraint on the character of the accretion flow Generalizing from these special cases and guessing that all black hole jets are made by a common mechanism, we suggest that the power for these jets derives from very close to the black hole, and possibly from the hole itself.

Another important observational development is the association of γ-ray sources with blazars (Hartmann et al 1996). Here it appears that ~ 1 GeV γ-rays are focused into narrow cones with opening angles $\sim 5^\circ$, presumably by relativistic motion of a scattering cloud of electrons and positrons. However, these γ rays can only be emitted outside the γ-ray photosphere or gammasphere where the opacity to pair production on the ambient X-rays falls below unity. In some sources, the gammasphere appears to be beyond $\sim 1000m$. The γ-rays are presumably created by electrons and positrons through inverse Compton scattering. However, these electrons and positrons cannot, themselves be the agency though which power is transported from the vicinity of the hole to the emission region because they are subject to a strong radiative drag. The only two viable means of carrying jet power all the way from $r \sim m$ to the emission region appear to be using protons and electromagnetic field (ie Poynting flux), eg Levinson & Blandford 1996. The latter choice seems to be favored, at least to this reviewer, because there does not seem to be any easy way to accelerate and collimate protons without invoking an even larger electromagnetic power.

ACKNOWLEDGEMENTS

I thank Pedro Avelino, Teresa Lago and their colleagues for the invitation to attend this meeting and financial support. Support under NSF grant AST 95-29170 and NASA grants NAG5-2837, NAG5-7007 is gratefully noted.

References

Balbus, S. A. & Hawley, J. F. 1998 RMP in press

Blandford, R. D. 1998a Proc. Ringberg Conference on M87 ed. H.-J. Röser & K. Meisenheimer Garching: MPG

Blandford, 1998b Accretion Processes in Astrophysics ed. S. S. Holt & T. R. Kallman New York: AIP

Falcke, H. & Melia, F. V. 1997 ApJ 479 740

Hartmann, R. C. it et al. 1996 ApJ 461 698

Hazard, C., Mackey, M. B. & Shimmins, A. J. 1963 Nature 197 1037
Herrnstein, J. *et al.* 1996 ApJ 468 L17
Levinson, A. & Blandford, R. D. 1996 ApJ 456 L29
Livio, M. 1997 Proc. IAU Colloq 163. ed. D. Wickramasinghe, G. Bicknell & L. Ferrario San Francisco : ASP
Magorrian, J. *et al.* 1997 AJ in press
Maoz, E. 1995 ApJ 447 L91
Morgan, E., Remiilard, R. & Greiner, J. 1997 ApJ 482 993
Narayan, R. & Yi, I. 1994 ApJ 428 L13
Nowak, M. A., Wagoner, R. V., Begelman, M. C. & Lehr, D. E. 1997 ApJ 477 L91
Salpeter, E. E. 1964 ApJ 140 796
Schmidt, M. 1963 Nature 197 1040
Tanaka, Y. et al 1995 Nature 375 659
Zel'dovich, Ya. B. & Novikov, I. D. 1964 Doklad. Acad. Nauk. SSSR 1964 158 811
Zhang, S. N., Cui, W & Chen, W. 1997 ApJ 482 L155

SOFT X-RAY ABSORPTION IN AGN SEEN BY BEPPOSAX/LECS

ASTRID ORR, A.N. PARMAR
Astrophysics Division, Space Science Department of ESA, ESTEC
P.O. Box 299, 2200 AG Noordwijk, The Netherlands

1. Introduction

The X-ray observatory BeppoSAX is a joint Italian and Dutch astronomy satellite launched in April 1996. It carries on-board a set of co-aligned narrow field spectroscopic instruments (NFI) which operate over an unprecedented wide energy band between 0.1 and 300 keV (Boella *et al.*, 1997a) and have imaging capabilities below 10 keV. In addition, a pair of wide field cameras (2–30 keV) are aligned perpendicularly to the NFI. Among the NFI, the Low and Medium Energy Concentrator Spectrometers are position sensitive gas scintillation proportional counters (GSPC), detecting photons in the range 0.1–10 keV (LECS) and 1.3–10 keV (MECS) respectively. Both the LECS and MECS have identical grazing incidence telescopes.

The LECS (Parmar *et al.*, 1997) was provided by the Space Science Department of the European Space Agency. It has an energy resolution of 8.8% FWHM at 6 keV and 32% FWHM at 0.28 keV. The angular resolution and effective area at these energies are 2.1′ and 9.7′, 50 cm^2 and 22 cm^2 respectively with a circular field of view of 37′ diameter. It achieves its extended low energy sensitivity thanks to a novel design which dispenses with the separate drift and scintillation regions in conventional GSPC's which would produce a significant degradation in low-energy performance. The LECS also includes an extremely thin entrance window which allows a good transmission of low energy photons.

The broad-band capabilities of BeppoSAX, as well as its spectral resolution and sensitivity down to 0.1 keV, make it particularly well suited to the study of active galactic nuclei (AGN). In addition, above 1.8 keV the effective area of the MECS (Boella *et al.*, 1997b) is larger than that of the LECS, with similar energy resolution. Therefore the combination of simultaneous data from both the LECS and MECS can bring strong constraints on spectral measurements between 0.1–10 keV.

AGN are known to be highly variable and to emit powerfully, from ra-

Astrophysics and Space Science is the original source of publication of his article. It is recommended that this article is cited as: *Astrophysics and Space Science* **261**: 253–256, 1999.

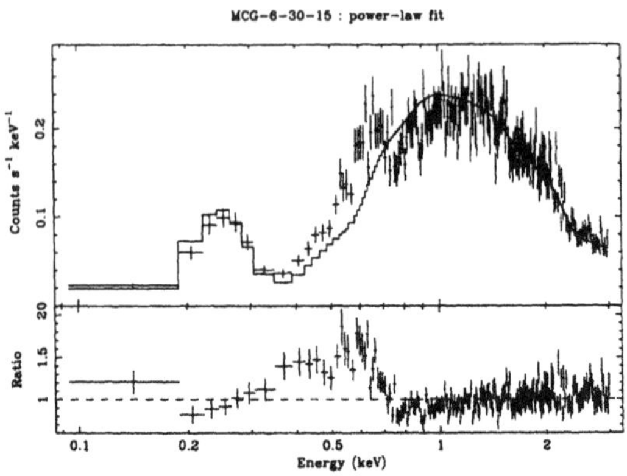

Figure 1. A power-law fit, including galactic neutral absorption, to the LECS spectrum of MCG-6-30-15. Large residuals due to a blend of O VII and O VIII K-edges are clearly visible below 1 keV. The lower plot shows the data-to model ratio.

dio waves to the most energetic γ rays. Their spectral energy distributions seem to consist of more than one component and the origin of these components is still a matter of debate. Due to strong photoelectric absorption by interstellar neutral gas in our Galaxy, it is especially difficult to measure the spectral shape of AGN between the Lyman limit at 912 Å and ~1 keV. At least two spectral components are likely to contribute in this energy range: an underlying non-thermal power-law continuum and the high energy tail of the "blue bump" (Walter, Orr *et al.*, 1994). Furthermore, with recent improvements in X-ray detectors, absorption and emission features due to both partly ionised, "warm", gas and neutral, "cold" line-of-sight material intrinsic to the source are being detected in a growing number of AGN (Reynolds, 1997; George *et al.*, 1998). Above 1 keV, broad Fe $K\alpha$ emission close to 6.4 keV (Nandra & Pounds, 1994) and an additional hard-continuum component (>10 keV), the "Compton hump" (Nandra *et al.*, 1990) blend in with the primary X-ray power-law. We discuss here BeppoSAX observations of two AGN whose X-ray intrinsic luminosities differ by a factor $\sim$ 1000 and which both display some of the spectral components listed above.

2. The Seyfert 1 galaxy MCG-6-30-15

The Seyfert 1 galaxy MCG-6-30-15 was observed during the Science Verification Phase (SVP) of BeppoSAX (Orr *et al.*, 1997; Molendi *et al.*, in prep.). An analysis of the LECS and MECS data reveals the presence of a complex warm absorber. Figure 1 shows that a simple power-law model with neutral absorption gives a poor fit to the spectrum below ~2 keV. The time averaged spectrum between 0.1–4 keV can be well described by a model composed of a power-law with photon index $\Gamma = 2.36 \pm 0.38$ and three absorption edges, for a $\chi^2 = 398.1$ and 394 degrees of freedom. Two edges have their energies fixed at the physical rest frame energies of the O VII and O VIII K-edges ($E_1 = 0.74$ keV, $E_2 = 0.87$ keV, $\tau_1 = 0.91 \pm 0.18$, $\tau_2 = 0.03 \pm 0.15$) and the third edge is at $E_3 = 1.12 \pm 0.10$ with $\tau_3 = 0.19 \pm 0.08$. The third edge is compatible with Ne IX absorption at 1.20 keV. An instrumental origin of these features can be excluded. In fact, a detailed study of the low energy calibration of the LECS (Orr *et al.*, 1997b) has shown that the maximum amplitude of fit residuals due to instrumental effects is less than 10 % of the response-folded model at a given energy above 0.5 keV and below 0.4 keV. Between 0.4–0.5 keV the maximum amplitude is <15%.

Flux variations of a factor of 4 occurred during the observation. In fact, in the first 7 hours of the observation the 0.7–1.5 keV count rate increased four-fold. The main results of a time resolved spectral analysis are:

- The continuum flux and slope as well as the warm absorber are variable on time scales shorter than 2×10^4 s.
- The variations in these two components are not always associated.
- The optical depth of O VII remains approximately constant throughout the observation.
- The optical depth of O VIII displays a significant change during the low state at the start of the observation. Thereafter it remains constant, even during large changes of continuum flux.

The dissimilar variability patterns of O VII and O VIII absorption have been observed previously in this source (Reynolds *et al.*, 1995; Otani *et al.*, 1996) by the Japanese X-ray satellite ASCA (Tanaka *et al.*, 1994). Multi-zone or stratified warm absorbers in photo-ionisation equilibrium have been suggested as a possible explanation (Reynolds *et al.*, 1995). The apparent lack of correlation of the optical depth of O VIII with changes in the continuum flux is different from what ASCA observed (Otani *et al.*, 1996) and may indicate that simple photo-ionisation equilibrium does not apply. This, for instance, can be the case if the electron density of the warm absorber medium is low (Reynolds & Fabian, 1995), thereby causing the recombination time-scale of the gas to be longer than the variability time-scale of the ionising continuum source.

3. The bright quasar 3C 273

The bright quasar 3C 273 was observed quasi-simultaneously by the BeppoSAX and ASCA satellites as part of dedicated program to cross-calibrate the detectors on a number of different X-ray astronomy missions (Orr *et al.*, in prep.). 3C 273 was observed during the BeppoSAX SVP. Simultaneous fits to the ASCA SIS0/SIS1 detectors (Tanaka *et al.*, 1994) and the BeppoSAX LECS data show a good agreement between the data sets below 4 keV. In particular there is no evidence for any systematic deviation in the common energy range 0.6–4 keV. The best model is composed of two power-laws ($\Gamma_{soft} = 6.5 \pm 1.7$; $\Gamma_{hard} = 1.66 \pm 0.02$), neutral absorption (galactic and excess) and an absorption edge ($E_{edge} = 0.58 \pm 0.02$ keV, in the source rest-frame; $\tau = 0.43 \pm 0.19$). The absorption at ~0.6 keV in the present LECS data set is the first unambiguous detection in 3C 273 of such a feature (Grandi *et al.*, 1997). The LECS is the only instrument on BeppoSAX capable of detecting this feature, and the SIS only detects the high energy ($\gtrsim$0.6 keV) side of the absorption spectrum. Therefore, the combination of SIS and LECS data provides a very good means of measuring the feature's overall spectral shape. No other absorption edges are apparent in the range 0.1–4 keV. When attempting to fit an absorption edge to a slight dip at ~1.3 keV (source rest frame) in the SIS spectra, the improvement in fit statistics is insignificant (less than 95 % significance using the F-statistic) and the upper limit to the optical depth is $\tau = 0.07$ (90% confidence for one parameter of interest).

A satisfactory identification of the single spectral feature must describe the spectrum self-consistently. The feature may be identified with an O II or an O III K-edge, which have physical rest frame energies of 0.58 and 0.6 keV, respectively. However, if the edge is O II the opacity from He I may be significant. Finally, an energy shift due to bulk motion of the absorber (and therefore another identification) cannot be excluded.

The authors wish to thank the BeppoSAX science team members. AO acknowledges an ESA Fellowship.

References

Boella G. et al. (1997a), *A&A Suppl.S.*, **122**, 299
Boella G. et al. (1997b), *A&A Suppl.S.*, **122**, 327
George I. et al. (1998), *ApJSuppl.*, **114**, 73
Grandi P. et al. (1997), *A&A*, **325**, L17
Molendi S. et al. (in prep.)
Nandra K. et al. (1994), *Nature*, **344**, 132
Nandra K., Pounds K. (1994), *M.N.R.A.S.*, **268**, 405
Orr A. et al. (1997), *A&A* , **324**, L77
Orr A. et al. (1997b), *BeppoSAX SDC technical report*, **TR-15**
Orr A. et al. (in prep. 1998)
Otani C. et al. (1996), *P.A.S.J.* , **48**, 211
Parmar A.N. et al. (1997), *A&A Suppl.S.*, **122**, 309
Reynolds C. et al. (1995), *M.N.R.A.S.*, **277**, 901
Reynolds C. & Fabian A. (1995), *M.N.R.A.S.*, **273**, 1167
Reynolds C. (1997), *M.N.R.A.S.*, **286**, 513
Tanaka Y. et al. (1994), *P.A.S.J.*, **46**, L37
Walter R., Orr A. et al. (1994), *A&A*, **285**, 119

THE ACTIVE CD GALAXY OF ABELL 2390

PEDRO AUGUSTO
Centro de Astrofísica da Universidade do Porto
Rua do Campo Alegre, 823, 4150 Porto, Portugal

The rich cluster Abell 2390 ($z = 0.23$) contains about 850 galaxies within 1 Mpc of its centre (Le Borgne et al. 1991; Yee et al. 1996) and one of the strongest cooling flows ever studied (Pierre et al. 1996). The central cD galaxy is surrounded by 17 galaxies within 50 kpc and by two, towards south, only 10 kpc away. This cD (the most luminous flat–spectrum radio source in a cooling flow cluster) was observed with the MERLIN and the VLBA at 5 GHz during a campaign to search for $90-300$ mas gravitational multiple images (Augusto et al. 1998). The combined map is presented in Figure 1. This source is classified as a compact symmetric object with a size of 0.8 h_{75}^{-1} kpc, thought to be a young radio galaxy. The structure on the northern radio jet may be due to precession as the radio source grows. This structure is most likely influenced by the dense environment surrounding the cD galaxy. ROSAT HRI X-ray data (Pierre et al. 1996) shows extended emission aligned with the shape of the outer optical cD 'halo' and with the main cluster direction at p.a. $\sim -60°$. However, the very central and brightest pixels of the X-ray data indicate a clear north-south orientation, the same orientation as the one of the radio double source although in a scale more than twenty times larger. The X-ray/radio correlation suggests that the radio emission may have a strong counterpart in the X-rays, perhaps superposed on the emission from the cooling flow. We have recently made EVN+MERLIN 18 cm observations in a long 4-hour run. These data may allow us to uncover all hidden extended structure of Figure 1. Spectral index maps will be produced to get estimates of ages for the shocked interstellar gas. Abell 2390 will be one of the first targets of the AXAF mission: the sub-arcsecond resolution images will test the X-ray/radio alignment. I wish to thank the Junta Nacional de Investigação Científica e Tecnológica for the grant Praxis XXI BPD 9985/96 and Alastair Edge for a valuable discussion.

Augusto, P., Wilkinson, P.N., Browne, I.W.A. (1998), *MNRAS*, subm.
Le Borgne, J.-F. et al. (1991), *Astron. & Astroph. Sup.*, **88**, 133
Pierre, M. et al. (1996), *Astron. & Astroph.*, **311**, 413
Yee, H.K.C. et al. (1996), *Astroph. J. Sup.*, **102**, 289

Astrophysics and Space Science is the original source of publication of his article. It is recommended that this article is cited as: *Astrophysics and Space Science* **261**: 257–258, 1999.

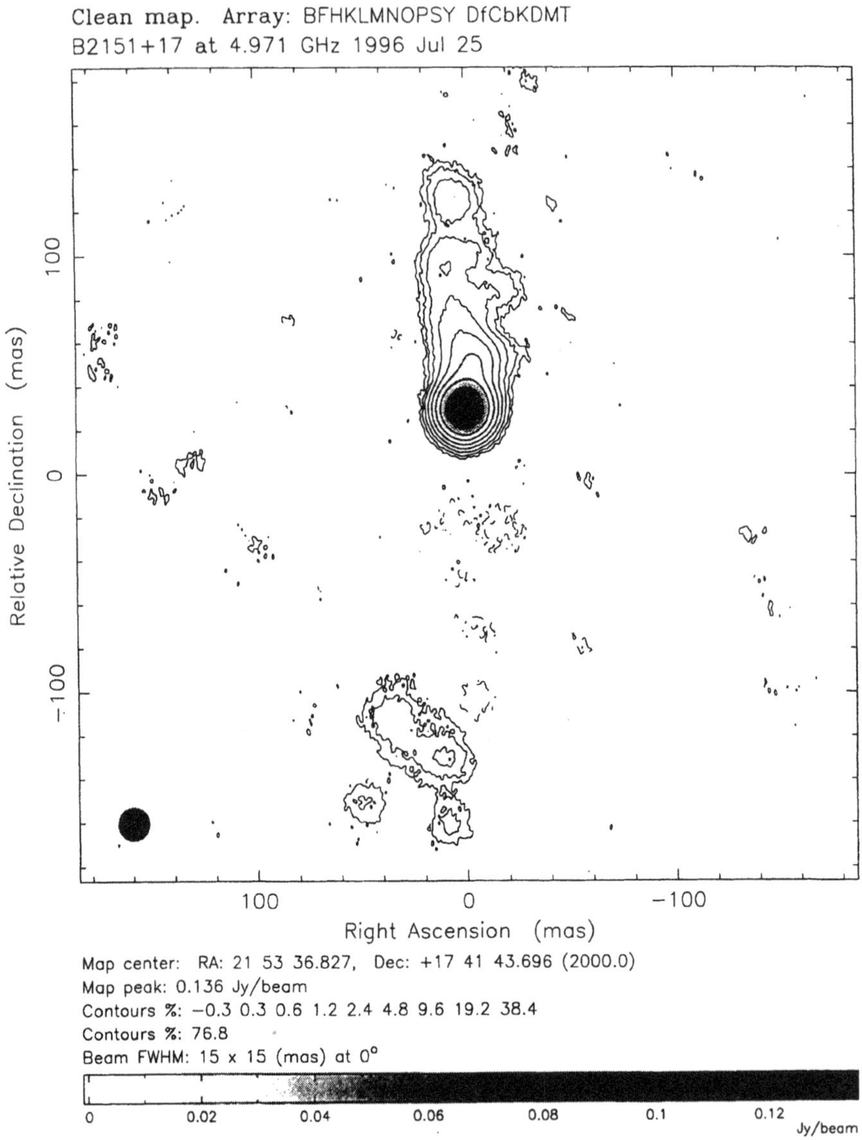

Figure 1. **The MERLIN+VLBA 5 GHz one-hour snapshot observations of the radio source B2151+174, identified with the cD galaxy of Abell 2390.**

THE 200 MJY SUBSAMPLE AT 1.4 GHZ

MERCEDES ESTEVES FILHO AND MARIA JOÃO MARCHÃ
University of Lisbon

The sources studied belong to the 200 mJy sample of optically bright flat spectrum radio sources (Marchã, Browne, Impey and Smith 1996). The main objective of the sample was to detect low luminosity *BL Lacs* and study any "natural" breaks in the properties of the 200 mJy objects. Polarimetric and spectroscopic studies enabled the division of the 200 mJy sources into 4 categories: *BL Lacs*; radio-galaxies with *Seyfert*-type spectra; GALs, galaxies with lines of equivalent width ≤ 20Å but whose properties did not enable us to infer a source of non-thermal continuum; 2 peculiar objects that do not fall into any of the above categories.

To complement the work already done on the 200 mJy objects, 18 of these sources were observed at 1.4 GHz with the VLA at various array configurations. The main objective was to investigate if the same systematic differences between categories were observed at radio frequencies as was observed with the polarimetric and spectroscopic data.

With that in mind, the following radio-parameters were measured for each source: the extended isotropic luminosity, L_{ext}; the R parameter defined as the anisotropic core lumniosity, L_{core}, divided by the extended luminosity; the 'radio-core polarization', m_{core}; depolarization in the core. The radio-morphologies were also studied taking into consideration the presence and intensity of cores, jets and halo-type extended emission. The 18 subsample objects were divided into 2 categories: one of *BL Lacs* and another of GALs where we include 2 *Seyfert*-type objects.

In terms of radio-morphology, the *BL Lacs* were predominantly one-sided or resolved cores while the non-*BL Lacs* were morphologically more heterogeneous, distributed almost equally between one-sided, two-sided and resolved radio-core morphology.

Using a Hubble constant of 75 km/Mpc, q_0=0 and spectral indexes of α=0 for the core emission and α=1 for the extended emission region, the extrapolated FR I/FR II extended luminosity break at the radio frequency of 1465 MHz is $L_{ext}=10^{31.0}$-$10^{31.5}$ erg s^{-1} Hz^{-1}. All except one *BL Lac*

Astrophysics and Space Science is the original source of publication of his article. It is recommended that this article is cited as: *Astrophysics and Space Science* **261**: 259–260, 1999.

have extended luminosity below the FR I/FR II break, while 12 out of the 15 non-*BL lacs* are also below the break, that is, these are objects with FR I-type extended luminosities. The only object presenting clear FR II-type extended luminosity is a *Seyfert*-type radio-galaxy with a core-jet-type morphology.

The R parameter, as defined by Orr and Browne (1982), is frequently used as a measure of the "beaming" of the object, that is of the alignment of the relativistic jet with respect to the observer. The objects are said to be aligned if Log $R \geq 0$. Analysis of the R distribution reveals that all the *BL Lacs* are aligned by the above criteria. One the other hand, 11 out of the 15 non-*BL lacs* are also aligned objects. However, the object with highest R parameter is a *Seyfert*-type radio-galaxy with a core-double lobe morphology.

Strong radio core polarization in sources is a good diagnostic of an extra source of non-thermal continuum. Most of the objects in the sample have radio-core polarizations between 1 and 4%. However, the objects with the largest radio-core polarizations of the sample, 11 and 9% respectively, are classified as GALs. Also, four of the subsample objects have no detectable radio-core polarization, two of them being GALs. However, no systematic difference between the radio-core polarizations of the *BL Lacs* and non-*BL Lacs* of the subsample can be seen.

The sources were also analysed in terms of depolarization, defined as the ratio between the radio-core polarization at 1385 MHz and 1465 MHz. Sources are said to be depolarized if the ratio is substantially different from one. This depolarization depends on underlying magnetic fields of the depolarizing medium and may be different among different types of objects. 5 out of 8 *BL Lacs* and 5 out of 10 non-*Bl Lacs* with measured polarization are not significantly depolarized. Their distributions are statistically indistinguishable so that no significant differences between depolarizing material can be established between the two categories.

Therefore, we conclude that, based on the 1.4 GHz data alone, the *BL Lacs* are not distinguishable from the rest of the subsample objects and we do not observe the systematic differences between the categories as is seen with the optical data of Marchã *et al.* 1996. However, we propose further observations of these subsample objects, namely follow-up VLBI observations that include polarization information.

References

1. M. J. M. Marchã, I. W. A. Browne, C. D. Impey and P. S. Smith, (1996), *MNRAS* **281**, 425–448
2. M. J. L. Orr and I. W. A. Browne, (1982), *MNRAS* **200**, 1067–1080

THE POPULATION OF KPC-SCALE FLAT-SPECTRUM RADIO SOURCES

PEDRO AUGUSTO
Centro de Astrofísica da Universidade do Porto
Rua do Campo Alegre, 823, 4150 Porto, Portugal

AND

PETER N. WILKINSON AND IAN W.A. BROWNE
Nuffield Radio Astronomy Laboratories
Jodrell Bank, Macclesfied, Cheshire, SK11 9DL, UK

Abstract. In this paper we present a subsample of 55 flat-spectrum radio sources dominated by ($\sim$ 100 mas) kpc-scale structure, selected from a parent sample of 1665 VLA sources. Most are core-jets and 23 are CSO/MSO candidates. Properties of the subsample are discussed.

1. Introduction

We have conducted a search for gravitational multiple imaging on $90 - 300$ mas scales, corresponding to lenses with masses on the range $10^8 - 10^9 M_\odot$ (Augusto, Wilkinson & Browne 1998). Starting from two VLA 8.4 GHz surveys with $\sim$ 200 mas resolution (the Jodrell-VLA Astrometric Survey (JVAS), Patnaik et al. (1992), and the Cosmic Lens All-Sky Survey (CLASS), Myers et al. 1995), we have selected a parent sample of 1665 flat-spectrum radio sources to search for sub-arcsecond lensing. This search has failed to reveal any new high-magnification lensed systems. The implications are described in Augusto et al. (1996) and Augusto & Wilkinson (1998).

The search has allowed the first systematic inspection of kpc-scale flat-spectrum radio sources. The compact/medium symmetric objects (CSO/MSOs), discussed in the end of this paper, are believed to be the precursors of the large classical double radio sources.

Astrophysics and Space Science is the original source of publication of his article. It is recommended that this article is cited as: *Astrophysics and Space Science* **261**: 261–264, 1999.

2. The Selection of the Subsample of kpc-Scale Radio Sources

We have made use of two surveys of radio sources produced with the VLA operating at 8.4 GHz containing a total of $\sim$ 4800 radio sources. One is the JVAS with $\sim$ 2300 sources and the other the CLASS, at the time of selection with $\sim$ 2500 sources (CLASS1). The JVAS contains all radio sources in the Northern Hemisphere with $S_5 \geq 200$ mJy, $\alpha_{1.4}^{5} < 0.5$ ($S_\nu \propto \nu^{-\alpha}$) and $|b^{II}| \geq 2\overset{\circ}{.}5$. The CLASS1 contains all radio sources with $35° \leq \delta \leq 75°$, $S_5 \geq 50$ mJy, $\alpha_{L-band}^{5} \leq 0.6$ and $|b^{II}| \geq 10°$. To produce the parent sample containing 1665 objects, we have 'cleaned' the JVAS+CLASS1 sample using the criteria: i) $S_{8.4} \geq 100$ mJy; ii) $\alpha_{1.4}^{5} < 0.5$; iii) $|b^{II}| \geq 10°$. This selection has preserved 1416 objects from JVAS and 249 from CLASS1.

Finally, to produce the subsample of 55 kpc-scale flat-spectrum radio sources (49 of which from JVAS), we have used a resolution criterion on the parent sample, selecting sources with sizes $90 - 300$ mas as described in Augusto et al. (1998). This size corresponds, at intermediate z, to $\sim$ kpc scales. All 55 sources were mapped with MERLIN at 5 GHz and about a fourth with the VLBA at 5 GHz. Currently, we are carrying out a survey with the EVN at 1.6 GHz of all the 55 sources, to study them in more detail.

3. Properties of the Subsample

The radio spectra (L to K band) of the 55 kpc-scale flat-spectrum radio sources follows a power-law for virtually all objects. There are no GHz-Peaked Spectrum sources in this subsample. This indicates a turnover at frequencies of a few hundred MHz, consistent with their kpc-scale structures. Furthermore, the distribution of spectral indices ($\alpha_{1.4}^{5}$) is different for the subsample ($< \alpha_{1.4}^{5} > \sim 0.3$) or the parent sample ($< \alpha_{1.4}^{5} > \sim 0.0$). This is also consistent with the resolution selection criterion.

As regards the 8.4 GHz fractional polarization distributions, they are the same for the subsample and the parent sample, with $<p>\sim 2\%$.

The Palomar Observatory Sky Survey (POSS) optical identifications of both the parent sample and the subsample, reveal differences in that there are proportionally more blue stellar objects (BSOs) in the parent sample than on the subsample. On the other hand, the latter contains three times more galaxies. The number of Empty Fields (EF) is about 30% on both.

The redshift distribution of the 19 sources in the subsample with such information has $<z>\simeq 0.7$. This distribution is clearly biased against the optically faint sources. The 16 EFs on POSS are probably objects at $z \sim 1$. When we finalize the spectroscopic study of our subsample, we will produce an *unbiased* redshift distribution.

4. The Two Classes of kpc-Scale Flat-Spectrum Radio Sources

We have identified two different classes of kpc-scale flat-spectrum radio sources. Their typical maps are presented in Figure 1. As suggested by their radio-morphology, we have called them the 'CSO/MSO' ($\sim$ 45% of the objects) and the 'core-jet' (55%) classes. The CSO/MSOs present radio structures typical of radio galaxies, although on much smaller scales. Half of the CSO/MSOs possess a dominating core. The core-jets present a nucleus plus one-sided jet. Roughly half of the jets are straight. The remaining jets are heavily distorted (bent by more than 90°).

The two classes present different 8.4 GHz polarization and optical properties. As regards polarization, there is a clear difference between the two: CSO/MSOs have lower fractional polarizations ($\sim$ 1%), sitting at the low polarization end of the distribution. The 'core-jet' class presents a long tail up to 11% polarization. As regards the POSS optical identifications, about 90% of the BSOs are 'core-jets'. We interpret these as members of a QSO/BLLac population, as suggested by the few identified BSOs. For the other class, it is noteworthy that about two-thirds of POSS galaxies are 'CSO/MSO' and the reverse is also true: about three-quarters of the CSO/MSOs are POSS galaxies or EFs. We interpret CSO/MSOs as galaxies in the optical. The EFs are most likely $m < 20$ galaxies.

The number of CSO/MSOs on $90 - 300$ mas scales is only $\sim$ 1.4% of the flat-spectrum radio population. This is to be compared with $\sim 5 - 10$% on $1 - 50$ mas from VLBI studies (eg. Taylor et al. 1996) to conclude that, somehow, the evolution of these 'young' radio galaxies goes through expanding their lobes at an early stage, before growing to eventually become the large radio galaxies we see today. When we have spectroscopic data for all the CSO/MSO candidates in our sample, we will be able to quantify this evolution more accurately.

Pedro Augusto acknowledges the Junta Nacional de Investigação Científica e Tecnológica for the grant Praxis XXI BPD 9985/96 and the hospitality of the Universidade da Madeira.

5. References

Augusto, P. et al. (1996), *Nuclear Physics B (Proc. Suppl.)*, **51B**, 118

Augusto, P. and Wilkinson, P.N. (1998), *M.N.R.A.S.*, in prep

Augusto, P., Wilkinson, P.N. and Browne, I.W.A. (1998), *M.N.R.A.S.*, submitted

Myers, S.T. et al. (1995), *Astroph. J. Letters*, **447**, L5

Patnaik, A.R. et al. (1992), *M.N.R.A.S.*, **254**, 655

Taylor, G.B. et al. (1996), *Astroph. J. Sup.*, **107**, 37

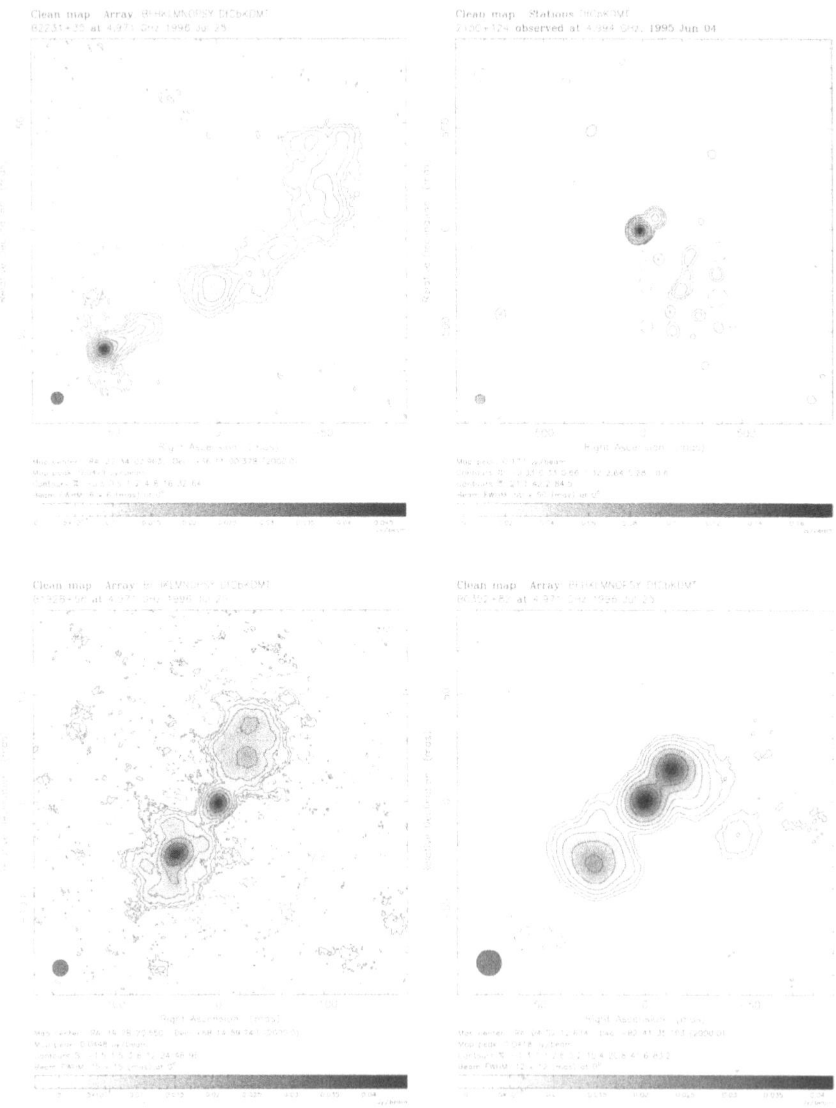

Figure 1. Three MERLIN+VLBA 5 GHz maps and one (top right) MERLIN 5 GHz map of typical kpc-scale flat-spectrum radio sources. On top, we present two examples of the class of 'core-jet' sources. On the bottom, we present two examples of the class of 'CSO/MSO' sources.

WHAT DO THE CBR FLUCTUATIONS TELL US?

A. BLANCHARD, J.G. BARTLETT, M. DOUSPIS, M. LEDOUR
Observatoire astronomique de Strasbourg,
11 rue de l'université,
67 000 Strasbourg
France

Abstract The detection of the fluctuations in the CMB by COBE has opened a new area of modern cosmology. The angular spectrum of these fluctuations potentially carries a lot of information, allowing for the possibility of determining accurately the cosmological parameters of the universe and giving constraint on scenario of the early universe. The detections of fluctuations on angular scale of the order of one degree are now numerous and allow one to start to put some constraints. The result presented here summarize our work on the question by the end of 1997. We first show that the Doppler peak is reasonably detected. The nature of the fluctuations has been investigated and found in agreement with a Gaussian distribution. Under the assumption of a critical density universe a large baryon fraction and a low Hubble constant are favored. Open models with $\Omega_0 \leq 0.4$ seems not adequate to describe the data.

1. Introduction

The quest for the values of the cosmological parameters has always been a central question in cosmology. It may even be considered as the most fundamental question. Since the discovery of the expansion law by Hubble, this activity has received a lot of effort, although it has become more clear and more clear that the measure of the geometry of the universe is a very difficult task in practice. The reason is that most of the classical tests rely on the determination of the observed properties of objects, like their luminosities, but who intrinsic properties are not known, like the intrinsic luminosities. Therefore one is often left with the solution of assuming that

Astrophysics and Space Science is the original source of publication of his article. It is recommended that this article is cited as: *Astrophysics and Space Science* **261**: 265–274, 1999.

the objects are standard candles, which implies that the properties of the objects remain unchanged over a period of time which should be longer than the age of the universe itself... A situation which is almost desperate. An alternative is the one offered by supernova, for which the typical time scale is much shorter than the age of the universe. It still remains to be proven that parameters which determine the intrinsic luminosity of a supernova do not evolved significantly over a period of the order of the Hubble time. The determination of the cosmological parameters from the CMB fluctuations offers a new potential way which is profoundly different from other techniques: it relies directly on the geometrical properties of the universe itself and on relatively simple physics, the evolution of the perturbations in the baryon-photon fluid during decoupling. There are few such tests, the evolution of the abundance of X-rays clusters being one of them (see the contribution by Sadat in these proceedings).

The precise measurement of the fluctuations of the CMB will allow one to confirm the gravitational instability picture and then to measure with a high level of precision , without ambiguity, the various cosmological parameters, like the Hubble constant, the baryonic content of the universe, its mean density, the cosmological constant, etc... This has motivated several experiments as well as the decisions to launch two space missions, MAP and Planck Surveyor. However, the present-day observations already allows us to draw some conclusions, which I will discuss shortly.

2. Basic principles

The CMB is a direct picture of the Universe as it was few hundred thousands years after the Big Bang (at least in the simple case where reionization was insufficient to blur up this picture). The existence of large-scale structures in the present day universe implies the existence of seeds, which are likely to have left some imprint on the CMB image. In the standard cosmological gravitational instability picture, the "passive fluctuations" case, these seeds are generated in the very early universe (essentially because we have not found any alternative way to generate them at a later epoch...). Because these fluctuations then exist on scales greater than the horizon, an inflation-like scenario for their generation is very likely. A possible alternative is the "active picture" (a terminology applying to both the fluctuations and to the people in the field...) in which the large-scale structure is generated from topological defects.

2.1. INFLATION

The pre-existence of seeds which lead to the formation of large scale structures had been postulated well before the invention of the inflationary scenario. However, Inflation are certainly a major breakthrough in modern cosmology: it allows well defined predictions for the type of fluctuations needed. For this reason Inflation makes very well defined predictions for the properties that the fluctuations of the CMB should possess (see the contribution by Liddle in these proceedings).

In its simplest version, inflation is extremely predictive and therefore extremely easy to falsify. Inflation predicts the value of the density parameter of the universe, $\Omega_0 \sim 1.0$, the spectrum of the primordial fluctuations, $n \sim 1.0$ (although inflation fails to predict its amplitude in a simple way) and their Gaussian nature. The power spectrum of matter fluctuations present at the epoch of structure formation depends on the nature of the dark matter (which has to be non-baryonic if primordial nucleosynthesis and the abundances of light elements have to agree). The reason is that the power spectrum of the primordial fluctuations is altered during its late evolution, something which is described by the tranfer function:

$$P_0(k) = T(k) * P_i(k)$$

2.2. PRESENT-DAY OBSERVATIONS

In the following, we will not try to directly test Inflation, because there is still not enough precision in present-day measurements of the CMB fluctuations to allow one to investigate the large variety of inflationary scenarios, but we try instead to examine what can be inferred from present-day observations. This subject is changing rapidly, as there are several experiments which have already been performed and for which data are in the analysis process. It might be that the conclusions drawn in this text will be out of date when it is published... or confirmed!

The fluctuations of the CBR on the celestial sphere can be decomposed over the spherical harmonics:

$$\frac{\Delta T}{T} = \Sigma a_{lm} Y_{lm}$$

Gaussian fluctuations can be characterised by their angular power spectrum:

$$C_l = < a^*_{lm} a_{lm} >$$

The shape of the C_l curve depends on the properties of the fluctuations which are assumed. At small l, the contribution is dominated by the so called Sachs-Wolfe effect. The measurement by DMR aboard the COBE

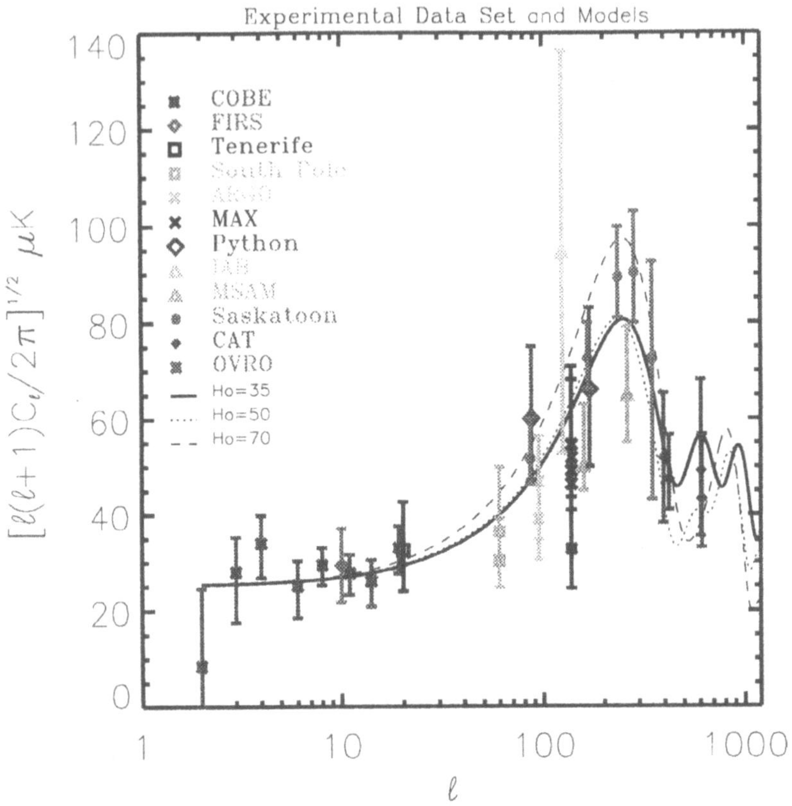

Figure 1. **Summary of present-day measurements of CBR fluctuations: theoretical models based on passive fluctuations predict the existence of a Doppler peak around $l \sim 250$. Observations on these scales provide a tentative evidence that this is actually the case.**

satellite has revealed the existence of fluctuations on these scales, and has furthermore shown that these fluctuations scale according to the inflation paradigm: $n \sim 1$. Furthermore these fluctuations are close to the level expected in the standard Cold Dark Matter picture, although a factor of 2 higher than predicted. This has been the fatal problem of a long list of difficulties... For CDM, it nevertheless remarkable that the level as well as the properties of the fluctuations were not far from the theoretical predictions.

2.2.1. *Doppler peak*

Beyond the existence of fluctuations at several scales, a general pattern expected for the fluctuations power spectrum is the existence of a so-called Doppler peak: the relative amplitude of the fluctuations should be locally maximum around $l \sim 200 - 400$, corresponding to angular scales of $\sim 1^o$. The reality of the Doppler peak in the data is critical for drawing any conclusion for cosmology. For instance, its actual presence would favour a inflationary model against a defect model, even if such conclusion is not yet definitive (Durrer & Sakellariadou, 1997).

There is no unique way to test for the possible existence of the Doppler peak. To this aim, we have analysed present-day data in the following way: we have fit a straight line to the data and evaluated the goodness-of-fit to the data. We are actually answering the question': are the data compatible with a straight line, with Gaussian uncertainties? A negative answer –as we got– is therefore indicative that either the data are not consistent with a straight line or the uncertainties are not Gaussian. The results of this analysis are commented on by Blanchard & Douspis (1997). It has been found that the existence of the Doppler peak (in the above sense) relies essentially on the Saskatoon experiment. Moreover, the existence of an uncertainty in the calibration of this experiment makes the conclusion a bit uncertain : only the new calibration by Leitch (1998), allows a firm conclusion. This was supplied by another test: we have estimated the best C_l model in order to fit the observations and ask the question: are the departures from the theoretical curve Gaussian distributed ? The answer we got was *yes*. (Using the same test, we were able to strongly reject a Poisson distribution with the same number of points.) The actual distribution is not expected to be Gaussian distributed: in fact, each experiment is likely to have its own distribution function which might be quite difficult to estimate in practice. However, it is still interesting to check whether the distribution could be regarded as Gaussian, probably meaning that nothing strange (or frightening) is happening: for instance if the uncertainties would be largely overestimated, one would find that the distribution is not Gaussian. With this indication that the departure from the best model is Gaussian, we may now average several experiments at similar l. We have done this average in 5 different bins in l. Such average values have also been done by others authors. This is extremely useful to understand in more detail the meaning of the likelihood estimation of the parameters.

This first step analysis is rich in consequences: the observed distribution of measured C_l, although quite noisy, does not reveal anything "strange". It is reasonable therefore to use these measurements for cosmological interpretation, like putting constraints on scenarios. For instance, the fact that the something like a Doppler peak is detected, together with the fact that the distribution of measurements is consistent with a Gaussian distribution is a serious case against the defect models. One should keep in mind, however, that the defect theory is not well understood and does not allow one to predict the C_l in a way as direct as in the case of inflationary models.

3. Parameter estimation

The exciting perspectives that CMB fluctuations experiments will offer are numerous: one first could test the validity of the general instability picture for structure formation. If the answer is positive, than the CMB fluctuations will provide a extraordinary source of information, both for the determination of the cosmological parameters and for theories of the early universe. This will also constitute a fundamental basis for early universe physics, with a detailed verification of inflation (see the contribution of A.Liddle in this proceedings). If the fluctuations are inconsistent with standard gravitational instability picture, but favour topological defect models, this could be an even more exiting result: CMB would then be a fundamental way to investigate the properties of these objects.

3.1. THE METHOD

Present-day data do not allow a detailed investigation. As we have already mentioned, the presence of a Doppler peak disfavours defect models, but the data are not good enough that such scenarios are fully ruled out. However, as we will see in more detailled way the detection of the Doppler peak, if real, allows one to draw some conclusion. Again, it should be stressed that such detection relies on the Saskatoon experiment. Clearly one would like to have a second experiment confirming this result. There are several experiments going which can soon give this critical information. Furthermore, the frequency range of the Saskatoon experiment is limited, so it has not be entirely demonstrated that an unidentified interstellar foreground is contaminating most of the data.

The general problem of estimating the cosmological parameters from CMB data has been widely discussed in the recent literature (See for instance Bunn, 1997). Let us first discuss briefly the general method. Let us assume that we have a pixelized map of the sky $\{x_i\}$, with a perfect experiment without any noise. The likelihood is written:

$$\mathcal{L} = \Pi p(x_i, \lambda) \tag{1}$$

where λ is the parameter describing the theory. In this espression, the x_i are given by the experiment and the λ are the parameter we want to estimate. The best parameter will be the one which maximise the likelihood. The quantity we may want to estimate is the angular power spectrum: C_l. Most experiments provide their estimate of the C_l, $\tilde{C}_l|_i$. We may then try to write the likelihood:

$$\mathcal{L} = \Pi p(\tilde{C}_l|_i, \lambda) \tag{2}$$

In such approach, we can consider the $\{\tilde{C}_l|_i\}$ as a compressed form of the data. For an ideal experiment (full sky, without noise), the C_l would be a particularly efficient way to compress the data: to test a Gaussian theory, there is no lost of information. Furthermore, the $\tilde{C}_l|_i$ would then be independent. For present day experiments the situation is not so simple. However, most experiments can be considered as independent. Furthermore, assuming that the distribution function of the experimental measurements can be considered as Gaussian, we can write the likelihood function as

$$\mathcal{L} = \Pi \exp\left(- \left(\frac{\tilde{C}_l|_i - C_l(\lambda)}{\sigma_i} \right)^2 /2 \right) \tag{3}$$

This procedure has been followed by Lineweaver et al (1997) and Hancock et al. (1997). It permits one to have a first look at the constraints that can be set from present day experiments.

3.2. RESULTS

The number of free parameters in the problem of anisotropies of the CMB, even in the class of Inflationary models, is quite large. Despite the availability of CMBfast, which has allowed these studies, we are still limited by the number of parameters that can be analyzed in detail. For this reason, we have investigated only a sub-class of these parameters: $\Omega_0, \Omega_b, H_0, n, Q$ (Q being the normalization of the anisotropies). The other main parameters which are neglected are possible reionization and the possible contribution of gravitational waves.

3.2.1. *The $\Omega_b - H_0$ diagram*

In this case, we kept $\Omega_0 = 1.$, the two remaining parameters are Q, n. We marginalized on Q and n, meaning that for each model we use the best n, Q. This allows us to determine the best $\Omega_b - H_0$. The confidence range can be determined by a standard method: for instance the contour $\Delta \mathrm{L} = 1.0$ can be projected to get the 68% confidence range on one parameter. One can also use the contour $\Delta \mathrm{L} = 2.3$ to find the 68% contour for the two parameters. Of course the confidence level holds only for truly Gaussian variables and is therefore only indicative in our case. It is interesting to examine the case $n = 1$ first. In this case, we obtain contours centred on a low value of H_0, but with Ω_b well consistent with nucleosynthesis. This constraint results both from the location and the height of the peak. Allowing n to vary allows a wider range of Ω_b, which is then essentially unconstrained. However, a low Hubble constant is still preferred, although the constraint is weak. The additional constraint on the baryonic content

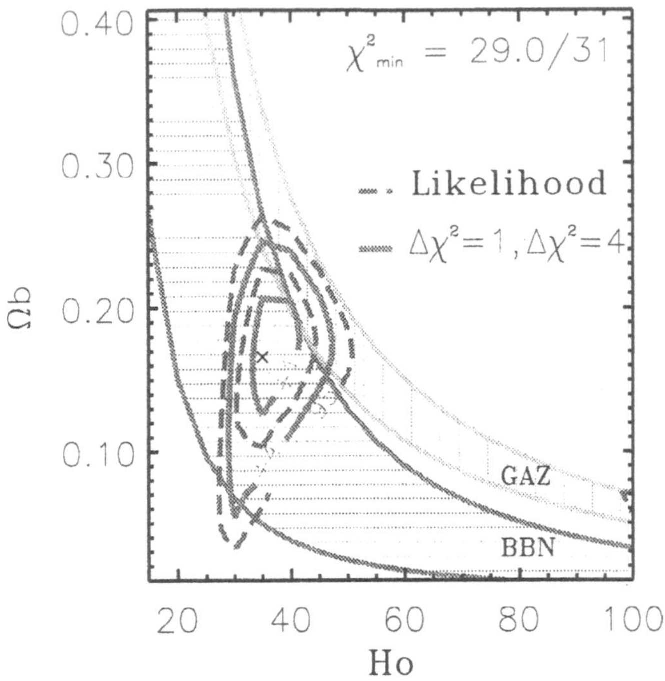

Figure 2. Constraints on critical density models ($\Omega_0 = 1$) in the $H_0 - \Omega_b$ plane inthe case $n = 1.0$. The best model can be estimated provide one assumes a likelihood function. Here we assume a likelihood function for independent Gaussian variables.

of the universe (from either the nucleosynthesis or the baryon fraction in clusters) still restricts the Hubble constant to a low value (in a $\Omega_0 = 1.$ universe) : $H_0 \leq 55$ km/s/Mpc.

3.2.2. *The flat case*

In the case of a flat geometry, allowing a non-zero cosmological constant, one still gets significant constraints in the λ_0, H_0 plane. Interestingly, the low Hubble constant preference relaxed moves up as the λ term becomes dominant: for a $\lambda \sim 0.7$, a Hubble constant as high as 70 is allowed. No interesting constraint is obtained on the cosmological constant itself.

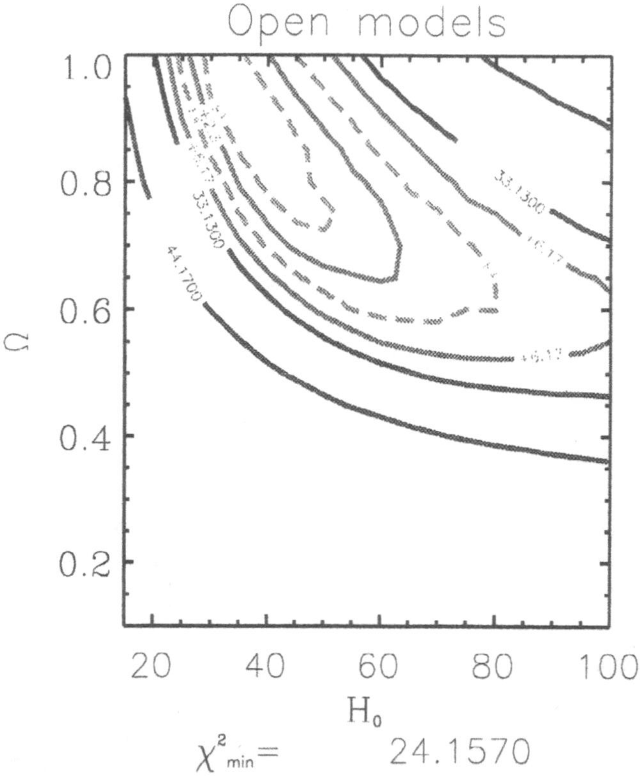

Figure 3. **Constraints on open models : the $H_0 - \Omega_0$ plane. Cosmological models with $\Omega_0 \leq 0.5$ are reasonably excluded. This is due to the absence of Doppler peak at the expected scale..**

3.2.3. *The open case*

The most interesting constraints are placed on the open cosmological models (with a vanishing cosmological constant). Following the same approach, it is possible to produce a diagram $\Omega_0 - H_0$, by marginalising the parameters Q, n, Ω_b. The result is presented on figure 3. One can see that models with $\Omega_0 \leq 0.5$ are disfavoured. In fact it can be shown that such conclusion hold even without taking the Saskatoon experiment into account. This conclusion is therefore quite robust.

4. Conclusion

The possible detection of the Doppler peak is reasonabily established by the Saskatoon experiment. This allows one to derive the first cosmological constraint that can be set from present day observations. A large region of the parameter space is already excluded, although it is still difficult to establish definitive conclusions on any single parameter, partly because of the intrinsic degeneracy of the problem. Furthermore, not all the parameters of the problem have been investigated. Nevertheless, we believe the tendencies we find should not be too sensitive to these simplifications.

In the Einstein-de Sitter case, a low value of the Hubble constant is favoured. This could be considered as a success of the proposal made by Bartlett at al. (1995), who emphasised the importance of CMB measurements for testing the possibility of a low Hubble constant. In the case of flat model with an additional cosmological constant, the constraint is not so strong, although a high Hubble constant (≥ 80) is disfavoured. Finally, the most interesting conclusion is probably that open models with $\Omega_0 \leq 0.5$ are put in difficulty by the CMB data. Taking with the evidence from the evolution of X-ray clusters (see the contribution of Sadat in this volume) this result gives serious indication for a high density universe, although the possibility of a non-zero cosmological constant is consistent with present-day observations of the CMB fluctuations.

References

Bartlett J.G., Blanchard A., Silk, J., Turner, M.S. (1995) *Science*, **267**, 980
Bunn, E.F. (1997) The Cosmic Microwave Background, NATO ASI Series, Series C, **Vol 502**, p. 135.
Blanchard, A., Douspis, M. (1997) http://astro.u-strasbg.fr/howfar/blanchard/lpp1.html
Durrer, R., Sakellariadou, M. (1997) *Phys.Rev.* , **Vol. no. D56** , pp. 4480–4493
Hancock, S., Rocha, G., Lasenby, A.N., Gutierrez, C. M. (1997) *MNRAS*, **294**, 1
Lineweaver, C. Barbosa, D., Blanchard, A., Bartlett, J. (1997) *A&A*, **322**, 365

CMB ANGULAR POWER SPECTRUM AND THE COSMOLOGICAL PARAMETERS

ANTÓNIO J. C. DA SILVA AND PAULO CRAWFORD
Centro de Física Nuclear da U. L. e Dep. Física da F.C.U.L., Av. Prof. Gama Pinto, 2 1699 Lisboa Codex, Portugal

In this paper we report the conclusions of a systematic analysis of the dependence of the amplitude, A_p, and position, l_p, of the primary acoustic peak in the CMB angular power spectrum with the cosmological parameters. Following the analysis made by Lineweaver et al in [2] for flat Cold Dark Matter (CDM) models, we obtain theoretical contour maps of A_p and l_p, in various regions of the parameter space, and discuss their behaviour in terms of the relevant cosmological parameters for Flat CDM, Λ-CDM and open CDM (with no Λ) models.

To compute the A_p and l_p diagrams for the matrices of models covering the desired regions of the parameter space, we use the Boltzmann code package CMBFAST (Seljak & Zaldarriaga [4] [5]) in conjunction with a subroutine (peakfinder) we developed, which full analyses the spectra in search for the positions and amplitudes of all peaks and valleys.

We have only considered the contribution of scalar perturbations to the power spectrum, with adiabatic initial conditions and no reionization phenomena. For the other relevant parameters needed by the CMBFAST to compute the C_l's we took a Helium fraction of $Y_{He} = 0.24$, CMB mean temperature $T_0 = 2.728\mu K$, neutrinos density $\Omega_\nu = 0$, 3 families of light neutrinos and normalization of the angular power spectrum given by the Bunn & White fitting formula (provided with the CMBFAST code).

For Flat CDM models we studied the $\Omega_b - h$ and the $\Omega_\Lambda - h$ plane for 3 discrete values of the spectral index n. For $n = 1$, we have obtained the same contour diagrams found by Lineweaver et al [2]. We have also constructed contour maps in the $n - h$ plane for 3 values of $\Omega_b h^2 = const.$ These diagrams are discussed in detail in the pre-print [1].

For open CDM models we considered the $\Omega_0 - h$ plane shown in the figure for 3 values of $\Omega_b h^2$ inside the region favored by the Big Bang Nucleosyntesis ($0.010 \leq \Omega_b h^2 \leq 0.026$). The solid lines enclosed by the dot contours are

Astrophysics and Space Science is the original source of publication of his article. It is recommended that this article is cited as: *Astrophysics and Space Science* **261**: 275–276, 1999.

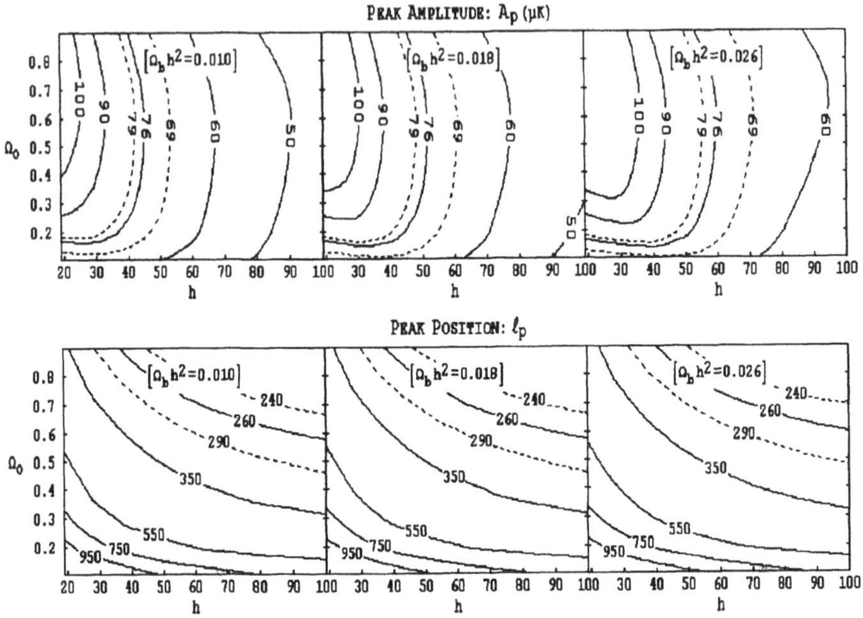

the amplitude and position of the best-fit model found by Lineweaver and Barbosa: $A_p = 76^{+3}_{-9}\mu K$; $l_p = 260^{+30}_{-20}$ [3] (see their comments on error bars).

From the analysis of these diagrams, and the ones obtained for flat CDM models, one can isolate the following general effects on l_p and A_p: - l_p increases with $\Omega_b h^2$, n and decreases with $\Omega_{cdm} h^2$, Ω_0; - A_p increases with $\Omega_b h^2$, n, Ω_0 and decreases with $\Omega_{cdm} h^2$. One must realise that these dependencies account only for the effect of each parameter considered "alone". Nevertheless the shape of the contours on all these diagrams result from the combination of these effects acting together. For example, in the figure, we see that for $h = const.$ and $\Omega_b h^2 = const.$, A_p is not monotonic with Ω_0. This happens because along these vertical lines Ω_0 and $\Omega_{cdm} h^2$ increase and produce competing effects on A_p. About Ω_b, Ω_{cdm} and h, they affect A_p and l_p whenever that implies a variation in $\Omega_b h^2$ and $\Omega_{cdm} h^2$. For flat CDM models A_p and l_p increase with Ω_b, and A_p is not monotonic with h. The effect of Ω_Λ was only analysed for flat models. See [1] for details. The authors wish to thank to D. Barbosa. This work was supported in part by JNICT, FMRH BM 3804/94 and by the project CERN/P/FAE/1164/97.

References

1. António J. C. da Silva and Paulo Crawford. 1998, pre-print in perparation.
2. Charlie Lineweaver et al. A&A, in press, asto-ph/9610133, 1997.
3. Charley Lineweaver and Domingos Barbosa. ApJ, in press, astro-ph/9706077, 1997.
4. Uros Seljak and Matias Zaldarriaga. ApJ 469:437-444,1996 (astro-ph/9603033).
5. Zaldarriaga, Seljak, and Bertschinger. ApJ, in press, astro-ph/9704265, 1997.

CONSTRANING Ω_0 WITH SUNYAEV-ZEL'DOVICH OBSERVATIONS

D. BARBOSA
Observatoire astronomique de Strasbourg, 11 rue de l'université. 67000 Strasbourg, France.
and
CAUP, Rua das Estrelas s/n, 4150 Porto, Portugal

AND

J.G. BARTLETT AND A. BLANCHARD
Observatoire Astronomique de Strasbourg, 11 rue de l'université. 67000 Strasbourg, France.

1. Introduction

As the largest virialized systems in the Universe, clusters of galaxies can be used as probes of the cosmological parameters. In fact, the evolution of the number of cluster of galaxies strongly depends on Ω_0 : in order to reproduce the observed abundance of X-ray clusters, open cosmological models require the presence of massive clusters at high redshifts. Recently, the discovery of two radio decrements in radio maps obtained by the VLA (Richard et al. 1997) and by the Ryle telescope (Jones et al. 1997) can have some implications for the density parameter Ω_o. If these decrements are indeed clusters of galaxies, then the lack of X-ray and optical counterparts can be explained by the existence of high-redshfit clusters. Because the number of massive clusters of galaxies at high redshift depends strongly on Ω_0, the observation of an even smaller number of such objects makes the viability of a critical Universe rather difficult (Oukbir & Blanchar 1992; Barbosa et al. 1996, Eke et al. 1996; Oukbir & Blanchard 1997).

2. The Sunyaev-Zel'dovich Effect

The same hot gas responsible for the X-ray emission of clusters of galaxies also produces a spectral distortion of the cosmic microwave background (CMB) by inverse Compton scattering of the incoming CMB photons (Sunyaev & Zel'dovich 1972). This results in an unique spectral signature, which makes a cluster appear as a decrement at wavelenghts longwards 1.34mm

Astrophysics and Space Science is the original source of publication of his article. It is recommended that this article is cited as: *Astrophysics and Space Science* **261**: 277–280, 1999.

in the mean sky brightness and as an excess at wavelenghts shortwards 1.34mm due to this diffusion of Rayleigh-Jeans photons towards the Wien part of the original CMB spectrum. The surface brightness of a cluster relative to the original CMB spectrum is expressed as the product of the SZ spectral function J_ν with the Compton y parameter, which is the integral of the electron pressure along the line-of-sight: $y \propto \int dl\ n_e T$. Integrating the surface brightness over the cluster solid angle and using the fact that there exists a relation between the X-ray temperature of a cluster and the virial mass -$T \propto M^{2/3}$ (Evrard et al. 1996), we have that $S_{sz} \propto \theta^2 n_e T l \propto f_{gas} M_{tot}^{5/3}(1+z) D_a^{-2}(z)$. More precisely (Barbosa et al. 1996) :

$$S_\nu = (8\,\mathrm{mJy}\,h^{8/3}) f_\nu(x) f_{gas} \Omega_o^{1/3} M_{15}^{5/3} [\frac{\Delta(z)}{178}]^{1/3}(1+z) D^{-2}(z), \qquad (1)$$

where f_{gas}, $f_\nu(x)$ and $D(z)$ are the gas fraction of a cluster and the dimentionless SZ spectral function and the dimentionless angular distance. For the X-ray flux of a cluster, we find that $f_x = \theta^2 n_e^2 T^{1/2} l (1+z)^{-4} \propto n_e f_{gas} M_{tot} T^{1/2} D_l^{-2}(z)$, D_l being the luminosity distance. It is clear that unlike the SZ surface brightness, the X-ray-flux suffer cosmological brightness dimming, thus making the SZ effect as the best way to probe high redshift clusters. Besides this difference, X-ray flux depends strongly on the core radius and profile of the intracluster medium, quantities with an evolution still poorly understood, while the SZ surface brightness **only** depends on the *total gas mass* and temperature of the cluster.

2.1. THE MASS FUNCTION AND THE NUMBER COUNTS

For our calculations we adopt the *Press-Schechter mass function* :

$$\frac{dn}{d\ln M} d\ln M = \sqrt{\frac{2}{\pi}} \frac{<\rho>}{M} \nu(M,z) \left(-\frac{d\ln\sigma(M,z)}{d\ln M} \right) e^{-\nu^2/2} d\ln M \qquad (2)$$

where $<\rho>$ is the comoving mass density of the Universe and $\sigma(M,z)$ is the matter power spectrum. The quantity $\nu = \delta_c D_g(z)/\sigma(M)$ gives the height of the over-dense regions collapsing at redshift z with a linear density contrast δ_c and is a function of the linear growth factor $D_g(z,\Lambda,\Omega_o)$. The appearance of D_g in the exponential of the PS mass-function makes it strongly dependent on Ω_o and the redshift distribution of clusters of a given mass is therefore only determined by the cosmological parameters (Oukbir & Blanchard 1997). With eq.1, we can convert the mass function into a distribution of clusters in SZ flux density and redshift and calculating the SZ number counts becomes a rather simple task. Fig.1 shows the redshift

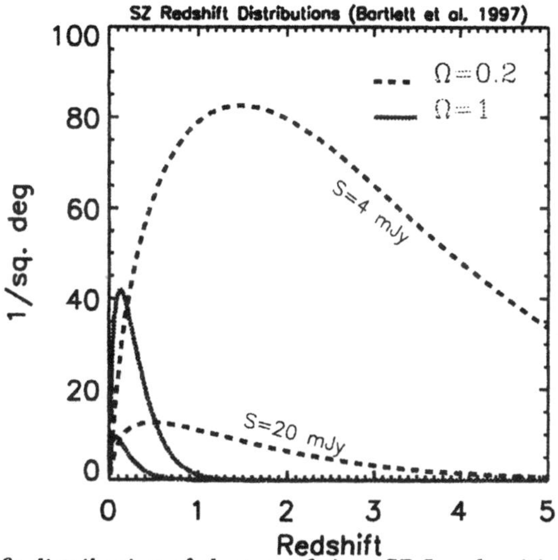

Figure 1. *Redsfit distribution of clusters of given SZ flux densities (4 and 20 mJy) for two cosmological models. The SZ flux densities correspond to the VLA and RT ones , when translated to a wavelength of* $\lambda = 0.75mm$.

distribution of clusters for two given SZ flux densities and for two cosmological models (a critical one and an $\Omega_o = 0.2$) at a wavelength $\lambda = 0.75$mm. We fixed the gas fraction to $f_{gas} = 0.06h^{-3/2}$.

3. Conclusions

While waiting confirmation of the true nature of the radio decrements with detections at other frequencies (at least the RT field seems to be contaminated by a double quasar), a critical Universe seems rather unlikely with the presence of these SZ sources (a quite different conclusion from the presentation of Rachida Sadat in these proceedings-see also Sadat et al. 1997). The existence of high-redshift SZ sources is natural in open Universes, thus making $\Omega_o = 0.2$ more likely (Bartlett et al. 1998). These results bring to light the importance of more SZ searches and show us the importance of the full sky coverage of the Planck Surveyor mission.

4. Acknowledgements

D.B. acknowledges support by the Praxis XXI CIÊNCIA-BD/2790/93 grant attributed by JNICT, Portugal.

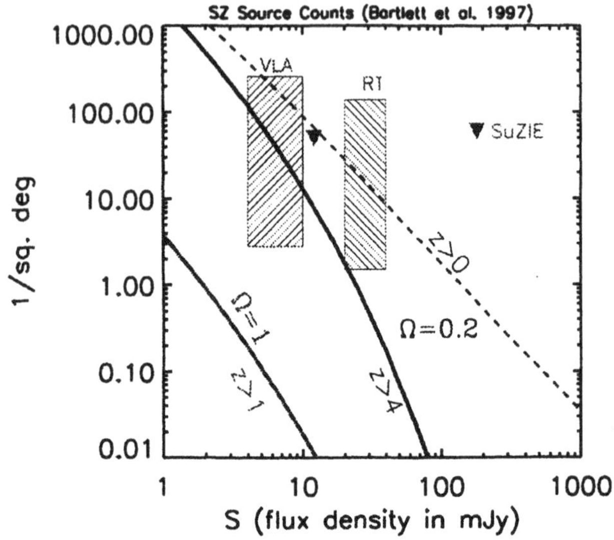

Figure 2. *SZ source counts with observational constraints, as a function of SZ flux density expressed at $\lambda = 0.75mm$. The two hatched boxes show the 95% one-sided confidence limits from the VLA and the RT; due to the uncertain redshift of the clusters, there is a range of possible total SZ flux density, which has for a minimum the value observed in each beam and a maximum chosen here to correspond to $z > 1$. From the SuZIE blank fields, one can deduce the 95% upper limit shown as the triangle pointing downwards (Church et al. 1997). We also plot the predictions of our fiducial open model ($\Omega = 0.2$) for all clusters (dashed line) and for those clusters with $z > 4$. The critical model has great difficulty explaining the observed objects even with a lower redshift cutoff of only $z > 1$; the actual limit from the X-ray data is stronger, but this would fall well off to the lower left of the plot. We assume $h = 1/2$.*

References

Bahcall, N.A., Fan, X., Cen, R., (1997), *ApJ*, **485**, L53

Barbosa, D., Bartlett, J.G., Blanchard, A., Oukbir, J., (1996), *A&A*, **314**, 13

Bartlett, J.G., Blanchard, A., Barbosa, D., (1998), *A&A*, **336**, 425

Eke, V.R., Cole, S., & Frenk, C.S., (1996), *MNRAS*, **282**, 263

Sunyaev, R. & Zel'dovich, Y.B., (1972), *Comm. Astrophys. Space Phys.*, **4**, 173

Evrard, A.E., Metzler, C., Navarro, J.F., (1996), *ApJ*, **469**, 494

Jones, M.E., Saunders, R., Baker, , J.C., Cotter, G., Edge, A., Grainge, K., Haynes, T., Lasenby, A., Pooley, G., Röttgering H., (1997), *ApJL*, **479**, L1

Oukbir, J., Blanchard, A., (1992), *A&A*, **262**, L1

Oukbir, J., Blanchard, A., (1997), *A&A*, **317**, 10

Press, W.H., Schechter, P., (1974), *ApJ*, **187**, 425

Richards, E.A., Fomalont, E.B., Kellerman, K.I., Partridge, R.B., Windhorst, R.A., (1997), *AJ*, **113**, 1475

Sadat, R., Blanchard, A., Oukbir, J., (1997), *A&A*, **329**, 21

INFLATION AND THE COSMIC MICROWAVE BACKGROUND

ANDREW R. LIDDLE
Astronomy Centre,
University of Sussex,
Brighton BN1 9QJ,
Great Britain

Abstract. I give a status report and outlook concerning the use of the cosmic microwave background anisotropies to constrain the inflationary cosmology, and stress its crucial role as an underlying paradigm for the estimation of cosmological parameters.

1. Introduction

For a long time now, inflation has been the leading paradigm for the origin of cosmological structures. This is largely due to its continuing success in confrontation with a wide range of observations, but also due in part to its theoretical simplicity compared to rivals such as cosmic strings, both in terms of making predictions for the perturbations and in the form (gaussian and adiabatic) of perturbations generated.

The interaction between observations and theoretical modelling of inflation plays a two-fold role in cosmology. The most-emphasized role is the possible use of observations, especially of cosmic microwave background anisotropies, to support or rule out the inflationary paradigm as the source of structures (see Liddle & Lyth 1993 for a review). Such a program may well offer the first glimpses of possible physics at very high energies. Much less has been said about the second role of inflation — that in providing a simple framework for structure formation, it is crucial in enabling the high-accuracy determination of more mundane cosmological parameters such as the Hubble constant and the density parameter. I shall focus particularly on this aspect towards the end of this review.

Astrophysics and Space Science is the original source of publication of his article. It is recommended that this article is cited as: *Astrophysics and Space Science* **261**: 281–290, 1999.

2. Inflation

Inflation is defined as any epoch of the Universe's history during which the scale factor $a(t)$ is accelerating. In fact, we have yet to prove conclusively that this is not happening at the present epoch, but here I am interested in whether such a behaviour might have happened in the Universe's distant past. I actually prefer to use the Hubble parameter $H = \dot{a}/a$ to write this in a somewhat different way

$$\ddot{a} > 0 \quad \Longleftrightarrow \quad \frac{d}{dt}\left(\frac{H^{-1}}{a}\right) < 0 \,, \tag{1}$$

where dots are of course time derivatives. The quantity H^{-1} is the Hubble length, the principal characteristic scale of an expanding Universe, and dividing it by a switches to comoving coordinates, i.e. the behaviour of relative to the expansion. In words, inflation is precisely the condition that the comoving Hubble length is decreasing with time. Since in comoving units all the objects just remain where they are as the Universe expands, and since at any epoch the Hubble length is a good estimate of how far light can travel during that epoch, this is telling us that inflation acts like a zoom lens, focussing in on an ever-tinier part of the initial region. That's why, for example, inflation can solve the flatness problem; even if curvature is important initially, when we zoom in on a tiny region the curvature becomes negligible, and enough 'zooming' can more than compensate for the subsequent increase of the comoving Hubble length after inflation ends.

Inflation's ability to generate large-scale perturbations is all down to the behaviour of the Hubble length, because it means that a given length scale may start well inside the Hubble radius, but finish up well outside it. Any irregularities existing at that time become 'frozen-in', unable to evolve. If we lived in a purely classical world, this would lead to a perfectly homogeneous Universe, simply because the assumption that the initial energy density be finite requires that there cannot be irregularities down to arbitrarily small scales.

Fortunately, we live not in a classical Universe but a quantum one, and the uncertainty principle implacably opposes inflation's attempts to make a perfectly smooth Universe. More so, it does this in a manner which is readily predictable for a given inflationary model. It turns out that quantum fluctuations give rise to two types of perturbations in the Universe, density perturbations and gravitational waves. These both take on a classical character once they are on scales well in excess of the Hubble length.

In order to give definite predictions for these, it is necessary to have a definite model for the inflationary expansion rather than just the hand-waving sketch I've given above. Here I want to stress that inflationary

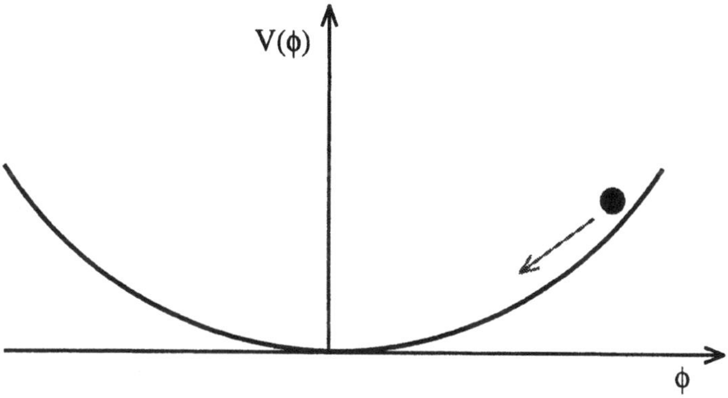

Figure 1. A schematic scalar field potential.

model-building has become quite a complicated and sophisticated occupation, and in presentations like this one necessarily gives an oversimplified picture. Bear that in mind, and I'll also remind you of it later.

The acceleration equation

$$\frac{\ddot{a}}{a} = -\frac{4\pi G}{3}(\rho + 3p) , \tag{2}$$

immediately tells us that we're going to need something a little out of the ordinary, namely $p < -\rho/3$. The standard way of achieving this uses a scalar field, the sort of thing which crops up all over the place in modern particle physics theories, especially when symmetry breaking is under consideration. The standard simplified picture, which does in fact cover a large fraction of the currently popular models, is that there is only one scalar field, and that it evolves classically by slow-rolling down a self-interaction potential $V(\phi)$ such as that shown in Figure 1. Either of these assumptions can be altered, giving more complicated models, but the bulk of my discussion will stay with the simplest case.

Such a homogeneous scalar field has effective energy density and pressure

$$\rho_\phi = \frac{1}{2}\dot{\phi}^2 + V(\phi) \quad ; \quad p_\phi = \frac{1}{2}\dot{\phi}^2 - V(\phi) , \tag{3}$$

and so the condition for inflation is satisfied as long as the potential dominates over the kinetic term. This will clearly happen if the potential is sufficiently flat; in fact, the flatness condition is very weak and so inflation is quite generic.

In principle, $V(\phi)$ is predictable from fundamental theories of physics. In practice there is as yet no clear guidance, and instead we treat it as a free function to be constrained by observations.

3. Describing the perturbations

If the inflationary expansion is rapid enough, physical conditions during inflation will change little between the origin of perturbations on the largest interesting scales and the smallest, so as a rough rule of thumb we expect that the perturbations will be nearly scale-invariant. For a long time, the perturbations were indeed taken to have that form. However, more recently the observations have reached such quality that the scale-invariance approximation is no longer an adequate description, and we have to do better. This is in no way a set-back (or climb-down) for inflation — it is an impressive success that the theory has done so well that we now regard small corrections to the initial picture as significant and observationally testable.

The breaking of scale-invariance will depend on the form of the potential $V(\phi)$ (it being the only input information), so we quantify this by defining two slow-roll parameters (Liddle & Lyth 1992)

$$\epsilon \equiv \frac{m_{\rm Pl}^2}{16\pi}\left(\frac{V'}{V}\right)^2 \quad ; \quad \eta \equiv \frac{m_{\rm Pl}^2}{8\pi}\frac{V''}{V}, \tag{4}$$

where prime means a derivative wrt ϕ, and $m_{\rm Pl} = G^{-1/2}$ is the Planck mass. Inflation requires that both these be less than one, in order to maintain dominance of the potential. The first measures the slope of the potential, the second its curvature.

The terminology I will use isn't very important; density perturbations are specified by $\delta_{\rm H}(k)$ and gravitational waves by $A_{\rm G}(k)$ where k is the comoving wavenumber. For scale-invariant spectra these are both independent of k; the power spectrum $P(k) \propto k\delta_{\rm H}^2$. Under the inflationary paradigm, these spectra are responsible for all the observed structures, with gravitational waves at best only significantly influencing large-angle microwave background anisotropies.

The formulation I'll describe is based on a perturbative approach, where we expand the (log of the) spectra in terms of log(wavenumber) about some scale k_*:

$$\ln \delta_{\rm H}^2(k) = \ln \delta_{\rm H}^2(k_*) + (n_* - 1)\ln\frac{k}{k_*} + \frac{1}{2}\left.\frac{dn}{d\ln k}\right|_* \ln^2\frac{k}{k_*} + \cdots, \tag{5}$$

and truncate at some level. For example

First term: Harrison–Zel'dovich spectrum (constant $\delta_{\rm H}$).

Second term: Power-law spectrum with spectral index n.

Third term: Includes scale-dependence of the spectral index.

Current observations require at least the second term and normally people stop there. I'll return to the third one later. The second term gives the power-law approximation, and the observables can be very nicely computed in terms of the slow-roll parameters (Liddle & Lyth 1992):

$$n = 1 - 6\epsilon + 2\eta \quad ; \quad n_{\rm G} = -2\epsilon \quad ; \quad \frac{A_{\rm G}^2(k_*)}{\delta_{\rm H}^2(k_*)} = \epsilon = -\frac{n_{\rm G}}{2} \,. \tag{6}$$

If slow-roll holds very well ($\epsilon \ll 1$, $|\eta| \ll 1$) we get scale-invariant density perturbations and negligible gravitational waves. Note also that although ϵ is positive by definition (the field always rolls downhill), η can have either sign and n can be greater or smaller than one. The final relation, giving the ratio of the spectra in terms of $n_{\rm G}$, is known as the consistency relation; it represents the inevitable 'entanglement' of the density perturbations and gravitational waves due to their common origin in the single function $V(\phi)$. In the unlikely event of it proving testable, it represents a very distinctive prediction of inflation.

4. COBE

The COBE observations have a very simple interpretation in terms of inflation. Because the COBE beam is so wide, it only probes scales larger than the horizon at the time of decoupling, so perturbations have not had time to evolve and are captured in their primordial form. In particular, the observed anisotropies do not depend on cosmological parameters such as the hubble constant h and the baryon density $\Omega_{\rm B}$.

COBE determines the perturbation amplitude extremely well, to about 10% accuracy, but is unable to distinguish the effect of density perturbations and gravitational waves. Its most useful application is to normalize the density perturbation spectrum; Bunn et al. (1996) obtained the result

$$\delta_{\rm H} = 1.91 \times 10^{-5} \, \frac{\exp\left[1.01(1-n)\right]}{\sqrt{1+0.75r}} \,, \tag{7}$$

using techniques described by Bunn & White (1995, 1997). Here $\delta_{\rm H}$ is the perturbation amplitude at the present Hubble radius, and $r = 12.4 A_{\rm G}^2/\delta_{\rm H}^2$ approximately measures the relative importance of gravitational waves and density perturbations in generating the anisotropies. The factor 12.4 comes from analytic evaluation assuming only the Sachs–Wolfe effect applies and perfect matter domination at last scattering; that the above expression contains the factor 0.75 indicates that this approximation fails at the tens of percent level on COBE scales.

COBE fixes the energy scale of inflation as (Bunn et al. 1996)

$$V_*^{1/4} = \left(6.6 \times 10^{16}\,{\rm GeV}\right) \epsilon^{1/4} \quad \pm 5\% \,, \tag{8}$$

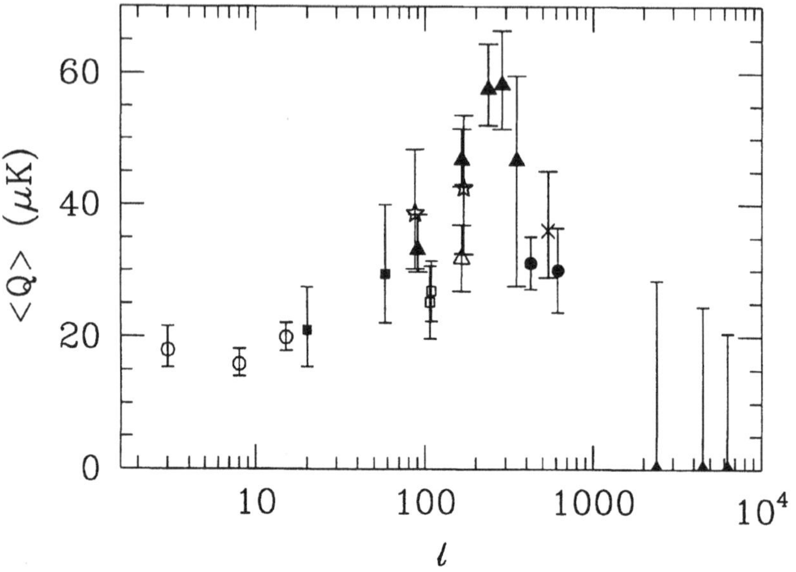

Figure 2. A compilation of measurements of microwave background anisotropies, by Martin White. The detections (1-sigma errors) are COBE (3 open circles), Tenerife and BAM (filled squares), Python (2 stars), Saskatoon (5 filled triangles), ARGO (2 open squares), MAX (open triangle), CAT (2 filed circles) and OVRO (cross). The 95% confidence upper limits are SuZIE, ACTA and Ryle. For a detailed discussion see Anthony Lasenby's contribution to these proceedings.

where $*$ indicates the value when the observed perturbations were generated. For a specific model ϵ is known and so this can be given exactly. Unless ϵ is tiny, which is perfectly possible in some models, the energy scale is around that expected of Grand Unification.

The spectral index n is only very weakly constrained by COBE, due to the limited range of scales sampled.

5. The current compilation

Nowadays, measured anisotropies go well beyond COBE, with a host of experiments reporting detections on a range of angular scales. Figure 2 shows a recent compilation by Martin White. The data are encouragingly compatible with inflationary preconceptions, but unfortunately at present don't allow us to say much more; the predictions on smaller angular scales (larger ℓ) depend on all the cosmological parameters and the observational errors are larger than one desires. The compilation does indicate a lower limit on n, which is somewhat model-dependent but is around 0.75, and at the moment that is its main inflationary implication.

6. Solving cosmology ...

The upcoming launch of the MAP and Planck satellites offers the prospect that cosmology could be more or less solved, in the sense that its most crucial parameters might be measured to a satisfyingly high accuracy (Jungman et al. 1996; Bond et al. 1997; Zaldarriaga et al. 1997). However, there are rather a lot of parameters which might come into play. Most attention has been focussed on the cosmological parameters, such as the Hubble constant h, total matter density Ω_0, baryon density $\Omega_{\rm B}$, cosmological constant Ω_Λ, hot dark matter density $\Omega_{\rm hdm}$ and the optical depth to the last-scattering surface $\tau_{\rm c}$. Some of these might be fixed by assumption, but as a very minimum h, $\Omega_{\rm B}$ and $\tau_{\rm c}$ must be determined observationally.

However, in this article I want to turn attention to the inflationary input. As I stressed at the beginning, this is absolutely crucial as a paradigm in cosmology. If we don't know the initial perturbation spectra, we have no chance of interpreting observed microwave anisotropies in terms of the cosmological parameters. This is potentially a very serious problem, since without guidance the perturbations could be free functions which need not be simple.

Fortunately, most models of inflation give very simple predictions, and are summarized in only two or three parameters, namely $\delta_{\rm H}$, n and perhaps r. However, scale-dependence of n is possible, and in accordance with the Taylor expansion of Eq. (5) it can be incorporated by inclusion of derivatives $dn/d\ln k$, $d^2n/d\ln k^2$ and so on. These represent extra parameters which must be determined from observations.

If extra parameters are introduced, then the determination of *all* parameters will deteriorate. We have estimated the extent of this deterioration (Copeland et al. 1997) for a configuration of the Planck satellite including polarized detectors. It is shown in Table 1.

We see that permitting scale dependence of n primarily only affects the determination of n, and not the other cosmological parameters, which is an encouraging conclusion. It suggests that if the modelling of inflation turns out to be over-simplified, the influence on parameter determination will not be too great. On the other hand, it is clear that n cannot be measured as well as has been claimed (e.g. Bond et al. 1997; Zaldarriaga et al. 1997) unless it is *assumed* that the spectrum is a perfect power-law.

7. Inflationary complications

Now, as promised, I turn to the question of possible complications to the inflationary modelling. First of all, one can ask whether we are in a position to compute the perturbations at the 1% or so accuracy level demanded by MAP and Planck. Finally, the answer to this appears to be 'yes'; the

TABLE 1. Estimated parameter errors (one-sigma) for the Standard CDM model, as extra scale-dependence is introduced.

Parameter	Planck 140 GHz channel with polarization		
$\delta\Omega_b h^2/\Omega_b h^2$	0.007	0.009	0.01
$\delta\Omega_{cdm} h^2/h^2$	0.02	0.02	0.02
$\delta\Omega_\Lambda h^2/h^2$	0.04	0.05	0.05
$\delta\tau_C$	0.0006	0.0006	0.0006
δn	0.004	0.04	0.14
δr	0.04	0.05	0.05
$dn/d\ln k$	–	0.006	0.04
$d^2n/d(\ln k)^2$	–	–	0.005

last problem (gravitational waves in open Universe models — see below) has been solved during the last year and there are no existing models in which the perturbations cannot be computed, at least through numerical integration of the relevant mode equations, to the required accuracy.

7.1. SINGLE-FIELD MODELS

So far we've been sticking to models with a single scalar field, with $V(\phi)$ kept as a free function. Normally the slow-roll approximation gives a very accurate analytic result (Grivell & Liddle 1996). However, a sufficiently complicated potential may lead to a failure of slow-roll severe enough that the slow-roll approximation is not good enough (Wang et al. 1997), and the numerical results are required. However, we have found (Copeland et al. 1998) that this need not be a bad thing; in particular, the near-failure of slow-roll makes it far more likely that the scale-dependence of n is observable, which allows one to determine more information about the inflaton potential than would otherwise have been available. Further, it would take a bizarre conspiracy for the slow-roll approximation to be failing, but yet for us to be oblivious to this. If things are going wrong to the extent that numerical computation is needed, we will know it.

The main worry in fact for the single-field models described so far is that many models [particularly those constructed under the currently-popular hybrid inflation strategy (see Lyth 1996 for a review)] predict a negligible level of gravitational waves. This limits seriously the amount of information that can be inferred about $V(\phi)$, which cannot even be determined uniquely if the gravitational waves cannot be detected.

7.2. MULTI-FIELD MODELS?

It is certainly possible that more than one scalar field can be dynamically important, and this can lead to a range of new phenomena. The perturbations may have an isocurvature component as well as the usual adiabatic one, and may even be non-gaussian. The perturbations can still be computed accurately, but now only on a model-by-model basis rather than via an all-encompassing formalism like that I've demonstrated for single-field models. This makes it much harder to 'guess' viable models from the observations.

However, calculational complications aside, a specific model of this type is as easy to exclude using observations as a single-field model.

7.3. OPEN INFLATION MODELS?

Open inflation models are an unfortunate late addition into the inflationary model zoo. Although dating back almost to the beginnings of inflation (Gott 1982), it is only relatively recently that they have been appreciated as a serious model. They rely on quantum tunnelling to generate an open Universe within an inflationary sea.

For a long time the perturbations in these models could not be computed (especially the gravitational waves), but finally the technology is in place (Sasaki et al. 1997; Bucher & Cohn 1997), and predictions can be obtained from them as readily as the more conventional models.

Mathematically they are much nastier models (for example the mode functions on hyperbolic geometry are very unpleasant indeed) than ones giving a flat spatial geometry. Fortunately, the spatial geometry is very readily measurable, even before the satellites go up, and hopefully these models will soon be consigned to the dustbin.

7.4. INFLATION NOT CORRECT?

If inflation is not in fact the correct theory for the origin of perturbations, this should be obvious from the observations. The inflationary prediction of passive, super-horizon perturbations is very distinctive, and leads to the familiar oscillations in the radiation angular power spectrum. If inflation is correct the observations should highly overdetermine the various parameters of the big-bang model, reassuring us that we are on the right track.

But here is a good point to stress once more the importance of inflation as a paradigm for the initial conditions. For example, it seems rather unlikely that predictions at the one percent level could come from the rival topological defect scenario in the forseable future, due to the horrendous non-linearities involved in the computations. Without accurate theoretical predictions for the anisotropies, huge extra uncertainties enter the cosmological parameter estimation game and severely dent one's ability to measure any of them.

8. Conclusions

My main point of emphasis has been to stress that the inflationary paradigm is a crucial underpinning of attempts to measure cosmological parameters from the microwave background. It provides a framework in which accurate predictions can readily be made, enabling the maximum to be squeezed out of quality observations. We are fortunate indeed that not only is the model theoretically appealing, but that it also stands in excellent shape when confronted with present data. We can only hope that in ten years time, when the large zoo of inflation models have been confronted with the new observations, that things may still look so good.

Acknowledgments

I was supported by the Royal Society. Thanks to Ted Bunn, Ed Copeland, Ian Grivell, Rocky Kolb, David Lyth and Martin White for their collaboration in parts of the work described here, and to Martin White for providing the data compilation shown in Figure 2.

References

Bond, J.R., Efstathiou, G. and Tegmark, M., 1997, Mon. Not. R. Astron. Soc. **291**, L33
Bucher, M. and Cohn, J.D., 1997, Phys. Rev. D **55**, 7461
Bunn, E.F., Liddle, A.R. and White, M., 1996, Phys. Rev. D **54**, 5917R
Copeland, E.J., Grivell, I.J. and Liddle, A.R., 1997, Sussex preprint astro-ph/9712028
Copeland, E.J., Grivell, I.J., Kolb, E.W. and Liddle, A.R., 1998, Sussex preprint
Gott, J.R., 1982, Nature **295**, 304
Grivell, I.J. and Liddle, A. R., 1996, Phys. Rev. D **54**, 7191
Jungman, G., Kamionkowski, M., Kosowsky, A. and Spergel, D.N., 1996, Phys. Rev D **55**, 7368
Liddle, A.R. and Lyth, D.H., 1992, Phys. Lett. B **291**, 391
Liddle, A.R. and Lyth, D.H., 1993, Phys. Rep. **231**, 1
Lyth, D.H., 1996, Lancaster preprint hep-ph/9609431
Sasaki, M., Tanaka, T and Yakushige, Y., 1997, Phys. Rev. D **56**, 616
Wang, L., Mukhanov, V.F. and Steinhardt, P.J., 1997, Phys. Lett. B **414**, 18
Zaldarriaga, M., Spergel, D. N. and Seljak, U., 1997, Astrophys. J. **488**, 1

STRUCTURE FORMATION WITH SCALAR FIELDS

PEDRO T.P. VIANA
Centro de Astrofísica da Universidade do Porto
Rua do Campo Alegre, 823, 4150 Porto, Portugal

AND

ANDREW R. LIDDLE
Astronomy Centre, University of Sussex
Falmer, Brighton BN1 9QH, United Kingdom

Abstract. We discuss structure formation models where a presently evolving scalar field behaves dynamically either as a decaying cosmological constant or as dust. We find that the first case provides a successful alternative to classical cosmological constant models, keeping all the characteristics that have made these some of the most able models in reproducing the available observational data, and with the added possibility of a different recent background evolution for the Universe. We allow the presence of spatial perturbations in the scalar field, and study their effect on the growth of adiabatic energy density perturbations in a cold dark matter component. We consider two alternatives for the scalar field potential: an exponential and that associated with a pseudo-Nambu Goldstone boson.

1. Introduction

The favoured ΛCDM models have an energy density Ω_Λ in the cosmological constant in the range 0.5 to 0.7 [1]. Though the preliminary data from the Supernova Cosmology Project [2], which attempts to determine the magnitude–redshift diagram of Type Ia supernovae, seemingly excluded these models by placing a 95 per cent confidence upper limit on Ω_Λ of 0.5, more recent results [3] have changed this picture dramatically, actually supporting a value for Ω_Λ around the interval given above.

Nevertheless, the available data is still scarce, and it would be useful to find structure formation models which imitate ΛCDM models, except for the background evolution of the Universe for $z < 1$. A way of introducing this extra freedom in the models is to substitute the cosmological constant by a dynamical quantity which decays with time (see [4] for references), and the most credible mechanism for obtaining such a term is to assume

Astrophysics and Space Science is the original source of publication of his article. It is recommended that this article is cited as: *Astrophysics and Space Science* **261**: 291–294, 1999.

the existence of a scalar field presently relaxing towards the minimum of its potential. It can also be argued that this situation may be more natural on particle physics grounds. For example, there is the well-known difficulty within quantum field theory to understand the very small vacuum energy density required by a cosmological constant.

The overall dynamics of the Universe in the presence of such a scalar field has been studied in detail before, and in general it has been assumed that the scalar field is always homogeneous and isotropic. However, as soon as one permits any form of time variation, general covariance immediately implies that spatial perturbations can also appear in the value of the scalar field. Though this does not change the global dynamics of the Universe, it may influence the growth rate of the matter density perturbations, being our main aim in this work to study this possible effect.

2. Evolution equations and parameter choice

We numerically integrated the differential equations describing the background and perturbation evolution since a redshift of 1100, at the time of electron-photon decoupling, when the cosmic microwave background radiation (CMBR) was produced. This period is well inside the matter dominated era for $\Omega_0 > 0.3$, and thus we considered only the dynamical contribution of two components: cold dust-like matter, and a scalar field, ϕ.

We derived two sets of perturbation equations [4], using two different gauge choices: the zero-shear gauge (ZSG) and the uniform-curvature gauge (UCG). For equivalent initial conditions, we should obtain the same results for perturbations whose lengthscale is today significantly smaller than the present Hubble radius, thus we used them as a check on the accuracy of the numerical integration.

We assumed that the initial amplitude in the scalar field perturbations is connected to that in the matter density perturbations via the adiabatic condition, as we would expect if the perturbations are generated according to standard models of inflation.

We considered two possibilities for the scalar field potential: an exponential potential of the form $V(\phi) = V_0 \exp(-\beta\phi)$, and the potential associated with a pseudo-Nambu-Goldstone-boson (PNGB) field, $V(\phi) = M^4[\cos(\phi/f) + 1]$. We chose two sets of values for the constants associated with each of the potentials. In models EXP1, with $\beta = \sqrt{40\pi G}$ and $V_0 = (0.025 \text{ eV})^4$, and PNGB1, where $M = 0.04\, h^{1/2}$ eV and $f = 1.25 \times 10^{18}$ GeV, the scalar field behaves dynamically like non-relativistic matter since $z \sim 100$. In models EXP2, for which $\beta = \sqrt{28\pi G}$ and $V_0 = (0.0025 \text{ eV})^4$, and PNGB2, with $M = 0.003\, h^{1/2}$ eV and $f = 1/\sqrt{8\pi G} = 2.4 \times 10^{18}$ GeV, the scalar field becomes dynamically important only very

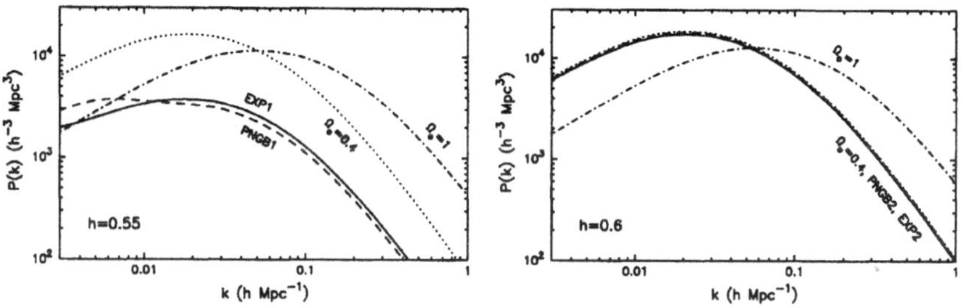

Figure 1. The power spectrum of the energy density perturbations in the matter component for the models, all normalized to COBE as described in the text.

recently, at $z \sim 1$, and is presently behaving as a slowly-decaying cosmological constant. In the first two models, we obtain an age for the Universe, t_0, around 12 Gyrs for $h = 0.55$, where h is the present value of the Hubble parameter in units of 100 km s^{-1} Mpc^{-1}. In the two other models, we chose to get $t_0 = 14$ Gyrs and $h = 0.6$. In all four models we have $\Omega_0 = 0.4$.

3. Results

In Fig. 1, we show the present-day power spectra, $P(k)$, for the matter density perturbations obtained for the four scalar field models that were chosen. They can be compared with what is expected for two models where no scalar field is present: a critical-density Universe and a flat Universe with $\Omega_0 = 0.4$. In these two comparison models it was assumed a scale-invariant primordial power spectrum for the energy density perturbations, $\Omega_{\rm baryon} = 0.016\,h^{-2}$ and either $h = 0.55$ or $h = 0.6$. The power spectra for the scalar field models were obtained by multiplying $P(k)$ for the shown cosmological constant models by the square of the ratio between the calculated present fractional amplitude of the matter density perturbation for any of the scalar field models and the same quantity for the comparison flat models. This procedure is correct since up to $z = 1100$ the background and perturbation evolution of any of the scalar field models is equivalent to that of a cosmological constant model which has the same values for Ω_0 and h as each individual scalar field model.

The results shown were obtained in the ZSG, but the results in the UCG coincide for values of k down to about $10^{-3}\,h\,\mathrm{Mpc}^{-1}$, and can therefore be regarded as gauge-independent for such k. For smaller k values the results will depend on the gauge used.

The critical-density and cosmological constant models are normalized to COBE [5]. The amplitudes of the other models, which have been computed relative to the latter, are correct up to a few per cent. The difference is due to a change in the line-of-sight contribution to the Sachs-Wolfe effect, which requires a full Boltzmann code for accurate computation.

The models EXP2 and PNGB2 yield a shape for $P(k)$ extremely close to that one obtains in a ΛCDM model where $\Omega_0 = 0.4$ and $h = 0.6$, though the amplitude is a few percent smaller. Given that this model is able to reproduce all the most reliable observational data presently available [1], we expect models EXP2 and PNGB2 to do as well. Because this work was done before the latest results on high-redshift Type Ia supernovae were published [3], models EXP2 and PNGB2, though easily fit the older data [2], have difficulty in reproducing the most recent one. However, due to the extra degrees of freedom scalar field models possess, it would not be difficult to find some which would keep all the characteristics of models EXP2 and PNGB2, while reproducing better the current supernovae results.

The situation is rather different for models EXP1 and PNGB1. In both cases, the shape for $P(k)$ is again very close to that one obtains in a ΛCDM model, this time with $\Omega_0 = 0.4$ and $h = 0.55$, model PNGB1 having the further interesting feature that the shape of $P(k)$ is altered for values of k below $0.01\,h\,\mathrm{Mpc}^{-1}$. However, this time the amplitude of the power spectra is only about 5 per cent of that for the cosmological constant model, which is very problematic from the point of view of forming the large-scale structures we today observe.

The results just described suggest that the presence of spatial perturbations in a presently-existing scalar field substantially affects the evolution of energy density perturbations in the matter component only if the scalar field has contributed significantly to the total energy density in the Universe for several Hubble times. If the scalar field is becoming dynamically important only now, the presence of spatial perturbations does not seem to have much effect.

Acknowledgements

This work was done while P.T.P.V. was at Sussex, supported by a PPARC fellowship. He is presently financed through grant BPD/11831/97 of the PRAXIS XXI program. A.R.L. is supported by the Royal Society.

References

1. Liddle, A.R., Lyth D.H., Viana, P. T. P. and White, M. (1996), *Mon. Not. R. Astr. Soc.*, **282**, 281
2. Perlmutter, S., *et al.* (1997), *Astrophys. J.*, **483**, 565
3. Perlmutter, S., *et al.* (1998), *Nature*, **391**, 51
4. Viana, P. T. P., and Liddle, A.R. (1998), *Phys. Rev. D*, **57**, 674
5. Bunn, E.F., Liddle, A.R., and White, M. (1996), *Phys. Rev. D*, **54**, R5917

ORIGIN OF STRUCTURE IN SQC

P. VARGAS MONIZ
DAMTP, University of Cambridge
Silver Street, Cambridge, CB3 9EW, UK

We will construct in this paper a model to establish *if* and *how* the inclusion of supersymmetry in a quantum cosmological scenario [1, 2] can lead to a scale-free spectrum of density fluctuations. Previous research in supersymmetric quantum cosmology (SQC)[1] has not provided quantum states which would represent: (*i*) a period of evolution from SQC towards a semi-classical stage, together with; (*ii*) identifying the existence of any quantum state associated to structure formation; (*iii*) followed by establishing how does conventional quantum cosmology harmonise into this picture; (*iv*) and hence, determining if a path from SQC physics down to a classical level can be consistently established. This report constitutes a response regarding (*ii*) above [6].

The action for our model is retrieved from the general action of N=1 of supergravity with scalar supermultyplets, as represented in equation (25.12) of reference [3]. Our background supersymmetric minisuperspace is constituted by the gravitational field, which is represented by a tetrad $e_\mu^{AA'} = e_\mu^a \sigma_a^{AA'}$ (in two-spinor notation) where [1] $e_{a\mu} = \mathrm{diag}\,(N(t), a(t)E_{\hat{a}i})$, In addition, we also have the gravitinos which must have the form (see ref. [1]) $\psi^A{}_i = e^{AA'}{}_i\bar{\psi}_{A'}(t)$, $\bar{\psi}^{A'}{}_i = e^{AA'}{}_i\psi_A(t)$, where $\psi_A, \bar{\psi}_{A'}$ constitute time-dependent spinor fields and $\psi_0^A(t), \bar{\psi}_0^{A'}(t)$ are Lagrange multipliers. A set complex scalar fields, $\phi(t), \bar{\phi}(t)$, and their fermionic superpartners, $\chi_A(t)$, $\bar{\chi}_{A'}(t)$ are also included. Finally, we choose a flat Kähler manifold.

As far as the perturbations about the background minisuperspace are concerned, we take the scalar fields as [4] $\Phi(x_i, t) = \phi(t) + \Sigma_{nlm} f_n^{lm}(t) Q_{lm}^n(x_i)$, together with its complex conjugate, where the coefficients $f_n^{lm}, \bar{f}_n^{lm}$ are functions of time. The fermionic superpartners are expanded as [5]:

$$\mathbf{X}_A(x_i, t) = \chi_A(t) + a^{-3/2} \Sigma_{mpq} \beta_m^{pq} \left[s_{mp}(t) \rho_A^{nq}(x_i) + \bar{t}_{mp}(t) \bar{\tau}_A^{mq}(x_i) \right], \quad (1)$$

[1]For a review on SQC see ref. [1].

Astrophysics and Space Science is the original source of publication of his article. It is recommended that this article is cited as: *Astrophysics and Space Science* **261**: 295–297, 1999.

together with its Hermitian conjugate, where $\rho_A^{mq}, \bar{\rho}_{A'}^{mq}, \tau_{A'}^{mq}, \bar{\tau}_A^{mq}$ are spinor hyperspherical harmonics on S^3.

After some suitable redifinitions[1] of the ψ^A and χ^A variables, the quantum supersymmetry constraints of the model can be constructed from the coefficients in $\psi_0^A, \bar{\psi}_0^{A'}$ in the Hamiltonian. We introduce a natural ansatz for the wave function of the universe, which has the form

$$\begin{aligned}
\Psi &= A + B\psi^C\psi_C + iC\psi^C\chi_C + D\chi^D\chi_D + E\psi^C\psi_C\chi^D\chi_D \\
&= A^{(0)}(a,\phi,\bar{\phi})\Pi_n A^{(n)}(a,\bar{\phi},\phi;f_n\bar{f}_n)\Pi_m A^{(m)}(a,\phi,\bar{\phi},s_m,t_m) \\
&+ B^{(0)}(a,\phi,\bar{\phi})\Pi_n B^{(n)}(a,\bar{\phi},\phi;f_n\bar{f}_n)\Pi_m B^{(m)}(a,\phi,\bar{\phi},s_m,t_m)\psi^C\psi_C \\
&+ C^{(0)}(a,\phi,\bar{\phi})\Pi_n C^{(n)}(a,\bar{\phi},\phi;f_n\bar{f}_n)\Pi_m C^{(m)}(a,\phi,\bar{\phi},s_m,t_m)\psi^C\chi_C \\
&+ D^{(0)}(a,\phi,\bar{\phi})\Pi_n D^{(n)}(a,\bar{\phi},\phi;f_n\bar{f}_n)\Pi_m D^{(m)}(a,\phi,\bar{\phi},s_m,t_m)\chi^C\chi_C \\
&+ E^{(0)}(a,\phi,\bar{\phi})\Pi_{n,m} E^{(n)}(a,\bar{\phi},\phi;f_n\bar{f}_n)E^{(m)}(a,\phi,\bar{\phi},s_m,t_m)(\psi)^2(\chi)^2 \qquad (2)
\end{aligned}$$

Expression (2) satisfies the Lorentz constraints associated with ψ_A, χ_A. The perturbation modes of the scalar fields and the fermionic partners do not couple to each other – see (2). In addition, the coefficients $s_m, t_m, \bar{s}_m, \bar{t}_m$ are taken as invariant under local Lorentz transformation to lowest order in perturbation [5]. Overall, this approach is fully satisfactory and we will extract a consistent set of solutions from the action of the supersymmetry constraints $S_A, \bar{S}_A$ on Ψ: $S_A\Psi = 0$ and $\bar{S}_A\Psi = 0$.

We then obtain a set of first-order differential equations. Among them we have:

$$\frac{\partial E}{\partial \bar{\phi}} + \frac{\phi}{2}\left[s_m\frac{\partial}{\partial s_m} - t_m\frac{\partial}{\partial t_m}\right]E + \frac{\partial E}{\partial \bar{f}_n} - 2a^2\Sigma_n(n-1)f_n E = 0 \qquad (3)$$

$$\frac{a}{\sqrt{3}}\frac{\partial E}{\partial a} - 2\sqrt{3}a^2 E + \frac{2}{\sqrt{3}}\phi\Sigma_m\left[s_m\frac{\partial}{\partial s_m} - t_m\frac{\partial}{\partial t_m}\right]E = 0\ , \qquad (4)$$

It is straightforward to obtain the following solutions:

$$E^{(0)} = \hat{E}_0^{(0)}\frac{e^{3a^2+\phi(2\lambda_6-\Omega_5)-\Omega_5\bar{\phi}}}{a^{\Omega_6}} \qquad (5)$$

$$E^{(n)} = E_0^{(n)}e^{-\lambda_7\bar{\phi}+\phi(2\lambda_8-\lambda_7)}e^{2\lambda_9\bar{f}_n+2a^2(n-1)f_n\bar{f}_n-(\Omega_7-\lambda_9)f_n+(\Omega_7-\lambda_9)\bar{f}_n} \qquad (6)$$

$$E^{(m)} = E_0^{(m)}e^{2\lambda_8\bar{\phi}-C_2\phi\bar{\phi}-\Omega_9\phi+\Omega_9\bar{\phi}}\tilde{E}, \qquad (7)$$

where $\hat{A}_0^{(0)} = A_0^{(0)}e^{3a_0^2}$, $A_0^{(n)}, A_0^{(m)}$, $\hat{E}_0^{(0)} = E_0^{(0)}e^{-3a^2}$ $E_0^{(n)}, E_0^{(m)}$ denote integration constants and $\tilde{A}$ and $\tilde{E} \sim s_{mp}$ or t_{mp}.

Charactheristic features of the no-boundary (Hartle-Hawking) solution are present in the bosonic coefficient E (see ref. [1, 4, 5]). This requires $|\Omega_6| \ll 1$ and the term $e^{-na^2 f_n\bar{f}_n}$, $(n \gg 1)$ in eq. (6) to dominate. It seems that the presence of supersymmetry *selects* a set of solutions where the Hartle-Hawking state is mandatory.

But do these results contribute to our understanding of the very early universe and if yes, how? The bosonic coefficient E implies $< f_n^{(1)} > \sim < f_n^{(2)} > \sim n^{-1} a_*^{-2}$. These constitute requirements[4] for a scale-free spectrum of fluctuations. Our results thus indicate that supersymmetry within a quantum description of the very early universe intrisically contains the relevant seeds for structure formation.

Within this context, the answer to those questions above is a *yes*, but where some caution is nevertheless required. SQC can indeed constitute an "observational" subject. This endorses supersymmetry as a mandatory component in any realistic analysis of a quantum universe. Further developments are expected to be reported in ref. [6].

Acknowledgments This work was supported by a JNICT/PRAXIS–XXI Fellowship BPD /6095/95 and is a summary of an Essay awarded an Honourable Mention in the 1997 Gravity Research Foundation competition. Finally, it was indeed a delight to attend this conference and special thanks go to P. Macedo and P. Avelino for their assistance, as well as to C. Martins for his assistance concerning the obscure LATEX features of KLUWER.

References

1. P. Moniz, Int. J. Mod. Phys. **A11** (1996) 4321 [Review–Report]
2. A.D.Y. Cheng and P.R.L.V. Moniz, Int. J. Mod. Phys. **D4**(1995) 189; Mod. Phys. Lett. **A11** (1996) 227; P. Moniz, Int. J. Mod. Phys.**A11** (1996) 1763; Gen. Rel. Grav. **28** (1996) 97; gr-qc/9604045; gr-qc/9606047.
3. J. Wess and J. Bagger, *Supersymmetry and Supergravity*, 2nd. ed. (Princeton University Press, 1992).
4. J.J. Halliwell and S.W. Hawking, Phys. Rev. **D31** (1985) 1777
5. P. D. D Eath and J.J. Halliwell, Phys. Rev. **D35** (1987) 1100
6. P. Moniz, *"Origin of structure in supersymmetric quantum cosmology"*, work in progress

AXIALLY SYMMETRIC COSMOLOGICAL MODELS WITH PERFECT FLUID AND COSMOLOGICAL CONSTANT

PAULO AGUIAR AND PAULO CRAWFORD
Centro de Física Nuclear e Departamento de Física da Faculdade de Ciências da Universidade de Lisboa

In this paper, we will discuss the behavior of some homogeneous but anisotropic models with axial symmetry, filled with a perfect and pressureless fluid (dust) and a non vanishing cosmological constant. For this, we will restrict our study to the the metric forms

$$ds^2 = -c^2dt^2 + a^2(t)dr^2 + b^2(t)\left(\frac{dv^2}{1-kv^2} + v^2d\phi^2\right), \tag{1}$$

with the two scale factors $a(t)$ and $b(t)$; k is the curvature index of the 2-dimensional surface and can take the values $+1, 0, -1$, giving the following three different metrics: Kantowski-Sachs, Bianchi I, and Bianchi III, respectively. Relevant Einstein equations for the metric (1), whith a perfect fluid and a cosmological term, Λ, content are then as follows:

$$2\,\frac{\dot{a}}{a}\frac{\dot{b}}{b} + \frac{\dot{b}^2}{b^2} + \frac{kc^2}{b^2} = 8\pi G\rho + \Lambda c^2, \tag{2}$$

$$2\,\frac{\ddot{b}}{b} + \frac{\dot{b}^2}{b^2} + \frac{kc^2}{b^2} = -8\pi G\frac{p^2}{c^2} + \Lambda c^2, \tag{3}$$

where ρ is the matter density and p is the (isotropic) pressure of the fluid.

If we consider a vanishing pressure ($p = 0$), which is compatible with the present conditions for the Universe, and integrate Eq.(3) we get

$$\frac{\dot{b}^2}{b^2} = \frac{M_1}{b^3} - \frac{kc^2}{b^2} + \frac{\Lambda}{3}c^2, \tag{4}$$

where M_1 is a constant of integration. In this models we have two Hubble constants, corresponding to the scales $a(t)$ and $b(t)$: $H_a \equiv \dot{a}/a \quad \dot{b}/b \equiv H_b$, respectively. Using the last equation we can rearrange it to get the folowing conservation equation

$$\Omega_M - \Omega_k + \Omega_\Lambda = 1. \tag{5}$$

Astrophysics and Space Science is the original source of publication of his article. It is recommended that this article is cited as: *Astrophysics and Space Science* **261**: 299–300, 1999.

Now, defining the dimensionless variable $y = b/b_0$, where $b_0 = b(t_0)$, is the angular scale factor for the present age of the Universe, and using Eq.(5) (taken for $t = t_0$), one may rewrite Eq.(4) as

$$\dot{y} = \pm H_{b_0}\sqrt{1 + \Omega_{M_0}(\frac{1}{y} - 1) + \Omega_{\Lambda_0}(y^2 - 1)}, \tag{6}$$

where the density parameters Ω and H_b with the zero subscript, denotes as before this quantities at the present time t_0. Substituting Eq.(4) into Eq.(2) and using analogous manipulations we get another conservation equation

$$\Omega_p - \Omega_M + 2\,\Omega_\Lambda = 2\,\frac{H_a}{H_b}, \tag{7}$$

Introducing other dimensionless variable $x = a/a_0$, Eq.(2) takes the form

$$\dot{x} = H_{b_0}\,\frac{\frac{\Omega_{M_0}}{2}(1 - \frac{x}{y}) + \Omega_{\Lambda_0}(-1 + xy^2) + \frac{H_{a_0}}{H_{b_0}}}{y\sqrt{\Omega_{M_0}(\frac{1}{y} - 1) + \Omega_{\Lambda_0}(y^2 - 1) + 1}}, \tag{8}$$

The behavior of $y(t)$ may be carried out looking for the y values where $\dot{y} = 0$. This analysis was made by Mariano Moles [1] for FLRW models, in great detail. Although we are considering anisotropic models, the Eq.(6) is exactly the same as the one obtained by [1, Moles] for the homogeneous and isotropic FLRW models. From Eq.(6) and Eq.(8) one obtains a differential equation in $\frac{dx}{dy}$ that can be integrated numericaly. With $H_{a_0} = H_{b_0}$, has three distinct possible integrations, one for each k value.

Although we are dealing with anisotropic models, we may admit that at a certain moment of time, which we can take as the present time t_0, the Hubble parameters along the orthogonal directions may be assumed to be approximately equal, $H_a \simeq H_b$. This hypothesis has been considered in [2, Henriques] for the case of a Kantowski-Sachs (KS) model. From this study was derived the conclusion that the classical tests of cosmology are not at present sufficient accurate to distinguish between a FLRW model and the KS defined in that paper, with ($H_{a_0} \simeq H_{b_0}$), execpt for small values of b_0.

Plotting the qualitative behaviour of scale factors for this three models we conclude that these models undergo isotropisation, except for the Kantowski-Sachs model ($\Omega_{k_0} > 0$) with $\Omega_{\Lambda_0} < \Omega_{\Lambda_M}$ and for the Bianchi III ($\Omega_{k_0} < 0$) with $\Omega_{\Lambda_0} < \Omega_{\Lambda_M}$.

An extended version of this paper may be found as an electronic preprint at bulletin board http://xxx.lanl.gov/abs/astro-ph/9801117.

References

1. Moles, M. (1991), *A. Journal* **382**, 369–376
2. Henriques, A. B. (1996), *Astrophysics and Space Science*,**235**, 129–140

COULD THE OBSERVABLE UNIVERSE BE INSIDE OF A BLACK HOLE?

ROSA DORAN AND PAULO CRAWFORD
Centro de Física Nuclear e Departamento de Física
da Faculdade de Ciências da Universidade de Lisboa

It is well known that the region inside of a Schwarzschild horizon has the same symmetries as a Kantowski-Sachs cosmological model. This possibility raised the question of whether our observable Universe could be viewed as the interior of a Schwarzschild black hole (SBH). The study of some important kinematic parameters such as expansion, shear, and deceleration were then carried out in order to find a possible answer to this question. In this paper we report this work which led us to conclude that the observable Universe could hardly be inside of a SBH.

Schwarzschild Metric for $r < 2M$

The Schwarzschild's metric (G = c= 1) can be rewritten in a more suitable coordinate system for the region $r < 2M$ [1]

$$ds^2 = -\left(\frac{2M}{t} - 1\right) dt^2 + \left(\frac{2M}{t} - 1\right)^{-1} dr^2 + t^2 d\Omega^2)$$

Some amazing results follow from the analysis of this metric if we take seriously the character of the t and z coordinates. For example we conclude that in the z direction, particles at rest, separated by a distance Δz, reach the singularity infinitely separated from each other, in a cigar type singularity. Particles at rest in this region follow geodesics. Furthermore, we can introduce a coordinate transformation that immediately allows us to study the Black Hole interior as a cosmological geometry.

Cosmological Interpretation

By introducing the following coordinate transformation $d\bar{t}^2 = dt^2 / (2M/t - 1)$ we can rewrite the metric as: $ds^2 = -d\bar{t}^2 + (2M/t(\bar{t}) - 1)\, dz^2 + t(\bar{t})^2 d\Omega^2$. This metric represents an anisotropic universe with a finite coordinate time

Astrophysics and Space Science is the original source of publication of his article. It is recommended that this article is cited as: *Astrophysics and Space Science* **261**: 301–302, 1999.

interval ($0 < \bar{t} < M\pi$). It has the same symmetries as the Kantowski-Sachs metric [2]. Now we can calculate the relevant kinematic parameters.

Kinematic Parameters

This model has an initial phase of expansion followed by a phase of contraction $\theta = (2M/t-1)^{1/2}(2t-3M)/\,(t(2M-t))$. Attention should be given to the fact that these study refers only to particles at rest which do not cross the horizon. In fact, taking into account the conserved quantities, it can be shown [3] that the particles coming from the exterior region can not be at rest. We can define a characteristic length scale R by $R^3 = AC^2$ where $A = (2M/t(\bar{t}) - 1)^{1/2}$ and $C = t(\bar{t})$. We have then two Hubble constants $H_a = M/t^2(2M/t-1)^{1/2}$, $\quad H_c = -(2M/t-1)^{1/2}/t$, showing that there is expansion in the z direction while there is contraction in the angular 2-D space. The study of the deceleration parameter gives an overall decelerating universe $q = (2(t-3M)^2/(-3M+2t)^2)$, and the shear tensor is as follows $\sigma = (3M-t)/(3t^3(2M-t))$, which shows that this metric has unequivocal anisotropic characteristics.

Conclusion

Although we are dealing with a spherically symmetric solution, these results show us that the geometry of the a SBHs interior region is unequivocally anisotropic since the shear is never sufficiently small to accomodate the present cosmological observations. Therefore it could hardly describe the observable Universe. What we should bear in mind, however, is that we are dealing with a model that is not a very realistic one. For instance, we could also consider the existence of a cosmological constant Λ. We know that the introduction of Λ will produce an effect of isotropization via an accelerated expansion (inflation). Therefore, it would smooth out the Kantowski-Sachs anisotropies and produce a more suitable model. This will be dealt with in other papers still in progress.

This work was finacially supported by grant BD 4345 PRAXIS XXI, from JNICT and by CERN/P/ FAE/1164/97 project.

References

1. Brehme, R.W., (1977) "Inside the Black Hole", Am.J.Phys., 45, 423-428.
2. Kantowski, R., Sachs, R.K., (1966) "Some Spatially Homogeneous Anisotropic Relativistic Cosmological Models", J.Math.Phys. 7, 443-445.
3. Doran, R., Crawford, P. (1998), work in progress

QUINTESSENTIAL COSMOLOGY

Novel Models of Cosmological Structure Formation

R.R. CALDWELL, R. DAVE, AND P.J. STEINHARDT
University of Pennsylvania
Department of Physics and Astronomy
Philadelphia, PA 19104 USA

Abstract.

We examine the possibility that a substantial fraction of the total energy density in a spatially flat Universe is composed of a time-dependent and spatially inhomogeneous component whose equation-of-state differs from that of baryons, neutrinos, dark matter, or radiation. In this lecture, we report on our investigations of the case in which the additional energy component, dubbed "quintessence", is due to a dynamical scalar field evolving in a potential. We have computed the effects on the background cosmological evolution, the cosmic microwave background (CMB) and mass power spectrum, finding a broad range of cosmologically viable models. We stress three important features of the quintessence or Q-component: the time evolution of the equation-of-state; the length-scale dependence of the speed of propagation of the fluctuations in the Q-component; and, the contribution of quintessence fluctuations to the CMB anisotropy spectrum.

1. Introduction

Our investigation into the missing energy in the Universe begins with the following line of reasoning. First, we consider that the Universe is spatially flat to very high precision, as a consequence of primordial inflation. Second, we note that the mass density in baryons and pressureless dust, cold dark matter (CDM), is much less than the critical density, according to recent measurements of large scale structure formation, large-angular scale microwave background anisotropy, and x-ray cluster luminosity and abundance [1-3]. Given these two conditions, there is a deficit in the cosmological energy density in order that the total sums to the critical value. A possibility for the missing energy that has been considered extensively

Astrophysics and Space Science is the original source of publication of his article. It is recommended that this article is cited as: *Astrophysics and Space Science* **261**: 303–310, 1999.

is that the remaining energy density is due to a cosmological constant, Λ. The ΛCDM cosmological scenario has certain advantages over the standard CDM scenario [2-5], owing primarily to the slower rate of growth of density perturbations and the lower mass fraction available to cluster. However, there is no compelling reason why the deficit must be made up by a constant vacuum energy density.

In this paper, we review some of our recent studies of an alternative possibility for the missing energy of the universe: a time-dependent and spatially inhomogeneous component with negative pressure. The negative pressure means that this component has an equation-of-state different from that of baryons, neutrinos, dark matter, or radiation. We refer to this fifth component as "quintessence." Quintessence is also distinct from a vacuum density or cosmological constant. A vacuum density is spatially uniform, whereas quintessence develops fluctuations. Both have negative pressure, but while the sum of pressure (p) and energy density (ρ) is precisely zero for the cosmological constant, the sum is positive and possibly time-varying for quintessence. In other words, quintessence is physically distinct from other forms of cosmological fluids usually considered.

Quintessence is introduced because, as we shall show, it is a logical possibility that appears to fit well all current observations. At this time, observations admit other possibilities, including ΛCDM. Each of the contending models has interesting and different implications for cosmology and fundamental physics. If quintessence is confirmed, it suggests a new realm of very low energy particles, fields and dynamics to be explored. Our purpose is to identify which quintessence models fit current data and which future cosmological observations can distinguish these models from alternative possibilities. In this sense, we show how near-future observations can probe macrophysics and microphysics simultaneously.

For this brief report, we focus on models for which the quintessence equation-of-state ($w \equiv p/\rho$) ranges between $-1 \leq w \leq 0$ for most of the evolutionary history. This allows both an older Universe than in a standard CDM scenario, and a relaxation of the classical distance-red shift constraints due to strong gravitational lensing, relative to the ΛCDM scenario [6-10]. In these regards, a quintessence scenario with cold dark matter (QCDM) has many advantages over other models.

Specifying w alone is not sufficient to define the model, as we also must provide the propagation speed of fluctuations in the quintessence or Q-component. In particular, we consider the quintessence to be described by a self-interacting scalar field, in which case the equation-of-state and speed of sound are automatically determined by the properties of the interaction potential. Extensions of the standard model of particle physics are replete with dynamical scalar fields, such as hidden sector fields, moduli, or pseudo-

Nambu-Goldstone bosons. Hence, there is no shortage of candidate models, in principle. One still needs to explain the coincidence that the quintessence component of the energy density is comparable to the matter density, even though the two change at different rates as the universe expands. However, one requires similar coincidences for models with vacuum density and open models. Whereas the curvature in open models or the vacuum density in ΛCDM models have no dynamical parameters to adjust to explain the coincidence, there is more freedom with quintessence. We are currently exploring this issue. In any event, it is fair to say that the quintessence possibility is no less attractive theoretically than the current viable alternatives, now that the standard cold dark matter (sCDM) scenario is ruled out.

Previous discussions of an energy component consisting of a dynamical, fluctuating cosmic scalar field have included fields evolving in a potential [6,11-15], or an energy component evolving according to a specific equation-of-state [16,17]. In the case of a scalar field, the detailed cosmic microwave background anisotropy and matter power spectrum were computed by Coble *et al* [12] for an ultra-light pseudo-Nambu-Goldstone boson with a cosine-type potential, and by Ferreira and Joyce [13,14] for the scaling solution of an exponential potential. In our work, we have gone beyond these specific examples to survey the range of imprints on the CMB anisotropy and large scale structure power spectrum [3,18-20].

In the limited space provided by these conference proceedings, we wish to emphasize three essential properties of quintessence: the effect of the time evolution of the equation-of-state; the length-scale dependence of the speed of propagation of the fluctuations in the Q-component; and, the contribution of quintessence fluctuations to the CMB anisotropy spectrum.

2. Effects of the Quintessence

The dynamical properties of quintessence result in cosmological predictions which differ from scenarios dominated by cold dark matter or a cosmological constant. To illustrate these properties, we focus on three particular QCDM models: 1) a Q-field with a constant $w = -1/3$ equation-of-state and $\Omega_Q = 0.6$; 2) an exponential potential, $V(Q) = M^4 \exp(-\beta Q)$, for which $w \sim -1$ at early times, but grows monotonically to $w(t_0) = -1/3$ with $\Omega_Q = 0.6$ by the present time t_0; 3) a quadratic potential, $V(Q) = \frac{1}{2}m^2Q^2$, with a very light mass, $m < H_0$, so that it has just begun to oscillate freely by today, with the instantaneous values $w(t_0) = -1/3$ and $\Omega_Q = 0.6$. These models share the feature that $w = -1/3$ and $\Omega_Q = 0.6$ today, but these parameters evolve differently in the past and future. Here we show that the CMB anisotropy and mass power spectrum for these models are distinguishable from sCDM or ΛCDM and, further, the model predictions can be distinguished from one another. This suggests that cosmological measurements can discriminate quintessence from vacuum and matter density, and even determine some aspects of the time-evolution of a Q-component.

2.1. EVOLUTION OF THE EQUATION-OF-STATE

Quintessence plays an important role in driving the expansion of the Universe when its energy density becomes an appreciable fraction of the total energy density. For a particular history of the equation-of-state, $w(a)$, the energy density is given by

$$\rho_Q(a) = \frac{3H_0^2}{8\pi G}\Omega_Q \exp\left[3\log\frac{a_0}{a} + 3\int_a^{a_0}\frac{d\tilde{a}}{\tilde{a}}w(\tilde{a})\right] \tag{1}$$

where a_0 is the present-day scale factor. This expression illustrates how ρ_Q depends on the evolution of w – the energy density in the Q-component may grow, decrease, or even oscillate. Trajectories of w and ρ_Q for the three demonstration models are shown in Figure 1. Clearly, quintessence behaves differently from CDM ($\rho_{cdm} \propto a^{-3}$) and Λ ($\rho_\Lambda$ = constant), leading to a different evolution of the cosmic expansion rate.

The evolution of fluctuations is also affected when the quintessence becomes a significant component of the total cosmological fluid. The departure from pressureless, dust-dominated expansion leads to a late-time, integrated Sachs-Wolfe contribution to the large angle CMB anisotropy. Furthermore, the rate of growth of the linear density perturbations in the CDM, determined by the evolution equation $\delta''_{cdm} + aH\delta'_{cdm} = 4\pi G a^2 \delta\rho_{cdm}$, is slowed by the effect of quintessence on the cosmological expansion rate. The simple equation

$$\frac{d\ln\delta_{cdm}}{d\ln a} \approx \left[1 + \rho_Q(a)/\rho_{cdm}(a)\right]^{-0.6}, \tag{2}$$

valid for scenarios in which the ρ_Q is negligible at radiation-matter equality, shows how the growth is slowed as ρ_Q approaches ρ_{cdm}. In order to achieve the same level of mass fluctuations, structure must form earlier in a QCDM than sCDM model, although the growth does not cut off as sharply as in a ΛCDM model with the same final matter density.

2.2. SCALE DEPENDENCE OF SOUND SPEED

An important aspect of quintessence, modeled as a dynamical scalar field, is that the propagation speed of perturbations is scale dependent. Let us decompose the scalar Q-field into spatially homogeneous and inhomogeneous pieces, $Q(\tau, \vec{x}) = Q_o(\tau) + \delta Q(\tau, \vec{x})$, where τ is the conformal time. Then

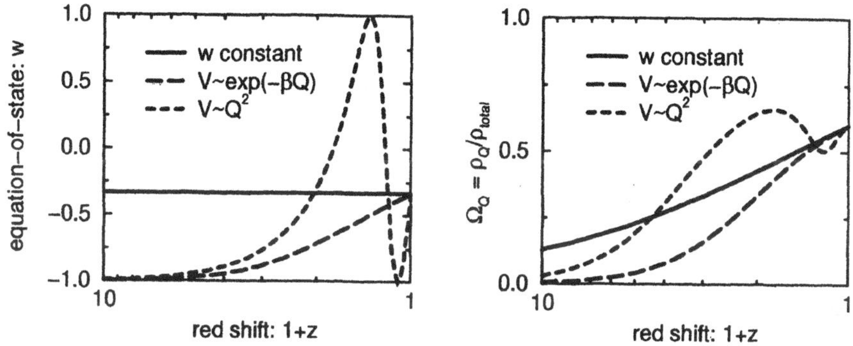

Figure 1. The evolution of the equation-of-state and energy density are shown for the three models discussed in the text. All have $w(t_0) = -1/3$ and $\Omega_Q = 0.6$ at the present time, t_0. However, the evolution and effects of the quintessence differ greatly: the Q-field oscillates in the quadratic potential, generating oscillations in ρ_Q; the Q-component with constant $w = -1/3$ becomes important at much higher red shift than in the case of the exponential potential.

we find that the Fourier transform of the fluctuating component obeys the wave equation

$$\delta Q_k'' + 2aH\delta Q_k' + (k^2 + a^2 V_{,QQ})\delta Q_k = -\frac{1}{2}h_k' Q_o' \tag{3}$$

where the prime denotes $\partial/\partial\tau$, the index k indicates the Fourier transform amplitude, and h is the trace of the synchronous gauge metric perturbation (see Ref. [21]). Let us consider the solutions which develop in response to the source term on the right hand side of the above expression. For the cosmologically interesting case in which $H^2 \gtrsim V_{,QQ}$, the evolution of the background field, Q_o, is friction dominated. We find that for $k^2 \ll a^2 V_{,QQ} \lesssim a^2 H^2$ the solutions decay in amplitude and oscillate with very slowly varying phase, $\arg(\delta Q_k) \sim a\tau\sqrt{V_{,QQ}}$. Perturbations propagate slowly, and density fluctuations accumulate on length scales $\lambda \gg 2\pi/\sqrt{V_{,QQ}}$. However, for $k^2 \gg a^2 V_{,QQ}$, $a^2 H^2$ the phase of the oscillatory solution, $\arg(\delta Q_k) \sim a\tau k$, rotates so rapidly that the time-averaged fluctuation, $\langle \delta Q_k \rangle$, vanishes. Because the density perturbation is linear in δQ, $\delta Q'$, then $\delta\rho_Q$ vanishes as well. Physically, the Q-field is so light it behaves relativistically, free-streaming on length scales $\lambda \ll 2\pi/\sqrt{V_{,QQ}}$. The dispersion term in (3) suggests a mode propagation velocity $v_k^2 = (1 + a^2 V_{,QQ}/k^2)^{-1}$ which is relativistic on small scales, with $v_k \to 1$, and non-relativistic on large scales, with $v_k \to 0$. The primary consequence of the scale dependence of the dispersion velocity is that, while the Q-field may develop density perturbations on very large scales, the quintessence models considered here do not clump on galaxy cluster scales and below. This means only baryons and

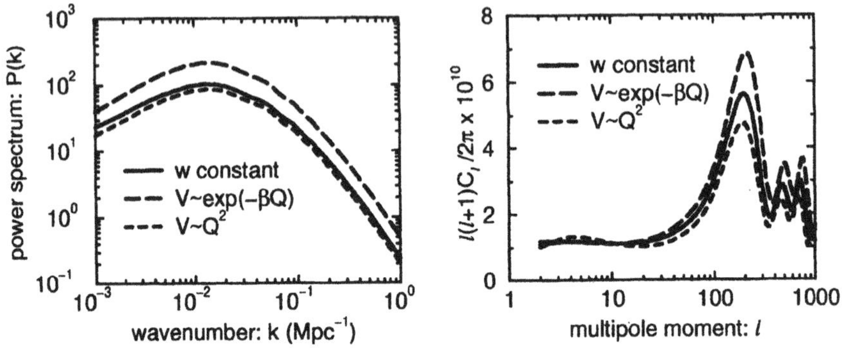

Figure 2. The mass power and CMB anisotropy spectra are shown for the three models discussed in the text. The COBE-normalized power spectra give $\sigma_8 = 0.58, 0.52, 0.83$ for the constant w model, quadratic and exponential potentials respectively. The CMB spectra also vary dramatically; note the notch near $\ell \sim 4$ in the spectrum for the quadratic potential, due to the oscillation of the Q-field at $z \sim 2$.

CDM contribute to the mass power spectrum. We illustrate in Figure 2 the power spectrum for a series of QCDM scenarios. The shape is affected only by the Hubble parameter and $\Omega_{cdm} = 1 - \Omega_Q$, and is insensitive to the microphysical properties of the quintessence. The overall amplitude, however, as determined via COBE normalization, is strongly dependent on the fluctuations in the Q-field, as discussed next.

2.3. ROLE OF FLUCTUATIONS

The fluctuations in the quintessence play an important role in the generation of large angle CMB anisotropy. We observed in the preceding subsection that the quintessence may cluster only on the largest scales, where the propagation velocity is non-relativistic. In order to determine the strength of the effect, we return to equation (3), where the disturbances in the Q-field are driven by the source term on the right hand side. Physically, the quintessence must respond to the changing gravitational field, h', of the perturbations in the CDM. The strength of the response is moderated by the kinetic energy of the homogeneous Q-component:

$$|Q_o'| = a\sqrt{(1+w)\rho_Q}. \tag{4}$$

A cosmological constant, an extreme type of quintessence with $w = -1$, has vanishing kinetic energy; it is "stiff" (being a constant) and does not respond to the local gravity. In the case that $w \neq -1$ and ρ_Q is a substantial fraction of the total energy density, the source term becomes important, turning on the fluctuations δQ. Hence, in QCDM scenarios in which the quintessence comes to dominate the energy density of the Universe at late times, the fluctuations $\delta\rho_Q$ will become important at late times and on large scales. In Figure 2 we show the CMB anisotropy for the three demonstration models, computed by integrating the Boltzmann equations from deep in the radiation era to the present [21,22]. Normalization to COBE, following the procedure of Bunn and White [23], disguises the dramatic effect of the quintessence on the large angular scales: even though the physics at the time of last-scattering, responsible for the acoustic oscillations, is unaffected by the quintessence, the high-ℓ amplitude varies widely among the QCDM models. In Figure 3 we show the CMB anisotropy in the case of a constant $w = -1/3$ QCDM model, with and without (achieved artificially by forcing $\delta Q = \delta Q' = 0$) the perturbations in the quintessence. One may interpret the rise at low-ℓ in the $\delta\rho_Q = 0$ spectrum as being due to the late integrated Sachs-Wolfe effect associated with the change in the background expansion rate only. The fluctuations clearly compensate for this change, when $\delta\rho_Q$ contributes to the perturbation spectrum.

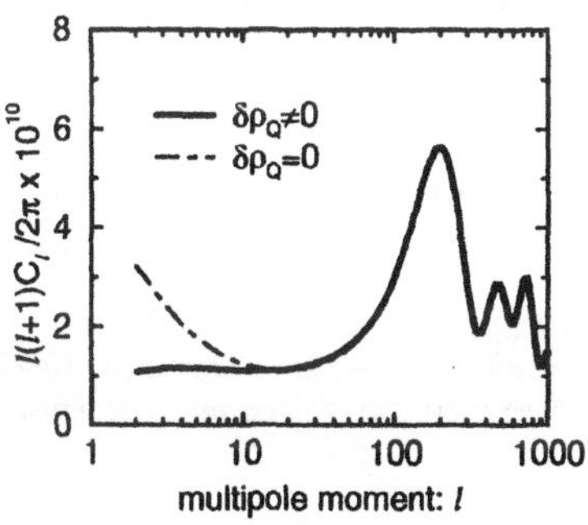

Figure 3. The CMB spectra for the constant $w = -1/3$ model (heavy, solid), and the same model with perturbations in the Q-component artificially turned off, $\delta\rho_Q = 0$ (light, dot-dashed), are shown. Fluctuations must develop in the Q-component, in response to the fluctuations in the surrounding baryons and CDM. The dramatic difference in the two curves demonstrates the importance of the contribution of the quintessence fluctuations to the anisotropy spectrum.

3. Discussion

We have examined the implications of "quintessence", a component of the energy density of the Universe, modeled here as a dynamical scalar field evolving in a potential. To demonstrate the cosmological effects, we have presented results for three different models. We have further surveyed a number of QCDM models, presented separately in the lecture, including constant w, exponential potentials, and both oscillatory and monotonically varying cosine, quadratic, and quartic potentials. Details of these models may be found in Refs. [18-20].

The three sample models shown here are part of spectrum which fit all current observations. A survey of best-fitting QCDM models is forthcoming [3]. The models result in an imprint on the CMB anisotropy and mass power spectrum that should be detectable in near-future experiments. Confirmation of any one of these models would indicate the existence of new, fundamental fields with profound implications for particle physics, as well as cosmology. A number of follow-up studies are under way, including: a quantitative analysis to determine how well future CMB experiments can resolve a Q-component and how well one can simultaneously resolve other cosmological parameters; numerical simulations of galaxy formation and evolution in the presence of a Q-component; and search for quintessential candidates in models of fundamental physics.

RRC thanks the organizing committee of the Non-Sleeping Universe Conference in the Centro de Astrofisica at the Universidade do Porto for an interesting and pleasurable meeting. This research was supported by the U.S. Department of Energy at Penn, DE-FG02-95ER40893.

References

1. Avishai Dekel, David Burstein, and Simon D. M. White, "Measuring Omega" in *Critical Dialogues in Cosmology*, ed. Neil Turok (World Scientific, 1997); and references therein.
2. J. P. Ostriker and P.J. Steinhardt, *Nature* **377**, 600 (1995); and references therein.
3. L. Wang, R.R. Caldwell, J.P. Ostriker, and P.J. Steinhardt, *in preparation*
4. L. Krauss and M.S. Turner, Gen. Rel. Grav. **27**, 1137 (1995).
5. Michael Turner, "The Case for ΛCDM", in *Critical Dialogues in Cosmology*, ed. Neil Turok (World Scientific, 1997); and references therein.
6. P.J.E. Peebles and Bharat Ratra, Ap. J. Lett. **325**, L17 (1988).
7. J. Frieman, C. Hill, A. Stebbins, and I. Waga, Phys. Rev. Lett. **75**, 2077 (1995).
8. V. Silveira and I .Waga, Phys. Rev. D **56**, 4625 (1997).
9. Joshua A. Frieman and Ioav Waga, astro-ph/9709063.
10. Paul J. Steinhardt, "Comments on How Forthcoming Progress in Cosmology Might Influence Fundamental Physics", in *Critical Dialogues in Cosmology*, ed. Neil Turok (World Scientific, 1997).
11. B. Ratra and P.J.E. Peebles, Phys. Rev. D **37**, 3406 (1988).
12. Kimberly Coble, Scott Dodelson, and Joshua A. Frieman, Phys. Rev. D **55**, 1851 (1997).
13. Pedro Ferreira and Michael Joyce, Phys. Rev. Lett. **79**, 4740 (1997).
14. Pedro Ferreira and Michael Joyce, astro-ph/9711102.
15. Pedro T.P. Viana and Andrew R. Liddle, Phys. Rev. D **57**, 674 (1998).
16. Michael S. Turner and Martin White, Phys. Rev. D **56**,R4439 (1997).
17. Takeshi Chiba, Naoshi Sugiyama, and Takashi Nakamura, Mon. Not. R. Astron. Soc. 289, 5 (1997).
18. R.R. Caldwell, R. Dave, and P.J. Steinhardt, *to appear* Phys. Rev. Lett. (astro-ph/9708069)
19. R.R. Caldwell and P.J. Steinhardt, *submitted to* Phys. Rev. D (astro-ph/9710062).
20. R.R. Caldwell, R. Dave, and P.J. Steinhardt, *in preparation*
21. C.-P. Ma and E. Bertschinger, Ap. J. **455**, 7 (1995).
22. U. Seljak and M. Zaldarriaga, astro-ph/9603033.
23. E.F. Bunn and M. White, Ap. J. **480**, 6 (1997).

WIGGLY COSMIC STRINGS

C.J.A.P. MARTINS
DAMTP, University of Cambridge
Silver St., Cambridge CB3 9EW, U.K.

Simulations [1, 2] of GUT cosmic strings [3] have established the existence of a significant amount 'wiggles' on scales much smaller than the correlation length. Given the limited dynamic range of these simulations, however, it is not clear under which conditions small-scale structure builds up indefinitely or eventually reaches a scaling solution like the large-scale network properties [4, 5, 6, 7]. Most of the work on cosmic strings so far was dealt with the structureless strings [8]. In particular, the first quantitative description of the complete evolution of these networks has been recently developed [9, 10, 11, 12, 13]. However, due to the limited number of degrees of freedom available, such models cannot realistically account for a number of cosmologically relevant phenomena, such as the build-up of small-scale structure.

One is therefore led to the introduction of further degrees of freedom on the string worldsheet [14, 15]. One of the simplest such models is the 'transonic' model with Lagrangian density $\mathcal{L} = -\mu_0\sqrt{1-\chi} = -\mu_0 w$, where $w^2 = T/U$. U and T are respectively the energy density in the locally preferred string rest frame and the local string tension, and this model has the equation of state $UT = \mu_0^2$. This arises in a 'smoothed' model of a wiggly string, that is a Goto-Nambu string containing a spectrum of small oscillations [16]. Here we report on the developement of an averaged description of the transonic model.

Consider a flat FRW background spacetime with the line element $ds^2 = a^2\left(d\tau^2 - \mathbf{dx}^2\right)$, the gauge choice $\sigma^0 = \tau$, $\dot{\mathbf{x}} \cdot \mathbf{x}' = 0$ (dots and primes respectively denoting derivatives with respect to the time-like and space-like coordinates on the worldsheet), and define $\epsilon^2 = \mathbf{x}'^2/1 - \dot{\mathbf{x}}^2$. Cosmic string motion in this background will be retarded by particle scattering; for a gauge string, the main contribution comes from Aharonov-Bohm scattering. Vilenkin [17] has shown that this can be described by a friction lenghtscale ℓ_f. Then the microscopic string equations of motion for this model can be

Astrophysics and Space Science is the original source of publication of his article. It is recommended that this article is cited as: *Astrophysics and Space Science* **261**: 311–314, 1999.

obtained from $\mathcal{L}$ by the usual variational method. Including this friction force [17] and defining the damping length as $1/\ell_{\rm d} = 2H + w/\ell_{\rm f}$ one finds

$$\ddot{\mathbf{x}} + \dot{\mathbf{x}}(1-\dot{\mathbf{x}}^2)\left[a\frac{w^2}{\ell_{\rm d}} + \frac{\dot{a}}{a}(1-w^2)\right] = \frac{w^2}{\epsilon}\left(\frac{\mathbf{x}'}{\epsilon}\right)', \tag{1}$$

$$\left(\frac{\epsilon}{w}\right)^{\cdot} + \left(\frac{\epsilon}{w}\right)\left[a\frac{w^2}{\ell_{\rm d}}\dot{\mathbf{x}}^2 + \frac{\dot{a}}{a}(1+\dot{\mathbf{x}}^2)(1-w^2)\right] = 0, \tag{2}$$

$$\frac{\dot{w}}{w} = (1-w^2)\left(\frac{\dot{a}}{a} + \frac{\mathbf{x}'\cdot\dot{\mathbf{x}}'}{\mathbf{x}'^2}\right). \tag{3}$$

The total energy of a piece of string is $E_{total} = \mu_0 a \int \frac{\epsilon}{w} d\sigma = a\int \epsilon U d\sigma$; part of this energy corresponds to the 'smoothed' string itself, while the rest is in the small-scale wiggles. The averaging procedure for the transonic elastic model closely follows what was done for the Goto-Nambu case [10]. The average of a generic quantity is defined giving more weight to regions with more small-scale structure. In particular, we will deal with the average RMS string velocity, v^2 and we define the renormalised string mass per unit length as $\tilde{\mu} \equiv E_{total}/E_{string}$. We distinguish between long or 'infinite' strings and loops and define the total energy density in a long string network as $\rho_\infty = \mu_0/L^2 = \tilde{\mu}_\infty\mu_0/\xi^2$ and the energy density in the 'smoothed' strings only as $\rho_{s\infty} = \mu_0/\xi^2$.

The crucial point in going from the above microscopic description to an averaged one in terms of quantities such as L, ξ, v or $\tilde{\mu}$ is the introduction of phenomenological parameters describing effects such as loop and small-scale structure production or gravitational radiation backreaction on the network [18, 19, 20, 6]. Comparisons with numerical simulations lead us to believe that this model does contain the essential physics of the problem, and can provide a reasonable description of wiggly string evolution. We define the fraction of the energy density transferred into loops per unit time as $(d\rho_{s\infty}/\rho_{s\infty}dt)_{toloops} = -\tilde{c}\tilde{\mu}_\infty v_\infty/\xi$. It is clear [1, 2] that small-scale structure enhances loop production. We model this by a simple linear dependence. As for the rate of transfer of the total energy in long strings to loops, we write $(d\rho_\infty/\rho_\infty dt)_{toloops} = -\lambda\tilde{c}\tilde{\mu}_\infty v_\infty/\xi$, where $\lambda(\tilde{\mu}_\infty)$ accounts for the fact that loops are preferentially 'chopped off' from regions containing more small-scale structure than average [1, 2]. Numerical simulations [1, 2] clearly show that such a dependence exists, although its is not clear what it is. We will again make the simplifying assumption of a linear dependence. There is a further phenomenological term, which will in fact be crucial. Whenever two strings inter-commute, kinks are produced [4, 5, 6], and this corresponds to energy being transferred from the strings to the small-scale wiggles. We will parametrise this as $(d\rho_{s\infty}\rho_{s\infty}dt)_{towiggles} = -\tilde{s}v_\infty/\xi$.

This being said, one can write the evolution equations for L, $\tilde{\mu}_\infty$ and v_∞ (the one for ξ is not independent of these) One finds that

$$\frac{2}{L}\frac{dL}{dt} = H\left(\frac{3\tilde{\mu}_\infty^2 - 1}{\tilde{\mu}_\infty^2} + \frac{2\tilde{\mu}_\infty^2 v_\infty^2}{2\tilde{\mu}_\infty^2 - 1}\right) + \frac{v_\infty^2}{\ell_{f_0}}\frac{\tilde{\mu}_\infty(1+\tilde{\mu}_\infty)}{2(2\tilde{\mu}_\infty^2 - 1)} + \tilde{\mu}_\infty^{3/2}\tilde{c}\frac{v_\infty}{L}, \quad (4)$$

$$\frac{1}{\tilde{\mu}_\infty}\frac{d\tilde{\mu}_\infty}{dt} = \left[k'\left(1 - \tilde{\mu}_\infty^{-2}\right) + \tilde{s} - \tilde{\mu}_\infty\tilde{c}(\tilde{\mu}_\infty - 1)\right]\frac{v_\infty}{\sqrt{\tilde{\mu}_\infty}L} \quad (5)$$

$$-(\tilde{\mu}_\infty - 1)\left[H\left(\frac{1+\tilde{\mu}_\infty}{\tilde{\mu}_\infty^2} - \frac{2\tilde{\mu}_\infty^2 v_\infty^2}{2\tilde{\mu}_\infty^2 - 1}\right) + \frac{\tilde{\mu}_\infty}{2(2\tilde{\mu}_\infty^2 - 1)}\frac{v_\infty^2}{\ell_{f_0}}\right] \quad (6)$$

$$\frac{dv_\infty}{dt} = (1 - v_\infty^2)\left[\frac{k}{\tilde{\mu}_\infty^{5/2}L} - \frac{\tilde{\mu}_\infty^2}{2\tilde{\mu}_\infty^2 - 1}v_\infty\left(2H + \frac{1+\tilde{\mu}_\infty}{2\ell_{f_0}}\right)\right]; \quad (7)$$

As in the Goto-Nambu case, comparison of the microscopic and averaged equations of motion for circular loops motivates the definition of the remaining phenomenological 'curvature' [10] parameters.

With our assumption $\lambda = \tilde{\mu}_\infty$, small-scale structure will in general reach a scaling regime (except for very large $\tilde{s}$). If $\tilde{s}$ is very small all small-scale structure can be removed and one ends up with a Goto-Nambu network; for intermediate values of $\tilde{s}$ the scaling value of $\tilde{\mu}_\infty$ can be significantly larger than unity. As long as there is a dependence of λ on $\tilde{\mu}_\infty$, the evolution does not depend strongly on the power chosen. We also find that the behaviour of L/t (that is, the total energy) is much less affected by the presence of small-scale structure than the behaviour of v_∞. This is in agreement with the numerical result that the evolution of small-scale structure does not affect substantially the large-scale evolution of the network [1, 2]. We will fit our parameters to values intermediate between those found by the Bennett & Bouchet [1] and the Allen & Shellard [2] simulations. Firstly we can obtain k and k' from our *ansätze*: $k_r \approx 0.39$, $k'_r \approx 0.25$ in the radiation era and $k_m \approx 0.72$, $k'_m \approx 0.43$ in the matter era. We can then use our averaged equations, under the assumption of full scaling, to find the remaining parameters, $\tilde{c}_r \approx 0.061$, $\tilde{s}_r \approx 0.030$, $\tilde{c}_m \approx 0.052$, and $\tilde{s}_m \approx 0.028$. Hence the two parameters are approximately equal in the two epochs. Note that we have $\tilde{c} \sim 2\tilde{s}$, indicating than in the scaling regime small-scale structure production through 'chopping' of loops is dominant.

We have studied the evolution of the network during the radiation-matter transition. Keeping $\tilde{c}$ and $\tilde{s}$ fixed at their radiation-era values, we are able to match the transition with an error slightly smaller than the one for our earlier Goto-Nambu model [10]. Note that we are matching the transition with a model containing only one free parameter, $\tilde{c}$. We can also establish a simple comparison with numerical simulations by studying the approach to scaling of a 'Vachaspati-Vilenkin' network. We study the evolution of radiation-era networks not subject to the friction force, and starting their evolution with $\tilde{\mu}_{\infty} = 1$ and $v_{\infty} \sim 0.6$ and compare with the corresponding Allen & Shellard [2] results. The agreement is rather good, although the timescale of the evolution of the network is slightly shorter in the numerical simulation.

Hence the new model successfully passed two very simple numerical tests. Further numerical testing will be needed, in particular to clarify the interpretation of our phenomenological parameters, but the usefulness of the transonic elastic model is already clear: by including additional degrees of freedom, one is able to provide a better physical description of the crucial evolution mechanisms.

This work has been done in collaboration with Paul Shellard, and was funded by JNICT (Portugal) under 'Programa PRAXIS XXI' (grant no. PRAXIS XXI/BD/3321/94). I would like to thank Pedro Avelino, Rob Caldwell and Proty Wu for useful discussions.

References

1. D. P. Bennett & F. R. Bouchet, Phys. Rev. **D40**, 2408 (1990).
2. B. Allen & E. P. S. Shellard, Phys. Rev. Lett. **64**, 119 (1990).
3. A. Vilenkin & E. P. S. Shellard, *'Cosmic Strings and other Topological Defects'*, Cambridge University Press (1994).
4. B. Allen & R. R. Caldwell, Phys. Rev. **D43**, 2457 (1991).
5. D. Austin, Phys. Rev. **D48**, 3422 (1993).
6. D. Austin, E. J. Copeland & T. W. B. Kibble, Phys. Rev. **D48**, 5594 (1993), and references therein.
7. G. R. Vincent, M. Hindmarsh & M. Sakellariadou, Phys. Rev. **D56**, 637 (1997).
8. T. W. B. Kibble, J. Mod. Phys. **A9**, 1387 (1976).
9. C. J. A. P. Martins & E. P. S. Shellard, Phys. Rev. **D53**, 575 (1996).
10. C. J. A. P. Martins & E. P. S. Shellard, Phys. Rev. **D54**, 2535 (1996).
11. C. J. A. P. Martins & E. P. S. Shellard, Phys. Rev. **B56**, 10892 (1997).
12. C. J. A. P. Martins, Phys. Rev. **D55**, 5028 (1997).
13. P. P. Avelino, R. R. Caldwell & C. J. A. P. Martins, Phys. Rev. **D56**, 4568 (1997).
14. B. Carter, in *'The Formation and Evolution of Cosmic Strings'*, G. W. Gibbons *et al.* (eds.), Cambridge University Press (1990).
15. B. Carter, in *'Formation and Interactions of Topological Defects'*, A. C. Davis & R. H. Brandenberger (eds.), Plenum Press (1995), and references therein.
16. A. Vilenkin, Phys. Rev. **D41**, 3038 (1990); B. Carter, Phys. Rev. **D41**, 3869 (1990); X. Martin, Phys. Rev. Lett. **74**, 3102 (1995).
17. A. Vilenkin, Phys. Rev. **D43**, 1060 (1991).
18. T. W. B. Kibble, Nucl. Phys. **B252**, 227 (1995); *Erratum:* **B261**, 750 (1986).
19. D. P. Bennett, Phys. Rev. **D33**, 872 (1986); *Erratum:* **D34**, 3592 (1986).
20. A. Albrecht & N. Turok, Phys. Rev. **D40**, 973 (1989).
21. E. P. S. Shellard & B. Allen, in *'The Formation and Evolution of Cosmic Strings'*, G. W. Gibbons *et al.* (eds.), Cambridge University Press (1990).
22. C. J. A. P. Martins & E. P. S. Shellard, *'Cosmic Wiggly Strings'*, submitted to Phys. Rev. **D**.

STRUCTURE FORMATION SEEDED BY COSMIC STRINGS

P. P. AVELINO
Centro de Astrofísica da Universidade do Porto
Rua do Campo Alegre 823, 4150 Porto, Portugal

E. P. S. SHELLARD AND J. H. P. WU
DAMTP, University of Cambridge,
Silver Street, Cambridge CB3 9EW, U.K.

AND

B. ALLEN
University of Wisconsin—Milwaukee, U.S.A.

We describe the results of high-resolution numerical simulations of structure formation seeded by a cosmic string network with a large dynamical range taking into account, for the first time, modifications due to the radiation-matter transition [1]. The resulting linear power spectrum of density perturbations is calculated with either cold or hot dark matter backgrounds and compared with the linear power spectrum inferred from various galaxy surveys [2]. Finally, we investigate the performance of cosmic string models in open universes and those with a non-zero cosmological constant. This direct numerical approach marks a considerable quantitative advance by incorporating important aspects of the relevant physics not included in previous treatments.

For flat models with $\Omega_\Lambda = 0$, strings induce an excess of small-scale power and a shortage of large-scale power, that is, this model requires a strongly scale-dependent bias (or even an antibias on small scales). This is not necessarily a fatal flaw on small scales because such excess power can be readily eliminated in a mixed dark matter model. On large scales the problem is less tractable and significant biasing is required. However, unlike inflation, defect models have never been wedded to an $\Omega = 1$ cosmology. The generalization to open or Λ-models tends to remove the excess small-scale power found in cosmic string models with $\Omega = 1$ and $\Omega_\Lambda = 0$, while also compensating for the shortage of large-scale power [3]. For open or Λ-cosmologies with $\Omega \approx 0.2$–0.3, the string + CDM power spectrum has a bias on large scales which is always close to unity and, overall, it is much

Astrophysics and Space Science is the original source of publication of his article. It is recommended that this article is cited as: *Astrophysics and Space Science* **261**: 315–316, 1999.

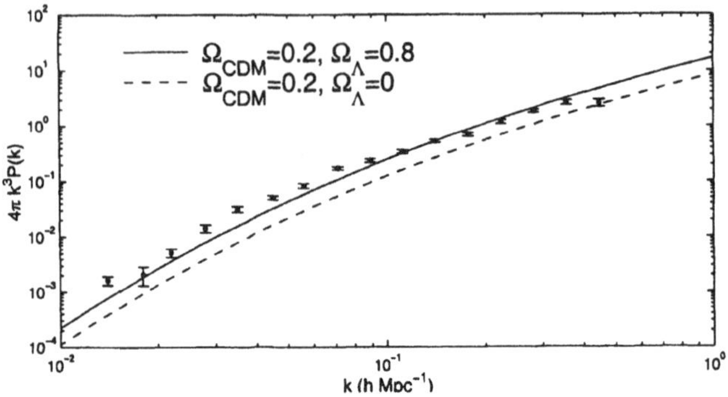

Figure 1. The linear power spectra induced by cosmic strings in CDM for different background cosmologies. Here, we use $h = 0.7$ and $G\mu_6 = 1.7$ [4]. The data points with error bars are the reconstructed linear spectrum by Peacock and Dodds [2].

less scale-dependent. We find that for $\Gamma = \Omega h = 0.1$–0.2, both $\sigma_{8(\mathrm{sim})}$ and the shape of the power spectrum induced by cosmic strings matches observations very well. The HDM power spectrum seems to require a strongly scale-dependent bias either on small or large scales, but we note that a high baryon fraction may help to increase small-scale power. Further investigation using a hydrodynamical code will be required to determine whether galaxies can form early enough. A key feature of all these string-induced power spectra is the influence of the slow relaxation to the matter era string density from the much higher radiation string density. Even by recombination in an $\Omega = 1$ ($h = 0.7$) cosmology, the string density is more than twice its asymptotic matter era value to which we normalize on COBE scales. This implies that the string model provides higher than expected large-scale power around $100h^{-1}$Mpc and below. Interestingly, this can also be expected to produce a significant Doppler-like peak on small angle CMB scales, an effect noted in ref. [4] and confirmed in ref. [5] using a simplified phenomenological model for cosmic strings. These results for the cosmic string scenario are encouraging.

We acknowledge funding from JNICT (PRAXIS XXI/BPD/9901/96), PPARC, NRS, ORS and Cambridge Overseas Trust. Simulations were performed on COSMOS, an SGI/CrayOrigin2000.

References

1. Avelino P. P., Shellard E. P. S., Wu J. H. P., Allen B., 1997, astro-ph/9712008.
2. Peacock J. A., Dodds S. J., 1994, MNRAS, **267**, 1020.
3. Avelino P. P., Caldwell R. R., Martins C. J. A. P., 1997, Phys. Rev. D **56**, 4568.
4. Allen B., Caldwell R. R., Dodelson S., Knox L., Shellard E. P. S., Stebbins A., Veeraraghavan S., 1997, Phys. Rev. Lett. **79**, 2624.
5. Albrecht A., Battye R. A., and Robinson J., 1997, astro-ph/9711121.

SUPERCONDUCTING STRINGS AND MAGNETIC FIELDS

C.J.A.P. MARTINS AND E.P.S. SHELLARD
DAMTP, University of Cambridge
Silver St., Cambridge CB3 9EW, U.K.

Observations show that our galaxy, and a good number of other spirals, possess magnetic fields with magnitude $B \sim 10^{-6}\,G$, on scales of several kiloparsecs (see [1] and references therein). No magnetic fields have been observed on larger scales, current observational bounds being about $B < 10^{-9}\,G$. The large coherence scales of these fields means that it is difficult to find mechanisms capable of creating them. Here we discuss the possibility of the galactic magnetic fields being generated by superconducting [2] cosmic strings, based on the quantitative evolution model of Martins and Shellard [3, 4], and a simple 'toy model' for the evolution of the superconducting currents [2].

It has been shown [1] that in order to explain the observed galactic magnetic fields one needs a seed field $B_s \geq 10^{-19}\,G$ on the comoving scale of a protogalaxy—such a field can then be amplified, by a dynamo mechanism, to the required value. Since protogalactic gravitational collapse enhances any frozen-in magnetic field, this seed field corresponds to an rms field $B_g \geq 10^{-22}\,G$ at the epoch t_g when galactic scales $d_g \sim 1\,Mpc$ fall inside the horizon.

Due to the statistical nature of a string network, analytic evolution methods must be 'thermodynamic': one must describe the network by a small number of macroscopic quantities whose evolution equations are derived from the microscopic string equations of motion. The first such model providing a quantitative picture of the complete evolution of a string network was developed by the present authors [3, 4]. It has two key quantities, the long-string correlation length L and the string RMS velocity, v^2, and also includes two 'phenomenological' parameters, a 'loop chopping efficiency' $\tilde{c}$ and a 'small-scale structure parameter' k. The loop size at formation, α, is a further parameter, but it turns out to be unimportant.

Recently, this has been extended with a 'toy model' for the evolution of the superconducting currents [2]. Assuming that there is a 'superconducting correlation length', ξ, which measures the scale over which one has coherent

Astrophysics and Space Science is the original source of publication of his article. It is recommended that this article is cited as: *Astrophysics and Space Science* **261**: 317–318, 1999.

current and charge densities on the strings, we can define N to be the number of uncorrelated current regions (in the long-string network) in a co-moving volume V. It is then straightforward to see how the dynamics of the string network affects N and obtain an evolution equation for it [2]. The only non-trivial issue is that of the dynamics of the currents themselves, and an 'equilibration parameter' f models that process.

We therefore obtain the following evolution equation for N

$$\frac{dN}{dt} = G\left(\frac{\ell}{\xi}\right)\frac{v_\infty}{\alpha}\frac{V}{L^4} - fHN\,, \tag{1}$$

where the 'correction factor' G is of order unity [2]; loops are assumed to form with a size $\ell(t) = \alpha(t)L(t)$ (see [4]). An equivalent description can be obtained in terms of N_L, the number of uncorrelated current regions per long-string correlation length.

Cosmic string-seeded magnetic fields should be coherent on the scale at which loops are being formed. Hence setting $L(t) = \gamma(t)t$ we find

$$B_g = \frac{2\pi e}{c^2 t_g^2 \alpha^{3/2}} \frac{N_L^{1/2}}{\gamma^2}\,, \tag{2}$$

and all we have to do is evaluate N_L and γ using our analytic model, while checking that at the nucleosynthesis epoch the corresponding magnetic. We find that that if equilibration is ineffective electroweak-scale strings just fall short of producing the required seed fields, $B_{seed} \sim 10^{-22}\,Gauss$, but GUT-scale strings can in the same circumstances produce such fields—all we require is an ineffective equilibration mechanism, $f \leq 0.5$. Also, there is almost no dependence on initial conditions in the GUT case, but such dependence persists for the electroweak-scale string network if equilibration is weak. this is because the GUT-scale string network is in the 'free' linear scaling regime at t_g, while the electroweak-scale network is in the friction-dominated Kibble regime (see [4] for a detailed description of these regimes).

Clearly, the outstanding issue, in this and other cosmological scenarios involving superconducting cosmic strings, is that of the importance of charge and current equilibration mechanisms on the strings, and a more detailed study will be required to resolve it.

References

1. C. J. A. P. Martins & E. P. S. Shellard, 'Galactic Magnetic Fields from Superconducting Strings', submitted to Phys. Rev. **D**.
2. C. J. A. P. Martins & E. P. S. Shellard, 'Evolution of Superconducting String Currents', Phys. Lett. **B**, in press (1997).
3. C. J. A. P. Martins & E. P. S. Shellard, Phys. Rev. **D53**, 575 (1996).
4. C. J. A. P. Martins & E. P. S. Shellard, Phys. Rev. **D54**, 2535 (1996).

THE PROMISE OF STRUCTURE FORMATION WITH COSMIC STRINGS IN AN OPEN UNIVERSE

P. P. AVELINO
Centro de Astrofísica da Universidade do Porto
Rua do Campo Alegre 823, 4150 Porto, Portugal

R. R. CALDWELL
Dep. of Physics and Astronomy, University of Pennsylvania
Philadelphia, PA 19104

AND

C. J. A. P. MARTINS
DAMTP, University of Cambridge
Silver Street, Cambridge CB3 9EW, U.K.

1. Introduction

The cosmic string scenario in an open universe has been developed – including the equations of motion, a model of network evolution, the large angular scale cosmic microwave background (CMB) anisotropy, and the power spectrum of density fluctuations produced by cosmic strings with dark matter [1]. We constructed a quantitative model of the evolution of the gross features of a cosmic string network in a dust-dominated, $\Omega < 1$ FRW space-time. We apply this model of network evolution to the results of a numerical simulation of CMB anisotropies induced by cosmic strings in a dust-dominated, $\Omega = 1$ FRW space-time [2], in order to estimate the cosmic string mass per unit length μ as a function of Ω. Finally, we consider the effects of the network evolution and normalization in an open universe on the large scale structure formation scenarios with either cold or hot dark matter (CDM, HDM).

2. Results and Conclusions

We extended the Martins and Shellard [3] model of quantitative string evolution to the case of an open, $\Omega < 1$ universe. We have found that with the onset of curvature dominated expansion, the long string energy density

Astrophysics and Space Science is the original source of publication of his article. It is recommended that this article is cited as: *Astrophysics and Space Science* **261**: 319–320, 1999.

and mean velocity decay rapidly. The resulting effect on the large angle CMB temperature fluctuations induced by cosmic strings is a lower level of anisotropy than in a critical, $\Omega = 1$ universe leading to a higher normalization of the cosmic string mass per unit length ($G\mu = 1.7\,\Omega^{0.3} \times 10^{-6}$ in an open universe with $\Lambda = 0$; $G\mu = 1.7\,\Omega^{0.05} \times 10^{-6}$ in a flat universe with a cosmological constant) consistent with all other observational constraints on cosmic strings, including the bound on a stochastic gravitational wave background arising from pulsar timing [4]. Finally, we computed the effect of an open, $\Omega < 1$ universe on the power spectrum of density fluctuations produced by cosmic strings with HDM and CDM. The string+HDM scenario for $\Omega < 1$ appears to produce too little power on scales $k \gtrsim 1\,\Omega h^2/\mathrm{Mpc}$. In a low density universe the string+CDM scenario is a better model for structure formation. We find that for cosmological parameters $\Gamma = \Omega h \sim 0.1 - 0.2$ in an open universe the string+CDM power spectrum fits the shape of the linear power spectrum inferred from various galaxy surveys. For $\Omega \sim 0.2 - 0.4$, the model requires a bias $b \gtrsim 1.5$ in the variance of the mass fluctuation on scales $8\,h^{-1}\mathrm{Mpc}$. In the presence of a cosmological constant, the spatially-flat string+CDM power spectrum requires a slightly lower bias than for an open universe of the same matter density. These results are consistent with those of ref. [5], in which a cosmic string network was used to compute the power spectrum of string induced perturbations. Similar results have also been recently reported in ref [6] for the case of cosmic strings in a ΛCDM cosmology. Finally, we point out that small-scale fluctuations generated by cosmic strings are non-gaussian, so that knowledge of $P(k)$ alone is insufficient to specify all the properties of the density field. We expect these non-gaussian effects to be even more pronounced in open models or flat models with $\Lambda \neq 0$ than in models with $\Omega = 1$ and $\Lambda = 0$.

Acknowledgments: P.P.A. is funded by JNICT (Portugal) under the 'Program PRAXIS XXI' (grant no. PRAXIS XXI/BPD/9901/96). The work of R.R.C. is supported by the DOE at Penn (DOE-EY-76-C-02-3071). C. J. A. P. M. is funded by JNICT (Portugal) under 'Programa PRAXIS XXI' (grant no. PRAXIS XXI/BD/3321/94).

References

1. Avelino P. P., Caldwell R. R., Martins C. J. A. P., 1997, Phys. Rev. D **56**, 4568
2. Allen B., Caldwell R. R., Dodelson S., Knox L., Shellard E. P. S., Stebbins A., Veeraraghavan S., 1997, Phys. Rev. Lett. **79**, 2624
3. Martins C. J. A. P., Shellard E. P. S., Phys. Rev. D **54**, 2535 (1996).
4. Caldwell R. R., Battye R. A., Shellard E. P. S., Phys. Rev. D **54**, 7146 (1996).
5. Avelino P. P., Shellard E. P. S., Wu J. H. P., Allen B. 1997, astro-ph/9712008
6. Battye R. A., Robinson J., Albrecht A. 1997, astro-ph/9711336

EVOLUTION OF SUPERCONDUCTING STRINGS

C.J.A.P. MARTINS AND E.P.S. SHELLARD
DAMTP, University of Cambridge
Silver St., Cambridge CB3 9EW, U.K.

Cosmic strings can in some circumstances behave as 'superconducting wires' carrying large currents and charges [1, 2]. Here we describe a generalization of the quantitative string evolution model of the present authors [3, 4] to the evolution of currents on long strings. We will use fundamental units in which $\hbar = c = k_B = 1$, and consequently $G = m_{Pl}^{-2}$.

A quantitative picture of the complete evolution of a string network (and the corresponding loop population) has been recently developed by the present authors [3]. This has two averaged quantities, the long-string correlation length L and the string RMS velocity, v^2, and includes two 'phenomenological' terms, a 'loop chopping efficiency' parameter $\tilde{c}$ and a 'small-scale structure parameter' k. Their evolution equations are

$$2\frac{dL}{dt} = 2HL(1+v_{\infty}^2) + v_{\infty}^2\frac{L}{\ell_f} + \tilde{c}v_{\infty}\,, \tag{1}$$

$$\frac{dv}{dt} = \left(1-v^2\right)\left[\frac{k}{L} - v\left(2H + \frac{1}{\ell_f}\right)\right]\,; \tag{2}$$

the explicit form of the 'friction lengthscale' ℓ_{f} will depend on the type of current involved.

We can extend this model by setting up a 'toy model' for the evolution of the currents, which for the time being we assume not to affect the long-string evolution [1]. One assumes that there is a 'superconducting correlation length' ξ, which measures the scale over which one has coherent current and charge densities on the strings. Associated with this we can define N to be the number of uncorrelated current regions (in the long-string network) in a co-moving volume V, and it is then fairly straightforward to see how the dynamics of the string network affects N [1]. The only non-trivial issue is that of the dynamics of the currents themselves. When regions with opposite currents inter-commute, new charged regions are created, setting up alternating currents. One expects electromagnetic processes

Astrophysics and Space Science is the original source of publication of his article. It is recommended that this article is cited as: *Astrophysics and Space Science* **261**: 321–322, 1999.

to make these currents die down, so that the charged region will eventually equilibrate with its neighbours. While it is beyond the present discussion to consider the complex backreaction processes causing this 'equilibration term', we will, as a first approximation, introduce it phenomenologically by assuming that after each Hubble time, a fraction f of the N regions existing at its start will have equilibrated with one of its neighbours, that is $(dN/dt)_{eq} = -fHN$. Presumably the only way to find out what f is will be by means of numerical simulations (in particular, we would expect it to be a model-dependent quantity), although some physical arguments can be used to constrain it.

However, setting this issue aside for the time being, we obtain the following evolution equation for N

$$\frac{dN}{dt} = G\left(\frac{\ell}{\xi}\right)\frac{v_\infty}{\alpha}\frac{V}{L^4} - fHN\,, \tag{3}$$

where G is a 'correction factor' G has the formof order unity; loops are assumed to form with a size $\ell(t) = \alpha(t)L(t)$ [3].

For what follows it is more convenient to introduce N_L, defined to be the number of uncorrelated current regions per long-string correlation length, $N_L \equiv L/\xi$ For neutral currents, analysis of the evolution equations reveals two types of behaviour. Firstly, if f is small (that is, equilibration is ineffective) then N_L grows without limit. In the particular case $f = 0$, we have $N_L \propto t$ once the long-string network has reached the linear scaling regime, meaning that $\xi \propto const.$ On the other hand, for $f > 1.88$ the late time behaviour is $N_L \propto const.$, $\xi \propto t$, that is the superconducting correlation length is scaling linearly just like the long-string correlation length; the scaling value of N_L has a fairly weak dependence on f, with smaller N_L's corresponding to larger f's as expected. Note that causality means that there is a maximum value allowed for f. We point out that if the cosmic strings interact with a plasma there will be a further contributing damping term, but one can show [1] that plasma effects are in general subdominant.

Our extended quantitative model allows reliable estimates of the currents on strings to be made at all times. Thus a more detailed analysis of some of the cosmological scenarios involving superconducting cosmic strings becomes possible. An outstanding example is the question of the abundance of vorton remnants.

References

1. C. J. A. P. Martins & E. P. S. Shellard, Phys. Lett. **B**, in press (1997).
2. E. Witten, Nuc. Phys. **B249**, 557 (1985).
3. C. J. A. P. Martins & E. P. S. Shellard, Phys. Rev. **D54**, 2535 (1996).
4. P. P. Avelino, R. R. Caldwell & C. J. A. P. Martins, Phys. Rev. **D56**, 4568 (1997).

COSMIC STRINGS WITH SELF-INTERACTING HDM

J. P. M. DE CARVALHO AND P. P. AVELINO
Centro de Astrofísica da Universidade do Porto
Rua do Campo Alegre 823, 4150 Porto, Portugal

1. Introduction

In recent work Atrio-Barandela and Davidson (1997), discussed the possibility that the dark matter could be constituted by light ($\sim$ 30 eV) self-interacting neutrino-like particles. They determined the linear power spectrum of density fluctuations generated by the present time in the context of primordial gaussian fluctuations and concluded that galaxy sized density perturbations could survive. Here we extend the work of Barandela and Davidson (1997) to the case of non-gaussian density perturbations induced by cosmic strings.

The evolution of radiation and non-relativistic interacting hot dark matter (IHDM) density fluctuations in the synchronous gauge is given by:

$$\begin{cases} \ddot{\delta}_h + \frac{\dot{a}}{a}\dot{\delta}_h - \frac{3}{2}\left(\frac{\dot{a}}{a}\right)^2(\Omega_h\delta_h + 2\Omega_r\delta_r) - c_s^2\nabla^2\delta_h = 4\pi(\Theta_{00} + \Theta_{ii}) \\ \ddot{\delta}_r - \frac{1}{3}\nabla^2\delta_r - \frac{4}{3}\ddot{\delta}_h = 0 \end{cases} \tag{1}$$

where $\Theta_{\alpha\beta}$ is the energy-momentum tensor of the strings, c_s is the adiabatic sound speed of IHDM and Ω_h and Ω_r express the densities in radiation and IHDM as fractions of the critical density. We are implicitly assuming that the strength of the neutrino-neutrino coupling is large enough for it to be a good approximation to treat the neutrino component as a perfect fluid. We obtained numerically the adiabatic sound speed of the self-interacting neutrinos, and fit the result by $c_s(a) = \frac{\sqrt{3}}{3}\left[1 + \left(\frac{a}{\alpha a_{eq}}\right)^2\right]^{-1/2}$, where $\alpha = 0.19$ is the best fit parameter.

Astrophysics and Space Science is the original source of publication of his article. It is recommended that this article is cited as: *Astrophysics and Space Science* **261**: 323–324, 1999.

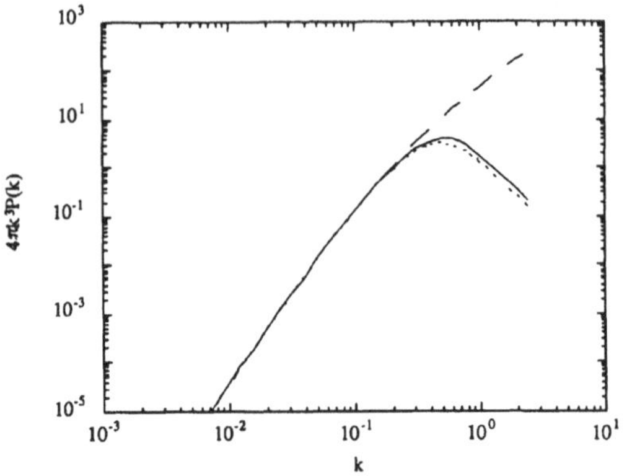

Figure 1. Comparison between the CMB-normalized linear power spectrum of cosmic string fluactuations with IHDM, CDM (dashed line) and HDM (doted line).

2. Results and Conclusions

In figure 1 we compare the power spectrum of the perturbations generated by cosmic strings for CDM, HDM and IHDM, after normalizing to the COBE-DMR observations (Allen el al. 1997), for $h = 1$ and $\Omega_0 = 1$.

We conclude that because gravitational instabilities can only be established on scales larger than the Jeans scale, small scale power is removed relative to the CDM case. In opposition to the case studied by Barandela and Davidson (1997), we note the absence of oscillations, which is due to the incoherent nature of the cosmic strings source (this was also found by Ferreira (1995) in the context of a purely baryonic universe). The gravitational force which the strings, long-lived seed perturbations, exert on IHDM drives sub-horizon pressure waves in a random way which prevents oscillations in the IHDM power spectrum from emerging. In fact, the results obtained for HDM and IHDM are very similar, the IHDM spectrum having a slightly larger amplitude on small scales.

Acknowledgments: P.P. Avelino is funded by JNICT (Portugal) under the Program PRAXIS XXI (grant nº PRAXIS XXI/BPD/9901/96). We thank CAUP for the facilities provided.

References

Allen B., Caldwell R. R., Dodelson S., Knox L., Shellard E. P. S., Stebbins A., Veeraraghavan S., 1997, Phys. Rev. Lett. **79**, 2624
Atrio-Barandela F., Davidson S., 1997, astro-ph/9702236
Avelino P. P., Caldwell R. R., Martins C. J. A. P., 1997, Phys. Rev. D **56**, 4568
Avelino P. P., Shellard E. P. S., Wu J. H. P., Allen B. 1997, astro-ph/9712008
Ferreira P. G., 1995, Phys. Rev. Lett. **74**, 3522

VORTONS: DARK MATTER FROM COSMIC STRINGS

C.J.A.P. MARTINS AND E.P.S. SHELLARD
DAMTP, University of Cambridge
Silver St., Cambridge CB3 9EW, U.K.

As first pointed out by Witten [1], cosmic strings can in some circumstances become superconducting. If a superconducting string loop has an angular momentum, it is semi-classically conserved, and it tries to resist the loop's tension. If the current is too large, charge carriers will leave the string in a burst of electromagnetic radiation, but otherwise it is possible that dynamically stable loops form.

These are called vortons (see [2] and references therein)—they are stationary rings that do not radiate classically, and at large distances they look like point particles with quantised charge and angular momentum. Their relevance comes from the fact that they provide very strong constraints on allowed particle physics models, since they behave like non-relativistic particles. Here we report on results of a recent study of the problem of vorton formation in the context of Witten's superconducting string model [2].

We have numerically studied the evolution of a number of loop solutions of the microscopic equations of motion for Witten's model, and from the results of this analysis parameters were introduced which characterise the loop's ability to evolve into a vorton state. Roughly speaking, loops which are always non-relativistic will become vortons, while those which are relativistic at some stage of their evolution will not. Note that the precise value of the 'cutoff' between what will be called the non-relativistic and relativistc regimes is model-dependent, and should be determined phenomenologically in each case.

Then, we have used the quantitative evolution model of the present authors [3] to determine the loop size at formation, and its generalisation for the evolution of the superconducting currents on the long cosmic string network [4] (based on the dynamics of a 'superconducting correlation length') to estimate the currents carried by string loops formed at all relevant times. When put together, these thus allow us to decide (according to the previously set criteria) whether or not each loop can become a vorton and calculate the corresponding density at any required epoch.

Astrophysics and Space Science is the original source of publication of his article. It is recommended that this article is cited as: *Astrophysics and Space Science* **261**: 325–326, 1999.

Based on our results, we don't expect any GUT-scale vortons to form at all. This is essentially because the friction-dominated epoch is very short for GUT-scale strings [3], so their currents and charges are never large enough to prevent them from becoming relativistic—and therefore liable to losses. Even if they did form, they wouldn't be in conflict with the standard cosmological scenario if they decayed soon after the end of the friction-domination epoch [3].

On the other hand, as we lower the string-forming energy scale we expect more and more efficient vorton production; in particular, at low enough energy scales, vortons will be a dark matter candidate. For example, for a string network formed around $T \sim 10^2\, GeV$ (typical of the electroweak phase transition) they can provide up to 6% of the critical density—this is in the best possible case in which all loops become vortons. Similarly, a string network formed at $T \sim 10^4\, GeV$ would provide a maximal vorton density equal to the critical density. This is therefore the strongest possible vorton constraint—it is based on the assumption that all loops form vortons. Naturally, realistic models are not expected to be fully efficient in producing vortons. One can therefore conjecture that the dark matter problem might be solved by a superconducting string network formed at an energy scale of $T \sim 10^5 - 10^6\, GeV$. We should point out that there are a number of super-symmetric models producing such network.

The analysis reported here can obviously be extended to other energy scales. One finds that in addition to the low-energy regime (which as we saw includes the electroweak scale) where vortons can be a source of dark matter and to an intermediate-energy range in which vortons would be too massive to be compatible with standard cosmology (thereby excluding these models), there is also a high-energy regime (of which the GUT scale is part) in which vortons don't form at all and therefore no cosmological constraints based on them can be set.

It is then curious (to say the least) that vorton constraints can be used to rule out cosmic string models in a wide range of energy scales, but not those formed around the GUT or the electroweak scales, where cosmic strings can be cosmologically useful.

References

1. E. Witten, Nucl. Phys. **B249**, 557 (1985).
2. C. J. A. P. Martins & E. P. S. Shellard, *'Conditions for Vorton Formation'*, submitted to Phys. Rev. **D**.
3. C. J. A. P. Martins & E. P. S. Shellard, Phys. Rev. **D53**, 575 (1996).
   C. J. A. P. Martins & E. P. S. Shellard, Phys. Rev. **D54**, 2535 (1996).
4. C. J. A. P. Martins & E. P. S. Shellard, *'Evolution of superconducting string currents'*, Phys. Lett. **B** in press (1997).

GENERAL RELATIVITY AS AN ATTRACTOR TO SCALAR-TENSOR GRAVITY THEORIES

JOSÉ P. MIMOSO AND ANA NUNES
Physics Dep. of Faculdade de Ciências de Lisboa,
Av. Prof. Gama Pinto, 2
1699 Lisboa CODEX, Portugal

Abstract. We investigate the conditions under which general scalar-tensor gravity theories relax towards General Relativity. We extend the work of Damour and Nordtvedt [2] by studying the effects of the inclusion of a cosmological potential term. When the universe is either radiation dominated or vacuum, we find that Einstein's gravity is indeed a cosmological attractor and, also, that the universe exhibits inflationary expansion. This latter feature provides another striking argument in favour of the inflationary paradigm, which in the present setting arises without the intervention of the usual inflaton field.

Scalar-tensor gravity theories are based on the following lagrangian [1]

$$L_\phi = \phi R - \frac{\omega(\phi)}{\phi}\, g^{ab}\phi_{,a}\phi^{,b} + 2U(\phi) + 16\pi L_m \quad , \tag{1}$$

where R is the usual Ricci curvature scalar of a spacetime endowed with the metric g_{ab}, ϕ is a scalar field, $\omega(\phi)$ is a dimensionless coupling function, $U(\phi)$ is a function of ϕ, and $\mathcal{L}_m$ is the Lagrangian for the matter fields. They provide the most natural generalization of general relativity (GR), and can be seen as a common denominator of most attempts at unifying gravitation with the other interactions. This happens, remarkably, in the low-energy limit of strings theories, in which the dilaton field exhibits the same role as that of the scalar field in Eq. (1). Thus the investigation of scalar-tensor gravity theories enables a generic, model-independent approach to the main features and cosmological implications of the unification schemes.

Recently Damour and Nordtvedt [2] have shown that general relativity is a generic attractor for massless scalar-tensor gravity theories, i.e. when

Astrophysics and Space Science is the original source of publication of his article. It is recommended that this article is cited as: *Astrophysics and Space Science* **261**: 327–330, 1999.

$U(\phi) = 0$. The importance of this result is best appreciated when one realizes that the evaluation of post-Newtonian effects of scalar-tensor theories on the solar system puts rather tight constraints on any departure from General Relativity, and that primordial nucleosynthesis also places similar bounds at a much earlier epoch [3]. Here we investigate whether any similar attractor mechanism exists for the case where $U(\phi) \neq 0$.

Consider the homogeneous and isotropic Friedmann-Robertson-Walker (FRW) universes characterized by the metric

$$ds^2 = -dt^2 + a^2(t)\left[\frac{dr^2}{1-kr^2} + r^2(d\theta^2 + \sin\theta^2 d\phi^2)\right] \quad , \qquad (2)$$

and assume the matter content of the universe to be a radiation fluid. Introducing conformal time η given by the differential relation $dt = a\, d\eta$, and the variables $X \equiv \phi\, a^2$, $Y \equiv \int \sqrt{\frac{2\omega+3}{3}}\, \frac{d\phi}{\phi}$, and $V(\phi(Y)) = U(\phi)/\phi^2$, the field equations become

$$(X')^2 + 4\,k\,X^2 - (Y'\,X)^2 = 4M\,X + \frac{4}{3}X^3 V(\phi) \qquad (3)$$

$$[Y'\,X]' = -\frac{2\,X^2\phi}{\sqrt{2\omega(\phi)+3}}\,\frac{dV(\phi)}{d\phi} \qquad (4)$$

where $Y' = \sqrt{(2\omega+3)/3}\,(\phi'/\phi)$ and M is a constant defined by $3M \equiv 8\pi G_0 \rho\, a^4$. By setting $M = 0$ we obtain the equations for vacuum.

From these equations (3,4), we immediately realize that $V(\phi) = \lambda_0$, where λ_0 is a constant, considerably simplifies the equations. This corresponds to $U(\phi) = \lambda_0\,\phi^2$, and we see that Eq. (4) reduces to that of the massless case ($U(\phi) = 0$) [4]. There is a first-integral $Y'\,X = f =$, where f is an integration constant. We see that $f = 0$ corresponds to general relativity since ϕ =constant is a solution.

We now study the likeliness of GR solutions $\dot{\phi} = 0$ or GR asymptotic solutions, that is $\dot{\phi} \to 0$, as η tends to some limit η_* in the particular case when $V(\phi) = \lambda_0 =$ constant, the equations decouple and become

$$(X')^2 + \mathcal{V}(X) \;=\; f^2 \qquad (5)$$

$$\sqrt{\frac{2\omega+3}{3}}\,\frac{\phi'}{\phi}X \;=\; f\,, \qquad (6)$$

where

$$\mathcal{V}(X) = -\frac{4}{3}\,\lambda_0\,X^3 + 4\,k\,X^2 - 4M\,X\ , \qquad (7)$$

that is, the first integral f behaves as an energy parameter for the conservative system with potential $\mathcal{V}(X)$ associated with the X variable.

The qualitative behaviour of the solutions $X(t)$ of equation (5) is easily determined as a function of the two independent parameters in the potential $\mathcal{V}(X)$. There are essentially two qualitatively different types of solutions: the one that we call $\mathcal{O}$, oscillatory, for which $X(t)$ is a periodic function, and the one that we call AGR, asymptotically GR, for which $X(t)$ is unbounded in both limits of its interval of definition. Clearly, the integral of equation (6) implies that, for most theories, these solutions will have $\phi' = 0$ asymptotically. We can obtain the asymptotic behaviour of $X(t)$ in the case of the AGR solutions. From (5) and (7) we have that, for $|X|$ large enough,

$$X \sim \frac{X(0)}{\left(1 \pm \sqrt{|X(0)|}\sqrt{\frac{|\lambda_0|}{3}}\,\eta\right)^2}, \tag{8}$$

that is $|X(\eta)|$ tends to infinity in a finite η time. It is then easy to see that this corresponds to the inflationary behaviour: $a \propto \exp\left(\sqrt{\phi_* \lambda_0/3}\; t\right)$, where ϕ_* is the asymptotic value of ϕ.

The following table characterizes the type of solutions of equation (5) in terms of the parameters and of the initial conditions $X(0), f$.

	$\lambda_0 < 0$	$0 < \lambda_0 < 3/(4M)$	$\lambda_0 > 3/(4M)$
$k = 0$	$\mathcal{O}$ if $f^2 < f_c^2$, $X(0) > X_c$ AGR otherwise	AGR	AGR
$k = 1$	$\mathcal{O}$ if $f^2 < f_c^2$, $X(0) > X_c$ AGR otherwise	$\mathcal{O}$ if $f^2 < f_c^2$, $X(0) < X_c$ AGR otherwise	AGR
$k = -1$	$\mathcal{O}$ if $f^2 < f_c^2$, $X(0) > X_c$ AGR otherwise	$\mathcal{O}$ if $f^2 < f_c^2$, $X(0) < X_c$ AGR otherwise	AGR

Here $f_c^2 = (4/3)X_c(kX_c - 2M)$, $X_c = (k + \sqrt{k^2 - M\lambda_0})/\lambda_0$.

Let us get back now to the complete system (3,4) with an arbitrary potential $V(\phi)$. Note that whenever ϕ_0 is a critical point of $V(\phi)$, $(X_{\phi_0}(\eta), \phi = \phi_0, \phi' = 0)$ is a solution of (3,4) with $X_{\phi_0}(\eta)$ a solution of the decoupled system (5,6) with $f = 0$ and $\lambda = V(\phi_0)$. The likeliness of GR solutions in this general setting depends on the stability of these $\phi' = 0$ solutions. As we shall see, this stability is determined by the behaviour of the function $X_{\phi_0}(\eta)$.

Let then $\phi(\eta) = \phi_0 + \varepsilon\,\varphi(\eta)$, where ε is a small parameter. Substituting in (3,4), linearizing about $\phi = \phi_0$ and keeping the first order terms, yields

$$\varphi'' + \frac{X'_{\phi_0}}{X_{\phi_0}}\varphi' + bX_{\phi_0}\varphi = 0\,, \tag{9}$$

where $b = 2\sqrt{3}\phi_0^2 V''(\phi_0)/\sqrt{2\omega(\phi_0)+3}$. The first necessary stability condition is that $X_{\phi_0}(\eta)\,V''(\phi_0) > 0$. On the other hand, we know by Liouville's formula that the volume $\Omega(\eta)$ in the (φ,φ') phase-space changes with time in accordance with $\Omega'(\eta) = \Omega(\eta)\,\alpha(\eta)$, where $\alpha(\eta) = -X'_{\phi_0}/X_{\phi_0}$. Thus we shall have exponential contraction, and hence stability, provided that X'_{ϕ_0}/X_{ϕ_0} is positive and bounded away from zero. Going back to the original coordinates ϕ and a, $X_{\phi_0} = \phi_0 a^2(\eta)$, we have

$$\frac{\Omega'}{\Omega} = -2\frac{a'}{a} = -2\dot{a}\,, \tag{10}$$

that is, the expansion of the scale factor a is an exact measure of the stability of the $\phi = \phi_0$ solutions.

So, summing up, a GR $\phi = \phi_0$ solution of the complete system (3,4) corresponding to a critical point of $V(\phi)$ is stable and will attract all nearby initial conditions if $X(0)\,V''(\phi_0) > 0$, and the corresponding solution $X_{\phi_0}(\eta)$ of the decoupled system is of type AGR and is an expanding branch. In other words, we find that an attractor mechanism indeed exists, when the universe is vacuum or radiation dominated. Futhermore, it is accompanied by a phase of inflationary expansion, thus providing another striking argument in favour of inflation [5].

Acknowledgements

The authors acknowledge the finantial support of the Fundação de Ciência e Tecnologia (grant PBIC/C/FIS/2215/95).

References

1. P G Bergmann (1968) *Int. J. Theor. Phys.* **1**, 25; R V Wagoner (1970) *Phys. Rev.* **D1**, 3209; K Nordtvedt (1970) *Astrophys. J.* **161**, 1059.
2. T. Damour and K. Nordtvedt, Phys. Rev. Lett. **70**, 2217 (1993); Phys. Rev. D **48**, 3436 (1993); T. Damour and A. Vilenkin, Phys. Rev. D **53**, 2981 (1996); Phys. Rev. Lett. **78**, 2288 (1997).
3. C. M. Will, Theory and Experiment in Gravitation (Cambridge University Press, Cambridge, England, 1993)
4. J.D. Barrow, Phys. Rev. D **47**, 5329 (1993); J. P. Mimoso and D. Wands, Phys. Rev. D **51**, 477 (1995).
5. J. Garcia-Bellido and D. Wands, Phys. Rev. D **52**, 6739 (1995)

HIGH REDSHIFT X-RAY CLUSTERS AND Ω_0

R. SADAT
Affiliation
Observatoire astronomique de Strasbourg, 11 rue de l'université. 67000 Strasbourg, France.
and
I.A.P. 98bis Bd. Arago, 75014 Paris, France

1. Introduction

Clusters of galaxies are undoubtedly the largest virialized systems in the universe, they offer a powerful tool in constraining cosmological parameters. For example, their present-day abundance has been used to constrain parameters such as the *shape* n and the *amplitude* σ_8 of the mass fluctuations as well the cosmological density parameter Ω_0. However this constraint is degenerate in σ_8 and Ω_0. Recently, Oukbir & Blanchard (1997; hereafter OB97), have nicely shown that the *evolution* of X-ray temperature distribution function *does depend strongly* on Ω_0, and claimed that the combined knowledge of the abundance of X-ray clusters together with the evolution of the luminosity-temperature relation would allow us to strongly constrain the cosmological density parameter independently of the power spectrum. I will present the result of our first attempt to apply this test to current X-ray data (Sadat et al. 1997).

2. Basic Recipe

The Press-Schechter (1974) (PS) simple analytical formalism allows us to calculate the comoving number density of dark matter halos of a given mass M. This approximation of the *mass function* seems to fit remarkably the simulations and then can be applied to clusters with present–day X-ray temperatures of $\sim 1-10$ keV. For more details on the derivation of PS formula see Bartlett 1997. To relate the *mass* to observable quantities, we prefer to deal with X-ray observations than with optical data because of the projection and contamination effects which may affect optical cluster observations. In principle, L_X is the simplest quantity to measure, nevertheless one must avoid to relate the *mass* to L_X, because of the strong

Astrophysics and Space Science is the original source of publication of his article. It is recommended that this article is cited as: *Astrophysics and Space Science* **261**: 331–334, 1999.

dependence of the gas emissivity on the spatial distribution of the gas in the core of the cluster which physics is not well understood. In contrast, X-ray temperature–mass relation is better understood and follows simple physics ($T_X \propto M/R$). The X-ray gas temperature is fairly well predicted by hydrodynamical simulations (Evrard et al. 1996). These authors have shown the existence of a tight relation between T_X and the mass: $T_X = (6.8h^{2/3}keV)M^{2/3}(1.+z)$. Thus, it is possible to use this simple relation to derive the expected X-ray gas *temperature* distribution function $\phi(T)$ from the *mass* function and then compare it to observations. Unfortunately, this *temperature* function is only available at $z = 0$, we do not yet have any information on the evolution of $\phi(T)$ with z.

2.1. EVOLUTION OF X-RAY CLUSTERS: A TEST FOR Ω_0?

Now that we have built a self-consistent modelling of X-ray clusters, one may address the following question: what one would expect from high-z X-ray clusters? Under some reasonable simplifications, the *mass function* can be written as (Blanchard et al. 1992):

$$N(M,z) = N(M,z=0)\frac{F\left(\nu_0 A(z)\right)}{F(\nu_0)}A(z). \tag{1}$$

and $N(M,z=0) = \frac{\bar{\rho}}{M^2}\nu_0\frac{d\log\sigma_0}{d\log M}F(\nu_0)$. where ν_0 being $\delta_{c,0}(z=0)/\sigma_o(M)$. F is well fitted by an $\exp(-\nu^2/2)$ independently of the power spectrum $P(k)$ (from numerical simulations) and A(Ω_0,z) is the growing rate. Equation (2) demonstrates clearly that the rate of cluster *evolution* with redshift z is mainly driven by the density parameter Ω_0, independently of the power spectrum index and σ_8. In open universe, we should expect more clusters at high z than in a critical universe. Consequently, the study of the *evolution* of the cluster number density, once normalized to present–day abundance, would strongly constrain Ω_0. However, the observation of such information that is – the *evolution* of the X-ray temperature distribution function– is far from being reached. The EMSS cluster redshift survey is up to now the largest sample of well-controlled X-ray selected clusters. This has been studied in detail by OB97 who found that in order to achieve a self-consistent modeling of the X-ray data, one needs to introduce a negative evolution in the L_X–T_X relation: $L_X \propto T_X^3(1+z)^\beta$ in order to match the data in an open universe (with β=-2.3), otherwise, i.e in the case of *no evolution* low Ω_0 models are simply ruled out. A recent analysis based on published clusters abundance has led to the same conclusion (Blanchard & Bartlett 1998, BB98) and Fig. 1. Therefore, the evolution of the L_X–T_X relation would allow us to disentangle between open and critical universe. Moreover, by iterating this analysis for different Ω_0, Sadat et al. (1997) have shown that the best fitting β is tightly related to Ω_0 accordingly to $\beta = 4. \times \Omega_0$ - 3.

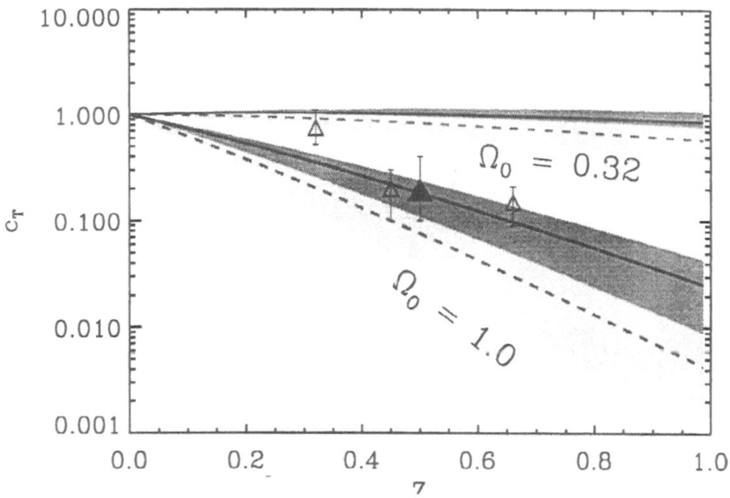

Figure 1. *Evolution with z of the relative number of clusters at a given T_X (from BB98). The solid and dashed lines correspond respectively to 4 keV and 6 keV clusters. Filled and (empty) triangles are 4 keV and (6 keV) clusters.*

3. Application of the test

Our sample results from a compilation of 57 high-z clusters among which $\sim$ 30 clusters at $z \geq 0.26$. Local observations have also been included. We added few clusters for which accurate mass estimate is available. We have derived the temperature, using simple scaling relations. In order to apply the test discussed above, that is to measurement of the evolution in L_X–T_X, we have fitted the following power law $C(z) = \alpha(1+z)^\beta$ to observational coefficient $C_i = L_i^{bol}/0.05T_i^3$. The result is presented in Fig. 2. While $\alpha = 1$ as expected, the best β value ranges in [0.–1.0] interval. This result shows no evidence of a significant evolution in the L_X–T_X relation in agreement with previous investigations.

3.1. WHAT DO THE DATA TELL US ON Ω_0 ?

As discussed in section 2.1, in low Ω_0 universe, a strong evolution is required in order to match the EMSS observations. Therefore, current data which show no significant evolution suggest that Ω_0>0.2. Now, we can go further and give an estimation of Ω_0 from the best fit value of β. Using our simple relation between β and Ω_0, one can directly derive the value of Ω_0 corresponding to the observed L_X–T_X evolution rate β: we find Ω_0=0.85 $\pm$0.2, although a full analysis of the errors still remains to be done.

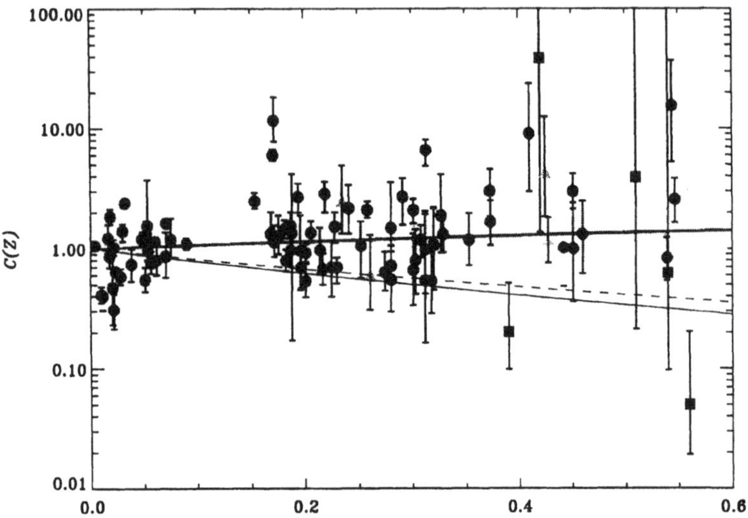

Figure 2. *Coefficient C_i versus z from Sadat et al. (1997). The error bars are 1σ. The thick line represents the best fitting power–law, the shaded area represents an estimate of the 90% confidence range. The prediction for an open universe is represented by the thin line.*

4. Conclusions

The evolution of the cluster temperature distribution function is crucial for cosmology, it represents a powerful test for Ω_0. The evolution of the L_X–T_X relation provides us with a useful variant of this test (OB97). We have reported the first attempt to apply this test to observations and found that –i) current X-ray data show no evidence of a strong evolution in the L_X–T_X relation, –ii) the lack of evolution is consistent with high Ω_0 universe, –iii) the evolution of the L_X–T_X relation has been quantified and suggests that $0.75 < \Omega_0 < 1$.

References

Bahcall, N.A., Fan, X., Cen, R., (1997), *ApJ*, **485**, L53
Bartlett, J.G., (1997), *From Quantum Fluctuations to Cosmological Structures*, D., Valls-Gabaud, M.A.Hendry, P.Molaro, K.Chamcham, *A.S.P. Conf. Ser.*, **vol. 126**, p 365.
Blanchard, A., Bartlett, J.G., *astro-ph/9712078*
Carlberg, R.G., Morris, S.L., Yee, H.K.C., Ellingson, E., (1997),*ApJ*, **479**, L19
Evrard, A.E., Metzler, C., Navarro, J.F., (1996), *ApJ*, **469**, 494
Oukbir, J., Blanchard, A., (1997), *A&A*, **317**, 10
Press, W.H., Schechter, P., (1974), *ApJ*, 187, 425
Sadat, R., Blanchard, A., Oukbir, J., (1997), *A&A* **329**, 21

NON-LINEAR GRAVITATIONAL CLUSTERING USING THE ZEL'DOVICH APPROXIMATION

MICHAEL JOSEPH CARR
Dunsink Observatory, Dublin 15, Ireland

AND

PEDRO PINA AVELINO AND PAULO MAURICIO DE CARVALHO
CAUP, Rua do Campo Alegre, 823, Porto 4150, Portugal.

Abstract. As an alternative to computationally expensive N-body simulations for gravitional clustering, the Zel'dovich approximation (ZA)was studied in $3D, 2D$ and $1D$. Plots of the density contrast were compared against linear theory and the exact solution. The ZA was found to perform very well in the linear regime, better than linear theory, and to give a good approximation well into the non-linear regime.

1. Introduction

We compare three different methods for evolving density fluctuations, for both positive and negative initial perturbations, namely the ZA, linear theory and the exact solution until shell crossing.

2. The Zel'dovich approximation and linear theory

In the absence of seeds and assuming the pertubations are irrotational to begin with and that the source term is irrotational then the ZA (Zel'dovich, 1970) can be written as $\left(\frac{\partial^2}{\partial t^2} + 2\frac{\dot{R}}{R}\frac{\partial}{\partial t} + 4\pi G\rho_b\Omega_c\right)\vec{\psi}_{\parallel} = 0$ where R is the scale factor, $\vec{\psi}_{\parallel}$ is the parallel componet of the comoving displacement, ρ_b is the background density and $\Omega_c = \rho^c/\rho_{crit}$ where ρ_{crit} is the critical density. The solution to the Zel'dovich equation for models with pressure $p = 0$, cosmological constant $\Lambda = 0$, and $\Omega = \Omega_c = 1$ in the absence of seeds, and with

Astrophysics and Space Science is the original source of publication of his article. It is recommended that this article is cited as: *Astrophysics and Space Science* **261**: 335–336, 1999.

initial condition $\dot{\vec{\psi}}(t_i) = 0$ is given by $\vec{\psi}(\vec{q},t) = \frac{3}{5}\vec{\psi}(\vec{q},t_i)\left(\left(\frac{t}{t_i}\right)^{2/3} + \frac{2t_i}{3t}\right)$. We choose the initial conditions such that the density contrast $\delta^c(t_i) = A\exp(-q^2)$ with $A \ll 1$. When $|\vec{\psi}|$ is small $\delta^c = -\nabla_{\vec{q}}\cdot\vec{\psi}$. Thus for spherical symmetry $\psi(\vec{q},t_i) = -1/q^2 \int_0^q Au^2\exp(-u^2)du$ in the linear regime up until shell crossing and $\delta^c(q,t) = \lim_{x\to q} \frac{x^3-q^3}{(x+\psi(x,t))^3-(q+\psi(q,t))^3} - 1$. Now taking the divergence of the Zel'dovich equation in the limit of small comoving displacements we obtain a linear equation for the evolution of δ^c. Using the same conditions as before the solution is $\delta_c(q,t) = \frac{3}{5}\delta_c(t_i)\left(\left(\frac{t}{t_i}\right)^{2/3} + \frac{2t_i}{3t}\right)$ with similar results holding for cylindrical and planar collapse.

3. Results and conclusions

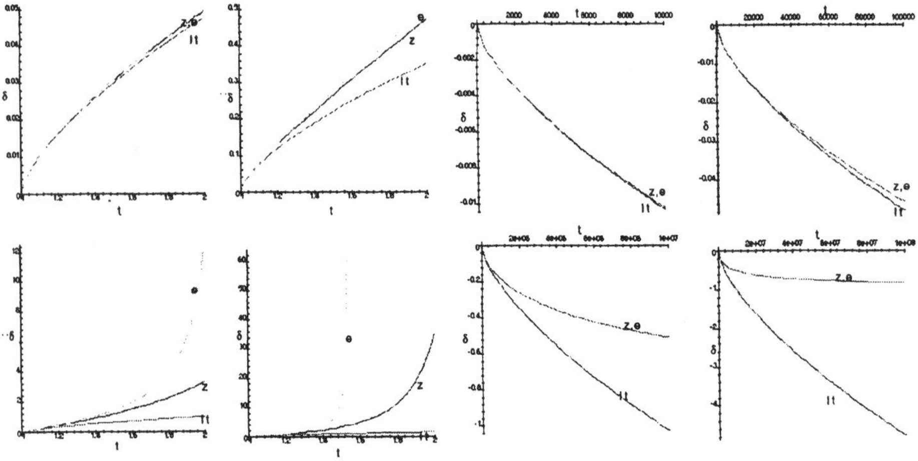

Figure 1. Plots of δ^c v t for linear theory, ZA and exact solution, for positive perturbations in $3D$ and negative perturbations in $1D$ respectively.

From our work we conclude the the ZA is more precise on going from spherical to planar collapse and from posive to negative perturbations. For positive perturbations the ZA, unlike linear theory gives $\delta(\vec{q},t) \to \infty$, as $t \to t_{zel} > t_{exact}$ but $t_{zel} \neq \infty$, where t_{zel} is the time required for the ZA to go to infinity, and for negative perturbations the ZA gives $\delta(\vec{q},t) \to -1$, as $t \to \infty$, unlike linear theory which gives $\delta(\vec{q},t) \to -\infty$, as $t \to \infty$.

References

Avelino, P.P. (1996) Phd thesis.
Carr, M.J. (1997) Msc. thesis.
Zel'dovich, Ya.B.,(1970) *A &A*, **5**, 84-89.

MEASURING TIME DELAYS FOR GRAVITATIONALLY LENSED VARIABLE QUASARS

F.P. PIJPERS
Theoretical Astrophysics Center
Aarhus University
Ny Munkegade
8000 Aarhus C
Denmark

1. Introduction

A common problem in astronomy is the determination of the time shift between two otherwise identical time series of measured flux from a variable source, in short the determination of a time delay. One example of where this problem occurs is in the determination of the Hubble constant from multiple images of gravitationally lensed variable quasars. This problem is very similar to the inverse problem problem of reverberation mapping of active galactic nuclei (AGN), and therefore the determination of time delays can also be seen as a restricted inverse problem.

2. Inverse problems

The problem of reverberation mapping comes from a view of the Broad Line Region (BLR) of AGN as gas clouds surrounding a variable continuum source. The gas clouds re-emit the radiation absorbed from the continuum source in spectral lines so that the time lag between the variation of the line emission and the continuum emission is a measure of the difference in path length to the observer and hence of the distance from the central source of the emitting gas clouds. Mathematically, the concept of reverberation mapping leads to the integral equation

$$L(t) = \int \mathrm{d}\tau \ \Psi(\tau) C(t-\tau). \tag{1}$$

Here L and C are the (velocity integrated) line flux of a broad line in the AGN spectrum, and the continuum flux respectively. The transfer function

Astrophysics and Space Science is the original source of publication of his article. It is recommended that this article is cited as: *Astrophysics and Space Science* **261**: 337–340, 1999.

is thus related to the distribution of clouds around the nucleus. Hence the problem is reduced to the inversion of the integral equation to obtain the transfer function Ψ as a function of time delay τ. If in equation (1) for $\Psi(\tau)$ a Dirac delta function is substituted, $\Psi(\tau) = \delta(\tau - t_0)$, equation (1) reduces to $L(t) = C(t - t_0)$. Conversely if two light curves are related by a time delay t_0 this is equivalent to equation (1) with a transfer function $\Psi(\tau) = \delta(\tau - t_0)$. In this paper the transfer function is therefore assumed to be essentially zero everywhere except at the unknown time-lag t_0 : $\Psi(\tau) = I\delta(\tau - t_0)$ where δ is the Dirac delta function.

Contrary to the problem of reverberation mapping where there is an assumed causal relationship between the variations in continuum and line flux, in the problem of time lag determinations the light curves are not distinguishable as driving or responding time series. The inversion method should reflect this lack of knowledge in a symmetric treatment of the two time series. The notation of equation (1) is slightly modified to express this :

$$F^{(b)}(t) = \int_{-\tau_{\max}}^{\tau_{\max}} \mathrm{d}\tau \; I\delta(\tau - t_0) F^{(a)}(t - \tau) \tag{2}$$

$$F^{(a)}(t) = \int_{-\tau_{\max}}^{\tau_{\max}} \mathrm{d}\tau \; \tfrac{1}{I}\delta(\tau + t_0) F^{(b)}(t - \tau) \tag{3}$$

The two equations both express that there is a time lag t_0 between the two time series $F^{(a)}$ and $F^{(b)}$ and therefore seem redundant. However both need to be used to ensure a symmetric treatment of the two time series in the algorithm. I is a constant to allow for a constant non-unity ratio between the two time series. Note that I can have any value larger than 0 and that the time-lag t_0 can be either positive or negative, since it is not known a-priori which light curve is leading and which is lagging. Just as in the application of the subtractive optimally localized averages method (SOLA) to reverberation mapping (cf. Pijpers & Wanders, 1994 : hereafter PW) the limits of the integration are set to a finite value. The reason for this is that for any measured time series its total extent is finite and therefore the range over which a time lag can be determined is limited to values within this range.

Equations (1) and (2,3) are idealized in the sense that finite sampling rates of the two light curves and finite measurement errors are not yet explicitly accounted for. Another problem can be that one or both of the light curves are contaminated by a foreground or background source.

3. A brief description of the algorithm

The strategy of the SOLA method in general is to find a set of linear coefficients which, when combined with the data, produce the value of the unknown convolved function under the integral sign for given value(s) of the time lag. In order to do this the time series under the integral sign is interpolated : to each measurement of the series outside of the integral sign is assigned a partial time series which is a section of the time series under the integral sign. As discussed by Pijpers and Wanders (1994) this means that to each measurement $L(t_i)$ the partial time series consisting of the ith measurement $C(t_i)$ and all n previous ones, $[C(t_{i-n}), C(t_i)]$, is assigned. These partial time series $[C(t_{i-n}), C(t_i)]$ for all i form the set of base functions for the SOLA method when it is applied to equation (1), and also (2,3). As Pijpers (1997) discusses, if the transfer function is narrow compared to the sampling rate or intrinsic time scale of variability of F one cannot hope to reconstruct its shape in an inversion. The width of the reconstructed transfer function will in this case be almost identical to the resolution given by the sampling of the time series (cf. PW and Pijpers 1994) rather than the real width. For simple time lag determinations this is not important, however, since only the single unknown time lag t_0 needs to be determined. To do this most conveniently one should combine the series of partial light curves of the time series under the integral sign into a function (the 'kernel') that is not the usual Gaussian (for estimates that are localized in time-delay around the peak of that Gaussian) but a linear function of τ. In this way the first moment of the transfer function is determined. The ratio of the first moment and the zero-th moment of the transfer function (which can be determined independently using the same technique) is the time-delay.

In order to ensure that the two time series are treated equally, the procedure is repeated after switching the role of the two series. As Pijpers (1997) demonstrates this can yield the time delay, the relative image magnification factor, and any contamination to the flux originating from for instance the foreground galaxy that is the lensing object. The former two are obtained from the mean of the determinations of the zero-th and first order moments of the transfer function. The latter (expressed as an offset and a drift) is obtained from the differences between the determinations when switching the role of the time series. In the absence of such contamination the determinations must be identical within the errors. Note that since the data is combined in a linear way, it is trivial to calculate the variance of the errors in the result when the variances of the errors in the observations are known.

4. QSO 0957+561

The largest homogeneous data set for the two optical images of the gravitationally lensed QSO 0957+561 is that reported by Schild & Thomson (1995). Since merging of datasets from different sources can cause spurious offsets and large error estimates it is best to use homogeneous datasets. Even so, it is immediately clear from the asymmetries obtained in both the values of I and of the time lags t_0 when reversing the role of the time series that some contamination of the time series is present. As reported by Pijpers (1997), after accounting for this contamination the magnification factor between the two images is $I = 1.35 \pm 0.03$ and the time delay $T_0 = 425 \pm 17$. From a careful analysis of the images of QSO 0957+561A, B combined with a tracing of the light deflection through a model potential to fit the positions and magnifications of the images it is possible to determine the gravitational potential of the lensing object. With this potential it is possible to determine the Hubble constant as long as the lensed source is variable. This is essentially a consequence of the light of the two images having traversed a different path through this gravitational potential. By measuring the time delay for a signal propagating at the speed of light an apparent separation of the images can be related to a physical distance which together with a measured redshift yields an estimate of the Hubble constant. Using the modelling of Grogin and Narayan (1996), a Hubble constant is deduced of $H_0 = 66 \pm 10$ km s^{-1} Mpc^{-1}.

5. Conclusions

A method developed to solve the inverse problem of determining time delays is described and it is applied to the time series measured for the double quasar QSO 0957+561. The resulting time delay is 425 ± 17 d. This leads to a best value for the Hubble constant of $H_0 = 66 \pm 10$ km s^{-1} Mpc^{-1}. An additional advantage of this method is that it determines the magnification ratio(s) of the images as well as the delay, and can detect contamination from foreground or background sources such as microlensing.

References

Grogin, N.A., Narayan, R., (1996), *ApJ*, **473**, 570

Pijpers F.P., 1994, in P.M. Gondhalekar, K. Horne, and B.M. Peterson, eds, Reverberation Mapping of the Broad-Line Region of Active Galactic Nuclei, ASP Conf. Ser. 69, San Francisco, 69

Pijpers, F.P. (1997), *MNRAS*, **289**, 933

Pijpers, F.P., Wanders, I. (1994), *MNRAS*, **271**, 183

Schild, R., Thomson, D.J. (1995), *AJ*, **109**, 1970

GRAVITATIONAL LENSES AND THE HUBBLE CONSTANT: PRESENT AND FUTURE

L.J. GOICOECHEA
Departamento de Fisica Moderna, Universidad de Cantabria
Avda. de los Castros, E-39005 Santander, Cantabria (Spain)
E-mail: goicol@besaya.unican.es

AND

E. MEDIAVILLA, A. OSCOZ, M. SERRA AND J. BUITRAGO
Instituto de Astrofisica de Canarias (IAC)
Via Lactea, E-38200 La Laguna, Tenerife (Spain)
E-mail: emg@iac.es,aoscoz@iac.es,mserra@iac.es,jgb@iac.es

1. Introduction

For a multiple QSO, the propagation time from the source to the observer varies from the image i to the image j, and this difference ($\Delta\tau_{ij}$) can be measured when the source is variable. In general, assuming a flat universe without cosmological constant, the parameter $\Delta\tau_{ij}$ x H_0 (H_0 is the Hubble constant) depends on the redshifts of the lens and the source, as well as the positions of the individual images and the source ($\vec{\theta}_i, \vec{\theta}_j, \vec{\beta}$), and the scaled surface potential Ψ_α at $\vec{\theta}_i$ and $\vec{\theta}_j$ (see, e.g., Blandford and Kundic 1996, Williams and Schechter 1997). The observations of multiple images of the same source are used to infer $\vec{\beta}$ and the adjustable parameters $\alpha \equiv (\alpha_1, ..., \alpha_p)$ that appear in the picture of the deflector, i.e., a lens model. From the lens model corresponding to the lens picture, $\Delta\tau_{ij}$, $\vec{\theta}_i$, $\vec{\theta}_j$ and the redshifts, one obtains H_0.

A golden system (which is suitable for determining H_0) must be a multiple QSO verifying some properties. The deflector cannot be dark or very faint (non-observable), and in order to model the gravitational potential, the lensing structure must be simple (e.g., an isolated galaxy). To infer α and $\vec{\beta}$, a relatively large number of constraints is also desirable. On the other hand, for obtaining $\Delta\tau_{ij}$, some typical scales of intrinsic variability of the source should be less than $\Delta\tau_{ij}$, and the absence of strong short

Astrophysics and Space Science is the original source of publication of his article. It is recommended that this article is cited as: *Astrophysics and Space Science* **261**: 341-344, 1999.

timescale microlensing is need. The microlensing events mask the intrinsic variability.

The best studied gravitational lens is the *double* QSO Q0957+561A,B. As a result of recent advances in observations and modelling, a ten percent measurement of H_0 is attainable. We discuss in detail this system (Sect. 2) and review the perspectives in a near future (Sect. 3).

2. Twin QSO (0957+561A, B)

For Q0957+561A, B, at optical wavelengths, there are epochs of a rapid intrinsic variability (a basic condition for measuring the time delay accurately) and epochs of calmness. They also appears evidences in favour of an important short timescale microlensing (Schild 1996), which masks the intrinsic variability. So, due to the double behaviour activity/calmness, the microlensing events, the gaps in the monitoring (three months per year), etc., it is difficult an accurate time delay determination based on a large dataset containing several years of observations.

The rough estimation $\Delta\tau_{BA} \approx T = 420$ days (Pelt *et al.* 1996; see also the paper by Pijpers in this volume) can be used for making nice datasets. A nice dataset contains an *active* light curve A during a period $[t_i, t_f]$ and the light curve B in the interval $[t_i + T, t_f + T]$. The microlensing should be weak and, on the other hand, the absence of gaps (each summer) is need. Very recently, using image A data in the band r (1994 Dec.-1995 May, which were kindly provided us by T. Kundic), and our image B data in the band R (1996 Feb.-July), we concluded that $\Delta\tau_{BA} = 424\pm3$ days (Oscoz *et al.* 1997). The A component was active (Kundic *et al.* 1995), and also, we have not found evidences of strong microlensing in the $r - R$ comparison (Oscoz *et al.* 1997, Goicoechea *et al.* 1998). A similar result (417±3 days; 2σ confidence level) has been derived by Kundic *et al.* (1997). They use light curves in the g and r bands. However, a reanalysis based on discrete correlation functions and two nice datasets, $A_r + B_r$ and $A_g + B_g$ (we basically exclude the photometric monitoring of image B in the interval 1995 Dec.-1996 Jan.), shows that 424 days is the most probable delay. For every fixed value θ (days), we construct the function

$$\delta^2(\theta) = (1/N)\sum_{i=1}^{N} S_i[DCC(\tau_i) - DAC(\tau_i - \theta)]^2, \qquad (1)$$

where DCC is the discrete $A - B$ cross-correlation function, DAC is the discrete $A - A$ autocorrelation function and $S_i = 1$ only when both DCC and DAC are defined at τ_i and $\tau_i - \theta$, respectively, and $S_i = 0$ otherwise. Then, we search for a minimum θ_0, such that $\theta_0 = \Delta\tau_{BA}$. In the two ($g - g$ and $r - r$) comparisons, it is inferred $\Delta\tau_{BA} = 424$ days (see Fig. 1).

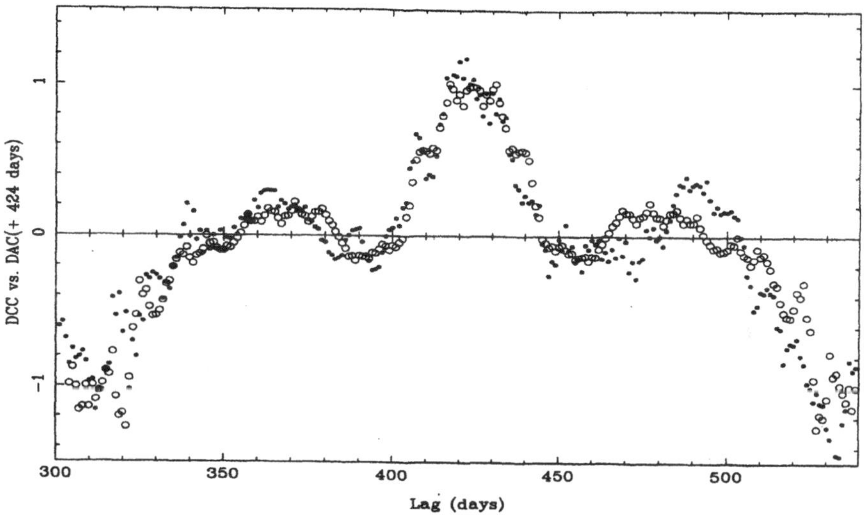

Figure 1. DCC (filled circles) vs. DAC (shifted by 424 days; open circles) in the $g-g$ comparison. The agreement is excellent (for details, see main text).

The main deflector is a cD galaxy at the center of a cluster. A good picture of the galaxy could be a King-type surface density profile with velocity dispersion σ_v and core angular radius θ_c, plus a point-mass (M_{bh}) at the centre of the King profile (Falco *et al.* 1991). Moreover, the gravitational effect of the cluster can be represented by means of a quadratic lens with a convergence κ and a shear γ with position angle ϕ. Grogin and Narayan (1996) used $\alpha \equiv (\sigma_v, \theta_c, M_{bh}, \gamma, \phi)$, $\vec{\beta}_1 \equiv (\beta_{1x}, \beta_{1y})$, $\vec{\beta}_5 \equiv (\beta_{5x}, \beta_{5y})$ as free parameters (Garrett *et al.* 1994, have fitted the A and B radio images with six Gaussian components, being A_1 and B_1 the respective core components and A_5-B_5 another jet components), since, in spite of the large number of constraints, there is a degenerancy in the convergence due the cluster. The whole lens model must include the nine parameters inferred from measurements of the lensed images as well as an estimation of κ derived from a direct measurement of the mass either in the cD galaxy or the cluster. A measurement of $\sigma_v(light)$, the 1D velocity dispersion of the luminous stars in the galaxy, allows to eliminate the cluster degenerancy.

Falco *et al.* (1997) derived $\sigma_v(light)$ = 279±13 km/s, and so, with the whole lens model and the time delay above mentioned, one obtains $H_0 = 66^{+15}_{-14}$ km/s/Mpc (2σ). This estimate is 10% accurate at 1σ (Oscoz *et al.* 1997). We however remark that a new lens model satisfying the constraints from radio mapping as well as new optical constraints deduced from *HST* observations is now in progress (Bernstein *et al.* 1997). Also, the approximation $\sigma_v = \sigma_v(light)$ must be reconsidered (Mediavilla *et al.* 1998).

3. Perspectives in a near future

The combined effort by several groups of astronomers will lead to the detailed analysis of a large number of gravitational mirages. For each system, if the deflectors are not dark, we must be able to obtain a promising set of constraints and some information on suitable pictures of the global lens (primary and secondary deflectors). Through a good picture and the constraints, one can infer a good lens model. Moreover, the system must be extensively monitored to get the light curves of the different images and deduce at least one time delay. This work could be difficult due to probable microlensing events, a weak source variability, an unsuitable sampling of the image light curves, etc. The technique for determining a time delay plays also a role. Even in the *old* Twin QSO there is an uncertainty (due to the methodology) of about one week, which is irrelevant in order to obtain H_0 from this system, but it could be dramatic in another multiply imaged QSO (e.g., the Triple QSO 1115+080$A_1 - A_2, B, C$).

Nowadays, from two individual systems (Twin QSO and Triple QSO), it is inferred a mean value of $H_0 \approx 60$ km/s/Mpc, in good agreement with other methods (see the paper by C. Frenk in this volume). However, new measurements of H_0 from gravitational lenses may surprise us. For example, when one measures H_0 via gravitational lensing, the influence of the large-scale structure is not taken into account. The effect introduced in the measurement of H_0 caused by large clusters and/or large voids, is a very interesting topic, which will be soon studied by our group.

4. References

Bernstein *et al.* (1997) *ApJ* **483**, L79
Blandford, R.D. and Kundic, T. (1996) *preprint (astro-ph/9611229)*
Falco *et al.* (1991) *ApJ* **372**, 364
Falco *et al.* (1997) *ApJ* **484**, 70
Garrett *et al.* (1994) *MNRAS* **270**, 457
Goicoechea *et al.* (1998) *ApJ* **492**, 74
Grogin, N.A. and Narayan, R. (1996) *ApJ* **464**, 92; **473**, 570
Kundic *et al.* (1995) *ApJ* **455**, L5
Kundic *et al.* (1997) *ApJ* **482**, 75
Mediavilla *et al.* (1998) *in preparation*
Oscoz *et al.* (1997) *ApJ* **479**, L89
Pelt *et al.* (1996) *A&A* **305**, 97
Schild, R. (1996) *ApJ* **464**, 125
Williams, L.L.R. and Schechter, P.L. (1997) *preprint (astro-ph/9709059)*

SUMMARY

R. D. BLANDFORD
Theoretical Astrophysics, 130-33 Caltech, Pasadena, CA 91125 USA

1. Introduction

This has been a most enjoyable and instructive "workshop" (in two senses of that word, as the lecture theatre was being constructed around us); enjoyable through the gracious hospitality of our hosts and instructive because the somewhat eclectic choice of topics meant that we all learnt something new and novel. In providing this brief, summary report, I hope that two flights of whimsy can be indulged. Firstly, like the biologists we are all supposed to become in the next century, we speak in three letter codes, (beginning with "NSU"). The use of acronyms is a sad reflection upon our impatient age, and may have the consequence of rendering our research inaccessible to outsiders, rather like the priests of medieval Oporto who spoke only in Latin. This is a conceit that we can ill afford. I have therefore taken the unusual step of providing translations. Secondly, I was struck by the strange title of this meeting, (which may have sounded less jarring in Portuguese). However, upon reflection, I think that it captures the essence of modern astronomy and cosmology. Our Universe is revealed as a lively and surprising place and I have chosen to develop this by adding some additional soubriquets.

2. GRB (Gamma Ray Bursts)

This has been good year for GRBs. Thanks to BATSE and its handlers, most of us, (especially those, including myself, who used to think that these were Galactic neutron stars), are now persuaded that we are witnessing the deaths of cosmologically distant stars [Meegan]. Furthermore, with the successful spectroscopy of GRB 970508, we now know that at least one burst really did come from $z > 0.835$. However, one is still left with the overwhelming impression that these are a heterogeneous class of sources,

Astrophysics and Space Science is the original source of publication of his article. It is recommended that this article is cited as: *Astrophysics and Space Science* **261**: 345–352, 1999.

like supernovae, that may contain a minority that are also Galactic, like the so-called soft repeaters. One reason for believing this is the burst duration distribution appears to be bimodal (although it should be recalled that the radio pulsar pulse distribution exhibited an irrelevant bimodality for a long while. Another clue concerns the "cyclotron" lines which still appear to be modestly convincing and have no natural place in a cosmological context. A third worry is that typical burst spectra show a characteristic photon energy $\sim m_e c^2$ a surprising conclusion if, as is popularly believed, the emission comes from relativistic outflows with Lorentz factors, $\mathcal{D} \sim 300 - 1000$).

Most current models would associate GRBs with current sites of star formation. It is then disturbing that the GRBs with good positions are not associated with luminous blue galaxies because we are told that we are seeing most of the star formation in the Universe. (This presupposes that the night sky is only as bright as the sum of known galaxies, a presumption which there may now be grounds to doubt.) One way out is to locate the sources at $z >> 1$. This increases the radiant energies to more than several percent of $M_\odot c^2$, approaching the binding energy of a pair of coalescing neutron stars. This, in turn, has stimulated alternative source models involving black hole binaries, "hypernovae", binary black holes and new physics [Bertolami]. The afterglows, have suggested their own puzzles. The first is that they are not obviously as compact as would be necessary to allow refractive or diffractive scintillation. The second concerns magnetic field amplification behind a relativistic blast wave, behind which, the magnetic energy density must increase by a factor $\sim 10^{12}$. This, for me at least, is a major shortcoming of existing models.

Whatever their nature though, GRBs surely signal a violent demise and I have to conclude that we live in *The Suicidal Universe.*

3. HDF (The Hubble Deep Field)

This meeting has presented a snapshot of contemporary research on galaxy formation [White]. Large scale redshift surveys are filling in the story of the evolution of galaxies from when the Universe was very young, less than ~ 1 Gyr old to the present day. At low redshifts, $z < 1$, we find a much larger incidence of irregular galaxies that show signs of merging and interaction.

Turning to redshifts $z \sim 3$, the remarkable discovery of large numbers of large galaxies has been a great breakthrough. Furthermore, these galaxies are turning out to be clustered, (to a degree that, somewhat surprisingly at first, does not challenge CDM models), and to expel gas (as may be relevant for the interpretation of quasar absorption lines). Finally, galaxies have been found with $z \sim 5$ where gravitational lensing effects are apparently quite significant.

Galaxies can be detected on the HDF with fluxes two orders of magnitude fainter than those for which direct spectroscopy is practical [Blandford]. There are a large number of the galaxies. Indeed there are ~ 100 billion of them on the sky, (Eddington's estimate in his popular writings of the number of galaxies in the Universe!). However, this is ~ 30 times larger than the product of the local density of bright galaxies and the comoving volume out to $z \sim 3$. We do not know where they all are and what are their powers. It seems quite likely that many of them are relatively local dwarfs that perhaps constitute the merging subunits in local interacting/merging galaxies. However, it also seems likely that many of them are also at high redshift as demonstrated by analyses of cluster gravitational lenses.

Elliptical galaxies are a puzzle. Recent simulations, in the framework of cold dark matter cosmogony, suggest that many ellipticals formed comparatively recently $z \sim 1$ from the merging of pairs of disk galaxies, for example. At first sight, this view seems quite at odds with the observations. The ellipticals observed at $z \sim 1$ on the HDF, are red, morphologically quite relaxed and isolated. They appear to be objects of great antiquity and are quite plausibly the descendants of the $z \sim 3$ galaxies. However, appearances can be deceptive. Disks at high redshift should be very small and a modest amount of star formation may go unnoticed in the light of the many old stars.

Returning to that stunning image of the HDF and recalling the shock and delight that I felt when I first saw it, I must say that we live in *The Overcrowded Universe.*

4. AGN (Active Galactic Nuclei)

Recent observations have amply vindicated the black hole model [Blandford]. Although we cannot say that we have proved that the "central dark objects" in galactic nuclei really are black holes, this must be true "beyond all reasonable doubt". The masses measured dynamically range from two million to three billion solar masses (and there are indications that the range can be extended at both ends). Knowing the black hole masses transforms interpretative research on accretion and is leading to much greater sophistication in fitting physical models to observational data.

There are several modes of accretion that have been reported here. Firstly, there is traditional disk accretion, in which the gas can cool and converts rest mass to radiant energy with an overall efficiency of $\sim 0.1c^2$. Secondly, there are Advection-Dominated Accretion Flows, or ADAFs, and these are relevant if there is viscous heating which goes primarily into the ions and these are coupled to the electrons just through Coulomb scattering. This results in a two-temperature, gas which cannot radiate if the accretion rate is sufficiently below the Eddington rate.

Although ADAFs are thought to be primarily steady, observations of sources accreting at much higher rates frequently exhibit quasi-periodic fluctuations of large amplitude. These require instabilities to occur in their central regions, presumably in their disks where, we now are fairly confident, magnetic stresses dominate the torques. This, is another great advance and encourages us to believe that we will one day have as good an understanding of these instabilities as we now do of, say, cepheid instability.

One way by which slowly accreting black holes in AGN can undergo violent outbursts is through stellar capture. When this happens to a normal star, there will be tidal disruption beyond the event horizon as long as the hole mass is less than $\sim 10^8$ $M_\odot$ [Ulmer]. However, capture of white dwarfs, by $\sim 10^5$ $M_\odot$ black holes can create degenerate rings of gas on relativistic orbits and these may have interesting observational consequences.

Another part of the black hole accretion process, and it is not clear yet quite how essential it is, is the formation of jets [Dennett-Thorpe]. In one extreme view these are the principal sink of angular momentum for the accreting gas, just as the solar wind is the main drain for the sun's spin. In the other, jets are occasional and incidental accoutrements to gas infall. We are getting some strong clues from recent observations. Firstly, there are now more sources like SS433 (eg GRS J1915+105) that create jets that appear to carry away more power than is associated directly with accretion. Furthermore, it is beginning to look as though, jet formation is not a stationary flow but more naturally described as episodic. This is apparent in GRS J1915+105, and the compact extragalactic radio sources. Returning to the extragalactic compact radio sources, there has been much recent attention on the kpc scale radio components which are found in abundance as "compact symmetric object" (CSOs)[Augusto]. A bold hypothesis has developed in which GHz peaked radio sources evolve into CSOs and then in to FR2's. The worry about this proposed evolution is that there appear to be far to few FR2's given the supposed ~ 1000 yr lifetimes of the CSOs.

More recently it is becoming apparent in the Herbig-Haro objects ejected by Young Stellar Objects. These now show evidence for correlated ejection in the two anti-parallel jets, demonstrating that the outflowing shocks are actually created in the nucleus rather than as independent instabilities in the two jets. It is interesting how observations of these very different types of jets are providing complementary information on the fundamental mechanisms for forming and propagating generic jets.

AGN, by their very nature are not creatures of great subtlety or refinement. They are powered by irruptive monsters that plunder and despoil their galactic hosts. For them, we have *The Ill-mannered Universe.*

5. CMB (Cosmic Microwave Background)

The study of microwave background fluctuations has come to occupy center stage in contemporary cosmology. Technically, this research has come of age [Lasenby]. After years of effort, complementary observations, from the ground, from balloons and from space have shown that these fluctuations really do exist on large angular scales at about the level predicted by contemporary (although not pre-CDM) theory, and as we learned here, on $\sim$ degree scales, the famous Doppler peak may have just been sighted through the clouds as well as the hinterland on the far side. Obviously caution is necessary; removing all the systematic effects is so hard. However, we are more than tantalized by the prospect of measuring almost all of the conventional cosmological parameters at the epoch of recombination, in the ancient Universe and giving ourselves the opportunity to compare them with their counterparts in the modern Universe as measured by classical cosmographic tests [Blanchard, Barbosa, Crawford].

On the theoretical front, it is claimed that matters are pretty well in hand. The Harrison-Zel'dovich spectrum is well understood to emerge from inflation and the deviations associated with tensorial fluctuations, (which may be measured through polarimetric studies), can, in principle, provide a quantitative verification of the theory of inflation itself. (Some of us are so old-fashioned that we are not prepared to endorse physical laws and principles on the basis of compelling theoretical imperatives alone!) The wherewithal to verify that the linear perturbations really are Gaussian exists, limited only by cosmic variance on large angular scales. A quite pretty example of just the sort of fluctuation one would hope not to see was presented here in the context of the CAT2 data.

The prospects for the future are great - a veritable alphabet soup of experiments comprehensively covering the frequency-wave number plane - BOOMERANG, VSA, CBI, VCA, MAP, PS and so on. It will all be necessary in order to have any chance at all of removing the foregrounds which are already looming ominously in the ground-based observational studies. There are high expectations for ~ 1 percent measurements of the fluctuations on the important scales and that, unlike with other cosmological endeavours, physical statistics will triumph over astronomical reality.

With all this talk of fluctuations either, just measured, or confidently expected from future observations, we should never lose sight of the remarkable and quite unprecedented overall uniformity of the microwave background. To the layman, unfamiliar with gravitational and quantum mechanical nuances, we must surely inhabit *The Boring Universe.*

6. LSS (Large Scale Structure)

We have had a lot of discussion of the details of cosmological fluctuation theory, as made manifest in both background radiation fluctuations and large scale structure [Liddle]. The kinematic content of inflation, as expressed by the enigmatic identity:

$$\rho + 3P < 0 \Leftrightarrow (H^{-1}a^{-1})' < 0 \qquad (1)$$

ensures that we must part and re-unite as long as we can find a "graceful" exit from the inflationary potential. Taylor expansion of the potential at exit, provides a means of constraining the character of the observed fluctuation spectrum, as mentioned above. There has been much talk of extended inflation, quintessence, scalar-tensor theories, supersymmetry and so on and there is much more freedom in physical theory than we can accommodate with cosmological models [Mimosa, Moniz, Viana, Caldwell]. Nonetheless, it is beginning to look as though the central notion of there having been an inflationary epoch is, at some level, falsifiable as a theory and it seems to me to be very important to try to achieve some consensus, as to what conceivable observational outcomes could rule out this explanation of the riddle of cosmic causality.

A quite different approach to the origin of large scale structure attributes it to cosmic defects, like textures and strings [Martins, Stebbins]. To this commentator, who remembers learning crystallography and metallurgy, this has always seemed entirely natural. However, as with any good theory, there are observable implications and these have been found wanting. At this meeting a much more confident understanding of the statistical mechanics of strings was reported. However, it seems unlikely that the theory can survive a confident measurement of the first Doppler peak and the establishment of ~ 100 Mpc structure in galaxy redshift surveys.

In addition to the anticipated microwave background observations, the observational future looks bright. Giant redshift surveys SDSS, 2DF will build upon the existing CFRS, LCRS and so on and provide a means of mapping and describing statistically the distribution of light in the Universe. The challenge is then to relate this to the distribution of mass. Here there are some disturbing trends. It appears that galaxies are quite well segregated in ways that relate to their ease of spectroscopic observation. In particular emission line galaxies are less luminous and more isolated than absorption line galaxies. Effects like these will have to be taken into account very carefully in analyzing these surveys.

A more direct approach to measuring the mass distribution is to measure weak gravitational lensing - a few percent, systematic distortions in the shapes of distant galaxies due to intervening large scale structure. Observing large areas of sky using wide field cameras provides a practical way to do this, although the systematic errors associated with the atmosphere, the telescope and the detector are challenges that still have to be squarely addressed before this becomes a reliable technique.

The history of predictions and observations of large scale structure is not, in my opinion, very pretty. There has been some degree of recidivism and denial of past error. However, out of all this may be emerging a robust, deductive path from the first $\sim 10^{-33}$s of the Universe to the present day. If this optimism turns out to be well-founded, and we should know within a decade (and maybe much sooner), then I think we can say that we inhabit *The Designer Universe.*

7. H$\omega\Lambda$ (The Hubble Constant, the Density Parameter and the Cosmological Constant.

There have been some excellent presentations on the local measurement of fundamental cosmological parameters and attempts to reconcile the contradictions between them [Frenk]. Firstly, according to many determinations, the Universe still appears to be younger than its oldest stars. However, three forms of help are in sight. Revisions of the distances to nearby clusters by the Hipparchos satellite may result in a reduction of globular cluster ages, (as measured by the main sequence turnoff), to $\sim 12 - 13$ Gyr. Secondly, the Hubble constant, as determined by Cepheids, appears to be falling from its initial high reported value and may not be inconsistent with values ~ 65 km s^{-1} Mpc^{-1} suggested by Type 1a supernovae and gravitational lenses [Pijpers, Goicoechea]. Thirdly, it now appears to be socially acceptable to discuss a low density Universe in which $t_0 \sim 0.9 - 1H_0^{-1}$ rather than $2/3H_0$ as in the Einstein-de Sitter case.

This brings us to the density measure Ω_0 and the kinematic acceleration parameter, q_0. Several lines of argument are now suggesting that $\Omega_0 \sim 0.3$, local dynamics, determinations of the baryonic content of rich clusters and the analysis of the evolution of their substructure and their incidence at high redshift when the Universe was less than half its present age [Sadat]. (AXAF should firm up a lot of this.) Again, Type 1a supernovae, treated as standard candles, are in rough agreement though this is in some sense, a combined measurement of q_0 and Ω_0.

This next raises the matter of Λ. The choice appears to be between an unbound, Λ-free universe and a flat Universe. Both require fine-tuning and I don't know which alternative is the less unpalatable, if one believes in inflation. (Expert opinion on this matter appears to be equally divided.) Gravitational lensing has been used to place rather discouraging upper bounds on Λ, but these are really statements about when a compensatory evolutionary effect is deemed unacceptably extreme. The differences between these two alternatives for the standard cosmographic measurements listed above are small but ultimately measurable, locally as, of course they may be from the microwave background.

However, for those of us who have watched confident claims of cosmological parameters come and go, like the Atlantic tides, I still worry that we live in *The Measureless Universe.*

Acknowledgements

Finally, to acknowledge the generous hospitality of our hosts, let us draw attention to *The Grateful Universe.* Obrigado e adeus.

SPRINGER NATURE

GPSR Compliance

The European Union's (EU) General Product Safety Regulation (GPSR) is a set of rules that requires consumer products to be safe and our obligations to ensure this.

If you have any concerns about our products, you can contact us on ProductSafety@springernature.com

In case Publisher is established outside the EU, the EU authorized representative is:

Springer Nature Customer Service Center GmbH
Europaplatz 3
69115 Heidelberg, Germany

Kolibri 360 GmbH
Ferdinand-Jühlke-Straße 7
99095 Erfurt, Deutschland
produktsicherheit@kolibri360.de